Cambridge Planetary Science Series

Editors: W.I.Axford, G.E.Hunt, T.O.Mutch

Interiors of the planets

T0296059

A. H. COOK

Jacksonian Professor of Natural Philosophy, University of Cambridge

Interiors of the planets

CAMBRIDGE UNIVERSITY PRESS

Cambridge

London   New York   New Rochelle

Melbourne   Sydney

CAMBRIDGE UNIVERSITY PRESS
Cambridge, New York, Melbourne, Madrid, Cape Town, Singapore, São Paulo, Delhi

Cambridge University Press
The Edinburgh Building, Cambridge CB2 8RU, UK

Published in the United States of America by Cambridge University Press, New York

www.cambridge.org
Information on this title: www.cambridge.org/9780521106016

First published 1980
This digitally printed version 2009

A catalogue record for this publication is available from the British Library

ISBN 978-0-521-23214-2 hardback
ISBN 978-0-521-10601-6 paperback

FOR ISABELL

'The works of the Lord are great, sought out of all them that take pleasure therein.'

From Psalm 111; carved over the entrance to the Cavendish Laboratory

CONTENTS

FOREWORD

The planets, which have always been objects of wonder and curiosity to those with the opportunity or need to lift their eyes to the heavens, now in our times shine with new and strange lights revealed to us by the far seeing instruments carried upon space craft. The Moon, Mars, Venus and Mercury all bear on their surfaces the crater scars of innumerable meteorites that have fallen upon them from the beginning of the solar system. The Earth alone has an active surface that has obliterated those scars. The fluid surface of Jupiter is in constant and vigorous motion, driven by heat flowing out from the interior or, it may be, brought to it by the ultra-violet radiation from the Sun or by the solar wind. The Medicean satellites of Jupiter now present to us strange and individual faces: would Galileo who first saw the mountains on the Moon or the spots on the Sun have been surprised by the eruption of sodium and sulphur from Io and the cloud of gas within which it moves, or by the strange stress patterns upon other of the satellites? Seeing these strange and varied faces of the planets, each apparently different from any other, who can forbear to ask, what bodies are these, how are they made up, that their appearances are so distinctive? Why are some active, and others apparently dead, some dry, and others thickly covered with atmosphere or ocean? We have indeed little to go on to answer those questions, just the sizes of the planets, the grosser features of the fields of gravity around them and the magnetic fields they possess. But within recent years, as we have learnt more by experiment and by theory of the behaviour of solids and liquids at very high pressures, it has become possible to supplement our knowledge of the planets with understanding of how their possible constituents might behave. That, in essence, is the theme of this book. I aim to explain how the mechanical properties of the planets are determined nowadays, to describe the behaviour of planetary materials at planetary pressures, and to combine the Newtonian physics of celestial mechanics with the quantum physics of highly compressed matter to establish the general constitution of the planets. No detailed explanation of the state of each planet can be expected, indeed I shall often emphasize the limitations upon our knowledge and understanding, but some connexions can be made between what we see of the surfaces and what there must be within, and, more speculatively, something may be said about how the system of planets came into being. But for all the great achievements of space research our understanding of the planets is still rudimentary, with many surprises no doubt yet to come, and my emphasis is upon the ways in which we approach the study of the planets, rather than on the results we have so far attained.

I am indebted to Mr. W. B. Harland and to Professor V. Heine for reading certain chapters in typescript and to many other colleagues for discussions on various topics of this book. I am most grateful to the staff of the Cambridge University Press for their outstanding help.

Note on the expression of planetary masses

The masses of all planets are derived from the acceleration of some object in the vicinity at a known distance. The fundamental quantity observed is thus the product, *GM*, of the mass and the gravitational constant. For example, the value of this product for the Earth is

$$GM_E = 3.986\,03 \times 10^{14}\,\mathrm{m^3/s^2}.$$

Values of such products are known to very high accuracy, that of the Earth, for example, to a few parts in a million. The constant of gravitation is, however, poorly known. In this book its value is taken to be

$$6.67 \times 10^{-11}\,\mathrm{m^3/s^2\,kg};$$

it has an uncertainty of a few parts in a thousand.

It follows that ratios of masses are well known, as are the accelerations to which they give rise, but densities expressed in kilogrammes per cubic metre have uncertainties of a few parts in a thousand, and that should be borne in mind in comparing estimates of planetary densities with laboratory data.

1

Introduction

1.1   The wonders of the heavens

The planets have been a subject of wonder to man from earliest recorded times. Their very name, the Wandering Ones, recalls the fact that their apparent positions in the sky change continually, in contrast to the fixed stars. Greek astronomers, Ptolemy particularly, had shown how the motions of the planets, the Sun and the Moon could be accounted for if they were all supposed to move around a stationary Earth, and in mediaeval times an elaborate cosmology was created, at its most allegorical, evocative and poetic in the *Paradiso* of Dante. The men of the Renaissance overthrew these ideas but provided fresh cause for wonder in their place. Placed in motion around the Sun by Copernicus, their paths observed with care by Kepler, the planets led Newton to his ideas of universal gravitation. Galileo, his telescope to his eye, showed that they had discs of definite size and that Jupiter had moons, the Medicean satellites, which formed a system like the planets themselves.

The discoveries of the seventeenth century settled notions of the planets for three centuries, but within that framework a most extraordinary flowering of the intellect attended the working out of the ideas of Newton. Closer and closer observation showed ever more intricate departures of the paths of the planets from the simple ellipses of Kepler, and each was accounted for by ever subtler applications of mechanics as the consequence of the gravitational pull of each planet upon its fellows. For some while the system of planets was tacitly or explicitly supposed to be closed, until William Herschel, in almost his first excursion into astronomy from his profession of music, saw from Bath an unknown planet – Uranus. Much later Adams predicted, a great achievement of dynamical theory, the existence of a further planet, and Leverrier found it. Now we know of yet a further planet, Pluto, and of the belt of asteroids between Mars and Jupiter. The major dynamical features of the motions

1

of the planets are now well understood, as are many minor features, among them the effects of the general-relativistic description of gravitation.

No doubt there are still discoveries to be made in orbital dynamics; they will come from the exploitation of modern developments in theory and, in particular, the execution of lengthy algebraic calculations by computer, and they will come also from the ever more precise measurements of the motions of the Moon and the planets and space probes that go close to them, by means of radio, radar and lasers. But these will probably be refinements; the heroic age of dynamical studies is almost closed, and other fields are now more productive. Early in this century, definite ideas about the internal constitution of the Earth began to develop, while estimates of the densities of the planets were available from dynamical investigations. It was realized that Mercury, Venus and Mars, as well as the Moon, must be in a general way similar to the Earth while the major planets, with much lower densities, were essentially different. In the 1930s the physics of the solid state and the quantum mechanics of materials at high pressures, rudimentary though they were by present day standards, were nonetheless adequate guides to thought about the nature of the planets, and the conjunction of dynamical studies on the one hand with the physics of condensed matter on the other is the theme to be developed in this book.

Space research has certainly contributed to the dynamical study of the planets, but rather in the way of refinement, for the observations of natural objects had led to considerable knowledge of the masses, densities, gravitational fields and moments of inertia of the Moon and a number of the planets before any space probes were launched towards them. The refinements have been valuable and have brought precision and simplicity to what previously may have been approximate and complex, but the great contributions of space research have been elsewhere, to the knowledge of surfaces and surface processes and of magnetic fields. We take it to be almost certain that the magnetic field of the Earth and the changes that the surface of the Earth undergoes are dependent on the internal state of the Earth and are driven by sources of energy within the Earth. The same is very likely true of the Moon and planets, but in different ways and to different degrees; surface features and magnetic fields no doubt contain clues to the nature of the interior, but we do not yet know how to unravel them; we do not understand the connexions in the Earth so how much less can we make use of them in studying the planets. We may nonetheless be fairly confident that it is here that the next major step in understanding the planets will be taken. The relation between the structures of the planets and the natures of the materials, so far as they can be unravelled by dynamical studies and understanding of the physics of condensed matter, is established as far as the major features are concerned. Some ideas of the limits of knowledge are also clear and further studies are, on the whole, likely to lead to refinements of present ideas but not to major changes. If major changes come, it will probably be as a result of understanding the way in which magnetic fields and surface processes are related to internal constitutions.

1.2 The system of the planets

Some of the principal facts about the planets are collected in Table 1.1. It gives the distance of each planet from the Sun, and its radius and mass, all in terms of the Earth's distance, radius and mass, it gives the mean density of each and it gives the period of spin about the polar axis. In Figure 1.1 the masses and densities are plotted against the distance from the Sun. The table and figure demonstrate the well-known division of the planets into two groups: the inner, terrestrial planets, relatively close to the Sun, of low mass and high density, and the outer, major planets, relatively far from the Sun and having high masses and low densities. The size and mass of Pluto, the outermost planet, are poorly known, but it is certainly smaller and denser than the other outer planets.

Table 1.1. *The system of planets*

	Distance from Sun (AU)[a]	Radius[b]	Mass[b]	Density (kg/m^3)	Spin period (d)
Moon		0.27	0.0123	3340	28
Mercury	0.387	0.382	0.055	5434	58.6
Venus	0.723	0.949	0.815	5244	243
Earth	1	1	1	5517	1
Mars	1.524	0.532	0.107	3935	1.026
Jupiter	5.203	11.16	317.8	1338	0.411
Saturn	9.539	9.41	95.1	705	0.428
Uranus	19.18	4.02	14.60	1254	0.96?
Neptune	30.06	3.89	16.78	1635	0.92?
Pluto	39.44	0.24	0.002	~1000	6.4

[a] The astronomical unit (AU) is the mean distance of the Earth from the Sun and is equal to 1.496×10^8 km.
[b] The radius and mass are given in terms of the Earth's radius (6378 km) and mass (5.977×10^{24} kg).

Any theory of the origin of the solar system must account for the sharp distinction between the two groups of planets. It is not my purpose to discuss the problem of the origin of the planets in any detail save that at the end I shall draw some conclusions about the origin of the planets from their present structure. The way in which the planets were formed does, however, have consequences for the chemical constitution of the planets, that is for the types of material of which we may suppose them to be made up, and hence for the physical properties of those materials at high pressure and temperature.

It is nowadays commonly supposed that the planets formed at an early stage in the history of the Sun, during the so-called Hiyashi phase, when the Sun was far more extended than it is now. Irregularities of density were brought about by the influence of a second star and led to condensations from which the planets formed. Many people currently favour a hot origin of the inner planets, that is to say, that they formed from the condensation of materials from a hot gas as opposed to accretion from a cloud of cold dust. It is therefore thought that the temperature of a planet immediately after formation would be the temperature of condensation. The distribution of temperature inside the planets is discussed in Chapter 9; here I am concerned to point out that theories of planetary origin entail

Figure 1.1. Masses, densities and distances of the planets. The mass of Pluto is about 0.002 that of the Earth.

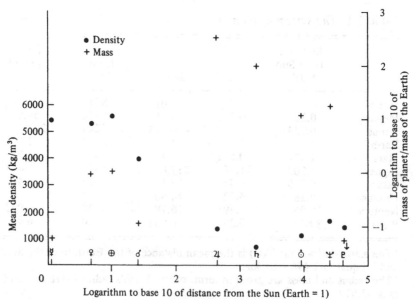

certain distributions. They also entail ideas about the compositions of the planets. In particular, if the planets condensed from a hot gas, then the different materials would condense in order of their boiling points, those with the highest points condensing first and so forming the innermost parts of the planets, followed by other materials in succession and leading to zoned structures for the planets. Quite detailed predictions of such sequences have been made, for the thermodynamic properties of materials that may form planets are known in some detail. Thus, if a mechanism for the formation of the planets is postulated, it may be possible to show that it entails a certain internal structure and a certain thermal history. Such an approach has been widely followed in much recent discussion, but it is not adopted here. The fact is that the origin of the solar system remains most obscure and while it may be plausible it is also surely hazardous to base our ideas of the internal structure of the planets on theories which of their nature do not admit of empirical verification.

A different approach is taken in this book. We start with the known dynamical properties of a planet, the size, mass, density and gravitational field, and ask what they imply for the internal distribution of density, and combine that with our knowledge of the properties of likely planetary materials to derive possible models of the internal structures. It may then be possible to make useful comparisons between the models derived in this way and those derived from theories of the origin of the solar system. No more will therefore be said about the origin of the planets until the final chapter when we look at the types of structure to which we shall have been led, and ask if they tell us anything significant about the origin of the planets. In brief, we are going to try to work back from observations of the planets, through models of their interiors, to criticisms of theories of origin, in contrast to going from theories of origin to models of structure which are constrained to fit the observed properties of the planets.

Seismological studies have enabled the internal structure of the Earth to be worked out in great detail, so that we know the density and elastic moduli as functions of pressure from the surface almost to the centre (Chapter 2). We have, in fact, empirical equations of state for the major constituents of the Earth and, by comparing them with equations of state found experimentally in the laboratory, it is possible to identify the chemical constituents of the Earth with some assurance. For no other body is that possible and in consequence other, indirect, evidence must be drawn upon to suggest the nature of the materials of the other planets. The densities of the major planets are so low that they must be composed largely of hydrogen and helium. Table 1.2 shows the densities of the condensed phases of the lighter elements at a pressure of 10^5 Pa,† together with their abundances in the solar system. Evidently the only abundant elements that could form the major planets are hydrogen and helium. The densities of the planets are of course much greater than the densities of hydrogen and helium at one atmosphere, but the pressures in the interiors of those planets are of the order of 10^{12} Pa or more and the compression of hydrogen and helium under such pressures is sufficient (Chapter 7) to account for the mean density of Jupiter and all but account for that of Saturn. Uranus and Neptune, with their greater mean densities but smaller size and so lower pressures, must have heavier elements, such as carbon, nitrogen and oxygen, mixed with the hydrogen and helium.

The increase of density through self-compression is much less in the smaller terrestrial planets, in which the greatest pressure is 3×10^{11} Pa at the centre of the Earth. Accordingly we must suppose that those planets are composed of materials with densities of 3000 kg/m³ or more at 10^5 Pa. There are three sources of evidence for the nature of such material. First, the surfaces of the Earth and the Moon are composed of silicates of aluminium, magnesium, sodium and iron and similar materials, having densities in the range of about 2500 to 3000 kg/m³. Secondly, meteorites are composed of similar materials together with free iron and nickel with densities of about 7000 kg/m³ (Table 1.3). Thirdly, there is the evidence of the internal structure of the Earth. Comparisons between the empirical equations of state of the different zones of the Earth and equations of state determined in the laboratory indicate that the outer zones of the Earth are composed predominantly of silicates of iron,

Table 1.2. *Densities of the condensed phases of the lightest elements at 10^5 Pa*

Element	Solar system abundance (Hydrogen = 1)	Density (kg/m³)
Hydrogen	1	89
Helium	10^{-1}	120
Lithium	2×10^{-9}	533
Beryllium	2.5×10^{-11}	1846
Boron	2.5×10^{-10}	2030
Carbon	5×10^{-4}	2266

† SI units are used throughout this book. 1 Pa (pascal) = 10^{-5} bar ~ 10^{-5} atmospheres.

magnesium and aluminium and possibly the oxides of iron, magnesium, aluminium and silicon at the greater depths, whilst the inner zone, the core, is composed mainly of iron diluted with some lighter material such as sulphur.

In all the compact bodies of the Universe, the inward attraction of their own gravitation would lead to condensation to an exceedingly high density were it not balanced by some pressure. Stars are hot and the balancing pressure is the radiation pressure of the thermal radiation flowing outwards. White dwarf stars have cooled down so that the radiation pressure is inadequate to balance self-gravitation and in consequence the density of the material has increased until the pressure of the degenerate gas of electrons in the ionized material balances the self-gravitation. Since the pressure depends little on temperature in a degenerate gas, white dwarfs may be considered to be cold. The pressures and temperatures in planets are much less than in stars, whether hot or white dwarfs, and the self-gravitation is balanced by the forces in solids and liquids which prevent their collapse. If the materials are unionized, the forces are the Coulomb forces between electrical charges in crystals; if the materials are ionized and metallic, the forces are those corresponding to the kinetic energy of the conduction electrons and the potential energy of the electrons in the field of the positive ions. In each case, the density is determined by the balance between internal repulsive forces, on the one hand, and internal attractive forces and external pressure, on the other; the repulsive forces arise mainly from the effect of the Pauli exclusion principle on states of electrons and from the kinetic energy of the

Table 1.3. *Composition of the surface of the Earth and chondritic meteorites*

	Crust of Earth	Average chondrite
SiO_2	0.587	0.380
MgO	0.049	0.238
FeO	0.052	0.124
Fe	—	0.188
FeS	—	0.057
Al_2O_3	0.150	0.025
CaO	0.067	0.020
Ni	—	0.013
Na_2O	0.031	0.010
K_2O	0.023	0.002
Fe_2O_3	0.023	—

electrons in metals. None of these forces depends greatly on temperature and so the planets may be considered to be cold bodies. Thermal vibrations lead of course, according to Debye theory, to additional energy and thermal expansion, but the coefficient of thermal expansion decreases rapidly with increase of pressure, so that to a high degree of approximation the densities of materials in the interiors of planets may be calculated as if they were at the absolute zero of temperature.

Let V be the specific volume of a substance, p be the pressure and T the temperature. Then

$$\frac{\partial}{\partial p}\left(\frac{\partial V}{\partial T}\right) = \frac{\partial}{\partial T}\left(\frac{\partial V}{\partial p}\right).$$

But $\partial V/\partial T = \alpha V$, where α is the coefficient of volume thermal expansion, and

$$\frac{\partial V}{\partial p} = -\frac{V}{K}$$

where K is the bulk modulus. Thus

$$\frac{\partial}{\partial p}(\alpha V) = -\frac{\partial}{\partial T}\left(\frac{V}{K}\right)$$

or

$$V\frac{\partial \alpha}{\partial p} - \alpha\frac{V}{K} = -\frac{1}{K}\alpha V + \frac{V}{K^2}\frac{\partial K}{\partial T},$$

that is

$$\frac{\partial \alpha}{\partial p} = \frac{1}{K^2}\frac{\partial K}{\partial T}.$$

The bulk modulus for many materials follows the approximate rule

$$K = K_0 + bp,$$

where K_0 is about 3×10^{11} Pa and b about 2; $\partial K/\partial T$ is about -1.5×10^7 Pa/deg for olivine. Thus, we find $\partial \alpha/\partial p = -1.7 \times 10^{-16}$/deg Pa at atmospheric pressure and somewhat less at 10^{11} Pa.

Thus the change of α over a range of 10^{11} Pa is -1.7×10^{-5}. But the value of α is about 10^{-5}; it may be inferred that α is negligible at pressures of the order of 5×10^{10} Pa or more.

1.3   Problems of inference

In all studies of the interiors of the Earth and the planets we are faced with having to derive the properties of the interior from observations made at the surface. In mathematical language, we wish to find a distribution of some property, density for example, as a function of radius. What we observe at the surface is some functional of the desired function; for example, the mass, which is equal to

$$\int_T \rho(r)\, d\tau,$$

where $d\tau$ denotes the element of volume and T the volume of the planet; or the moment of inertia, which is equal to

$$\tfrac{2}{3}\int_T r^2\rho(r)\, d\tau.$$

If we are concerned to determine the elastic moduli as functions of radius, then more complex functionals are involved, namely the times of travel of elastic waves from one point on the surface to another, or the periods of various modes of free oscillation of the planet.

The determination of the desired functions from the observed functionals is the key problem of geophysics. It is of the essence of the subject. Quite generally the number of functionals that can be observed is finite and so the detail with which the functions can be estimated is limited. The study of the optimum ways of determining the unknown functions is known as *inverse theory* (Backus, 1970a, b, c; Backus and Gilbert, 1967, 1968, 1970) and has revolutionized our understanding of what can be learnt about the interior of the Earth from observations at the surface.

Seismology, carried out in a systematic way with a worldwide network of instruments and frequent large natural earthquakes, has provided an immense quantity of data from which quite detailed knowledge of the interior of the Earth has been derived. Comparable data are lacking for any of the other planets and are very inadequately represented for the Moon. We are therefore faced with the problem of trying to learn what we may about the planets from the values of two functionals, at the most, namely the mass and moment of inertia, which can be obtained by dynamical analysis without landing space craft on the planet. It may be expected that with such a dearth of information little can be learnt of the interior. The problem has been considered by Parker (1972), who has shown how certain limits may be placed on the models that may be constructed.

Given just the radius, mass and moment of inertia of a planet, the number of parameters by which a model may be characterized is limited. Essentially there are two types of model we may use: one in which the planet is divided into two zones and in which we attempt to determine the radius of the division between them, together with the densities in the two zones; or the other in which there is a continuous distribution of density specified by two parameters. The former is appropriate to the terrestrial planets, for we know the Earth is divided into two major zones, while the density in any zone changes relatively little with pressure. Of course the model is only a first approximation, for, with only the mass and moment of inertia to go on, we cannot determine finer subdivisions nor variations of density within zones. The second type of model may be more suitable for the major planets if, as suggested above, they are composed predominantly of a substance of uniform composition.

The simple models are still indeterminate if only the radius, mass and moment of inertia are given. Consider the model with two zones. Let the radius of the planet be a and the radius of the inner zone be a_1. Let α be the ratio a_1/a.

Let the densities of the inner and outer zones be respectively ρ_2 and ρ_1 and let the mean density of the planet be $\bar{\rho}$. We then have for the mass

$$\tfrac{4}{3}\pi a_1^3 \rho_2 + \tfrac{4}{3}\pi (a^3 - a_1^3)\rho_1 = M = \tfrac{4}{3}\pi a^3 \bar{\rho}.$$

If we let $\rho_2/\bar{\rho} = \sigma_2$, and $\rho_1/\bar{\rho} = \sigma_1$, this may be written

$$\alpha^3 \sigma_2 + (1 - \alpha^3)\sigma_1 = 1. \tag{1.1}$$

We suppose we also know the mean moment of inertia, I. Then

$$\tfrac{8}{15}\pi a_1^5 \rho_2 + \tfrac{8}{15}\pi (a^5 - a_1^5)\rho_1 = I.$$

Let us define an *inertial mean density*, $\gamma\bar{\rho}$, by the relation $I = \tfrac{8}{15}\pi a^5 \gamma\bar{\rho}$. Note that

$$\frac{I}{Ma^2} = \tfrac{2}{5}\gamma$$

and that, for a body of uniform density, $\gamma = 1$.

We then have the second equation in the form

$$\alpha^5 \sigma_2 + (1 - \alpha^5)\sigma_1 = \gamma. \tag{1.2}$$

It is evident from the form of these equations that given γ from observation the most we can do is obtain a relation between σ_1, σ_2 and α; if a value of a_1, for example, is chosen, ρ_1 and ρ_2 are determined. There are, however, some limits on the range of variables, as is discussed in Appendix 1.

Evidently, if we wish to select a particular model as in some sense the preferred one, we must have some *a priori* principles on which to make

our choice. The general approach which, following Parker (1972), I adopt
is to choose that model which maximizes some function of the density.
For example, one might maximize the variance of the density

$$\int (\rho - \bar{\rho})^2 \, d\tau,$$

subject to the conditions that the mass and moment of inertia have their
observed values.

Suppose we wish to maximize some integral I subject to the conditions
(1.1) and (1.2). Then, by the procedure of undetermined multipliers, we
find the unconstrained maximum of

$$I + \lambda_1 [\alpha^3 \sigma_2 + (1 - \alpha^3) \sigma_1 - 1] + \lambda_2 [\alpha^5 \sigma_2 + (1 - \alpha^5) \sigma_1 - \gamma]$$

with respect to variations of α, σ_1 and σ_2.

How is I to be chosen? We can either attempt to set a limit on one or
more of α, σ_1 or σ_2, or we may attempt to estimate the best values of these
parameters. Parker was concerned with the first problem. He wished to
find the *least* value of σ_2 consistent with the mass and moment of inertia
and showed that then the integral I must be

$$\left(\int_0^a |\rho(r)|^p \, dr/a \right)^{1/p},$$

where p is allowed to go to infinity, for, as that happens, I becomes the
greatest maximum of $|\rho|$.

Parker then finds the simple result that

$$\rho(r) = \rho_0 \quad \text{for } \lambda_1 r^2 + \lambda_2 r^4 > 0$$
$$\rho(r) = 0 \quad \text{for } \lambda_1 r^2 + \lambda_2 r^4 < 0.$$

ρ_0 is the least value of the maximum density, i.e. the value of ρ_2, and, for
it, ρ_1 is zero.

Parker shows that

$$a_1/a = (\tfrac{5}{2} I/Ma^2)^{1/2},$$
$$\sigma^2 = \rho_0/\bar{\rho} = (\tfrac{2}{5} Ma^2/I)^{3/2}.$$

This is essentially a very simple result: it says that the least value of σ_2 is
that which corresponds to the outer zone having zero density. Any finite
value of σ_1 leads to a smaller value of α and a greater value of σ_2. It is of
value in specifying the maximum information we can obtain from the
mass and moment of inertia alone.

If we wish to select some best value of the model parameters, we must
invoke considerations of *a priori* probability. A principle which, following

developments in information theory, is now often followed is to choose that model which has the most random variation of parameters consistent with the observed functionals. The reason for making that choice is that then the model contains the least extra information extraneous to the data: the effect of *a priori* hypothesis is reduced to the minimum. By analogy with thermodynamics an *entropy* is defined and maximized. Here we follow a suggestion of Rubincam (see Graber, 1977) and define the entropy corresponding to a distribution of density to be

$$-\int_{\mathrm{T}} \sigma \ln \sigma \, \mathrm{d}\tau,$$

where σ is $\rho/\bar{\rho}$.

The problem of maximizing the entropy subject to the constraints of the observed mass and moment of inertia, both for the two-zone model and for a continuous distribution, is discussed in Appendix 1.

The limits set by Parker's rules are not in general very stringent and, if we do no more than follow them or the prescription of maximum entropy, we in fact ignore other *a priori* information we have, not indeed certain information about a particular planet, for we have already exhausted that, but general, somewhat probable, ideas based on what we know of other planets and the behaviour of materials. We have already drawn on such general ideas in choosing to look at two-zone models of the terrestrial planets. If we consider solely an individual planet we have no grounds for making that or any other choice, but we observe that the Earth has the major division into core and mantle and we note (Chapter 5) that there is probably a similar division in the Moon and so we are led to think that the other planets of broadly similar size and density may be similarly constituted. The conclusion may be wrong, but it has a high *a priori* probability of being right in our present state of knowledge.

We shall therefore attempt to construct models of the terrestrial planets by analogy with the Earth and the Moon, making them as close to the Earth and the Moon as the observed properties allow. Not only will zoned models be adopted, but the physical properties of the materials will be chosen to be as close as possible to those of the Earth and the Moon. However, analogies must be followed with caution: we already know (Chapter 5) that the materials in the Moon are not identical with those in the corresponding parts of the Earth and we therefore have a hint that materials may show some variation from planet to planet even though the planets are grossly similar.

When we turn to the major planets it is clear, for the reasons set out already, that the analogies with the terrestrial planets fail, for the densities of the major planets are too low for them to be composed of similar materials. The materials of the terrestrial planets, metals and metal silicates and oxides, are too complex and admit of too great a variety in composition for realistic theoretical calculations to be made of their physical properties (Chapter 4). On the other hand, Jupiter appears to be made of so nearly pure hydrogen that it is reasonable to choose as a first approximation a model consisting of cold hydrogen compressed under its self-gravitation. The equation of state of hydrogen, especially in the metallic form it assumes at high pressures, can be calculated with considerable assurance (Chapter 7) and it therefore seems that a secure theoretical basis exists for models of Jupiter and probably Saturn. Uranus and Neptune do not agree with hydrogen models and essentially more complex structures are required (Chapter 8).

1.4   The interest of the internal structures of the planets

No one will ever visit the depths of the planets, not even the Earth, and observe their nature directly. What then are the reasons for studying them? In the first place, the problem itself is of great intellectual, indeed human, interest. What must these bodies be like inside for them to be as we find them from the outside? How should we account for the differences and similarities between them? The attempt to understand the planets is a major challenge to the human mind.

The study of the planets raises problems of physics. Often in the history of physics, astronomical observations have led to experimental and theoretical studies which have extended in major ways our understanding of physics. Some of the fundamental observations in atomic physics are based on astronomical observations. The same is also true of the planets, where condensed matter is subject to pressures until recently unobtainable in the laboratory; indeed the highest ones are still unobtainable. The observed properties of the Earth and the need to understand the behaviour of hydrogen in its metallic form have contributed to important developments in understanding of the condensed state of materials. The reason why astronomy has been and continues to be so effective a stimulus to physics is that conditions are encountered there far outside the limited range of laboratory experiment, so that phenomena occur that would never have been anticipated from laboratory studies alone. The same is true of planetary physics; a wider range of conditions leads us to consider behaviour and structures which might never otherwise have been suspected. In a very real sense, all physics, especially the physics of condensed matter, is subsumed in solutions of Schrödinger's equation, but the possible solutions are so various that we need a guide to what they might be and which are the most interesting. Experiment and observation provide that guide and they also provide the solutions themselves when theory fails on account of the complexity of the systems. Such is the situation for planetary materials at high pressures.

Mankind has always tried to understand the way in which his home, the Earth, and its neighbours in space have come into existence. As mentioned above the problems are at present intractable, but attempts have been made to produce models of the evolution of the solar system based on current understanding of the evolution of a star from a condensing cloud of gas. Any such predictions should account for the planets as they are: their orbits, their masses and sizes, the rates at which they spin and their compositions. It has already been seen that the outer planets must consist primarily of hydrogen while the Earth, at least, and the other terrestrial planets probably, consists of metals and their silicates and oxides. Theories of the origin of the planets must predict such compositions. They must also predict core and mantle for the Earth and a zoned structure for the Moon; is it possible to set even further constraints upon theories of planetary origin from inferences that may be drawn about the interiors of other planets? To that question I return at the conclusion of this book.

The models of the planets so far discussed, and those which are the prime concern of this book, are concerned with distributions of density with radius and the closely related equations of state – the variations of density with pressure – and it has been argued that to a first approxima-tion planets may be taken to be cold and the temperature within them ignored. This is only a first approximation although it seems that it gives a good basis from which to develop more complex models. That more complex models are needed is evident from the existence of magnetic fields around the Earth, Jupiter, Mercury and Saturn, and from the surface irregularities of the Earth, the Moon and Mars which show that the surfaces have undergone changes of internal origin (as distinct from meteoritic bombardment) in the course of their histories. Both phenomena imply relative motions of material within the planets. They show that the planets must contain sources of energy to drive the motions and that they cannot be in hydrostatic equilibrium. The idea of cold planets in hydrostatic equilibrium under their gravitational self-attrac-tion is thus inadequate, and we are led to ask whether it is possible to learn anything about the departures from hydrostatic equilibrium or about the sources of energy by studying the magnetic field or the surface features of those parts of the gravitational field which are incompatible with hydro-static equilibrium. From the point of view of the human condition, this is a study of great concern, for human life exists by reason of the nicely adjusted departures of the Earth from hydrostatic equilibrium, sufficient to permit land to emerge from the waters, but not so great as to lead to extreme and inhospitable conditions (see Cook, 1979).

The plan of this book is as follows. Chapter 2 is devoted to a summary of our knowledge of the internal structure of the Earth. It has already been observed that through seismology we have a very detailed know-ledge of the properties of the materials of the Earth; properties that may be compared with theory and experiment and that must form the basis for attempts to understand the other terrestrial planets and the Moon. Chapter 3 is concerned with the ways in which the dynamical data for the planets may be derived from observation. The next three chapters deal with the Moon and the terrestrial planets. Chapter 4 discusses the properties of the materials of which they are most probably made, Chapter 5 is devoted to the Moon and Chapter 6 to Mars, Venus and Mercury. In Chapter 7 we take up the study of the major planets with the theory of the structure and behaviour of hydrogen and helium at high pressures and in Chapter 8 the theoretical results are applied to models of the major planets. Chapter 9 deals with magnetic fields, surface proper-ties, the departures from hydrostatic equilibrium and how all these may be related to internal structure. Finally, in Chapter 10, threads are drawn together, especially the implications for theories of the origin of the planets.

2

The internal structure of the Earth

2.1 Introduction

The Earth, as will appear, is not typical of the planets. It is the largest of the inner planets, it is the only one on which active tectonic development of the surface appears to be going on at present and, so far as we know, it has the most complex structure. Yet it is the only one which can be studied in detail; from it we may derive empirically equations of state of the materials of the inner planets; and the methods that have been used to study the structure of the Earth are those we should like to use, but are inhibited from using by the difficulties of observing, in the investigation of the other planets. For these reasons it is helpful to preface an account of the methods used to study the planets and of the results that have been obtained with a review of the way in which the Earth is examined and what has been discovered.

Our knowledge of the internal structure of the Earth comes by two routes. In the first place the mass, size and density of the Earth provide a rough idea of the overall composition and of the central pressure, while the value of the moment of inertia shows that the density increases strongly towards the centre. Naturally a wide range of models could be constructed to fit just three facts and so it is necessary to turn, in the second place, to seismology to provide more detailed information. Seismologists study the times of travel of elastic waves through the body of the Earth or round the surface and they study the periods with which the Earth vibrates as a whole. In general, rather complex calculations are required to construct a model of the interior of the Earth from seismological data and, furthermore, seismological data by themselves do not suffice for the construction of models: the models must also give the right mass and moment of inertia for the actual radius. At the same time, rather simple arguments from seismic observations lead to the conclusion that the Earth is divided into two major parts: an outer solid shell, or mantle,

16

and an inner liquid denser core. The division into core and mantle, the core having a radius of just over half the radius of the Earth as a whole, is the major feature of the internal structure of the Earth, so let us consider the evidence for it in more detail.

2.2 Evidence for a core and mantle

The size and mass of the Earth are very well known. The means by which those of the Moon and planets are found will be discussed in Chapter 3; the methods used for the Earth can be quickly summarized here. In former times, the size of the Earth was found from measurements of the lengths of arcs over the surface combined with astronomical measurements of the angles subtended by the arcs at the centre of the Earth; the equatorial radius and the polar flattening had to be estimated simultaneously. At the same time, measurements of the attraction of gravity were made over the surface and from them the equatorial value of gravity and the variation from equator to pole were estimated. By ignoring the flattening of the Earth, the mean value of gravity is given by

$$g = \frac{GM}{a^2},$$

where M is the mass of the Earth, a the mean radius and G the constant of gravitation.

Although simple in principle, the procedure is not straightforward in practice because the surface is irregular and gravity varies greatly from place to place. Much more satisfactory results are now obtained from the observations of the motions of artificial satellites. Consider a satellite in a circular orbit far enough away that it is not affected by the drag of the atmosphere. Let its distance from the centre of the Earth be R and its angular velocity in its orbit, n. Then by Kepler's third law

$$n^2 R^3 = GM,$$

while D, the measured distance of the satellite from the surface of the Earth, is

$$D = R - a.$$

The observations on a single satellite do not suffice to determine both a and M because R is also unknown, but if observations are made to a second satellite (which might be the Moon) then a and GM can be found.

There are complications. The orbit of a satellite is not circular, nor even just a Kepler ellipse with the Earth at one focus: it is perturbed by the attraction of the Sun and by the variation in the Earth's gravity corresponding to the polar flattening, but both effects can readily be calculated.

Another method is to observe the acceleration of a space probe as it travels away from the Earth after the motors have been shut down. The instantaneous velocity relative to the Earth can be found from the Doppler shift of a radio transmission from the space vehicle as received on Earth, and the acceleration may be found from the distance from the surface of the Earth as obtained by integrating the velocity. Once again, the radius of the Earth is unknown but, because the acceleration is found for a range of distances from the surface, both the mass of the Earth and the radius of the Earth can be derived.

The precision of both methods is high and the accuracies of current values of the radius and mass are a few parts in a million. However, it should be noted that the observations yield·values of the product GM and not of the mass separately; the uncertainty of the product is a few parts in a million but the uncertainty of G is a few parts in a thousand so that the value of the mass is also uncertain by a few parts in a thousand. Needless to say the radius and mass of the Earth are better known than those of any other planet. Where possible, similar methods using space probes or artificial satellites are employed to find the radius and mass of the Moon and planets, but they are not available for all planets and in any case the conditions for observation are less favourable than for the Earth.

Turn now to the polar flattening and the corresponding variation of gravity. Because the gravitational potential, V, satisfies Laplace's equation outside the Earth, it may be written as a series of spherical harmonics, functions of radial distance, r, from the centre of mass of the Earth, of co-latitude, θ, and of longitude, λ. Thus

$$V = -\frac{GM}{r}\left[1 - \sum_{n=2}^{\infty} \left(\frac{a}{r}\right)^n J_n P_n(\cos\theta)\right.$$

$$\left. + \sum_{n=2}^{\infty} \sum_{m=1}^{n} \left(\frac{a}{r}\right)^n (C_{nm}\cos m\lambda + S_{nm}\sin m\lambda)P_n^m(\cos\theta)\right].$$

Here $P_n(\cos\theta)$ is a Legendre function and $P_n^m(\cos\theta)$ is an associated Legendre function (Whittaker and Watson, 1940) and J_n, C_{nm} and S_{nm} are numerical coefficients. It is usual to choose the numerical factor in $P_n^m(\cos\theta)$ such that the integral of the square of an harmonic term over the surface of the unit sphere is 4π and values of C_{nm} and S_{nm} are most often given for that normalization. The $P_n(\cos\theta)$ are also sometimes normalized in the same way, but often the convention is that

$$\int_{-1}^{+1} [P_n(\cos\theta)]^2 \, d\cos\theta = \frac{2}{2n+1}.$$

The theory of the motion of a satellite in such a potential is now highly developed (see Chapter 3) but, so far as the principal features of the internal structure of the Earth are concerned, it is only the second zonal harmonic term, that with the coefficient J_2, that concerns us. That term is related to the polar flattening by the formula

$$f = \tfrac{3}{2}J_2 + \tfrac{1}{2}m.$$

Here f is the ratio $(a-b)/a$, where a is the equatorial and b the polar radius of the Earth, and m is the ratio of centrifugal to gravitational acceleration at the equator; if ϖ is the spin angular velocity of the Earth,

$$m = \frac{\varpi^2 a^3}{GM}.$$

The polar and equatorial radii and the flattening have to be carefully defined. The actual rough solid surface of the Earth is not what is involved, the radii and flattening are those of the surface on which the combined potential of gravity and rotation is constant and equal to the value it has over the open oceans; the surface is thus mean sea level where that can be seen and it is the continuation of that surface in continental regions.

The effect of the second zonal harmonic in gravity on the orbit of an artificial satellite is particularly simple. Suppose the orbit lies in a plane which makes an angle i with the plane of the equator (Figure 2.1). Let the plane of the orbit intersect the equatorial plane in the line NN'; because one focus is at the centre of the Earth, NN' passes through the centre. N as drawn is called the *ascending node* of the orbit; through it the satellite passes from south to north. The position of N is measured by the angle Ω subtended between it and a reference direction at the centre of the Earth; the reference direction is usually the vernal equinox or *first point of Aries*, Υ.

To a first approximation, the effect of the second zonal harmonic is that the position of the node moves steadily backwards along the equator at the rate

$$\dot{\Omega} = -\frac{3}{2}\left(\frac{a}{a_s}\right)^2 nJ_2 \cos i,$$

where n is the mean angular velocity of the satellite in its orbit and a_s is the semi-major axis of the orbit (Chapter 3).

Let P be the perigee of the orbit, the position at which the satellite is closest to the centre of the Earth. The position of P is measured by the angle ω drawn round the orbit from the ascending node. To a first approximation the effect of the second zonal harmonic in gravity is to cause the perigee to move at the steady rate given by

$$\dot{\omega} = \frac{3}{4}\left(\frac{a}{a_s}\right)^2 nJ_2(5\cos^2 i - 1).$$

The formulae for $\dot{\Omega}$ and $\dot{\omega}$ by no means give the whole story: they are the dominant parts of the motions, but for practical application more detailed results are required.

As a result of many observations of many artificial satellites, the value of J_2 is known to be 1082.65×10^{-6}. The importance of J_2 for the study of the interior of the Earth and the planets is that it is related to the moments of inertia. Let A, B and C be the principal moments of inertia, of which C is the polar moment, the largest. Then by McCullagh's theorem (Cook, 1973)

$$J_2 = \frac{C - \frac{1}{2}(A+B)}{Ma^2}.$$

Figure 2.1. Geometry of satellite orbit.

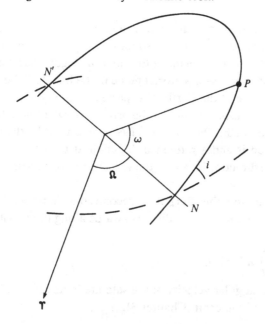

A and *B* are very nearly identical for the Earth and to a good approximation

$$J_2 = \frac{C - A}{Ma^2}.$$

J_2 is proportional to the difference of moments of inertia and some other relation between C and A is required if each is to be obtained separately. The required relation is provided by the luni-solar precession of the Earth. As a consequence of the torque exerted by the Sun and the Moon on the equatorial bulge of the spinning Earth, which behaves as a gyroscope, the direction of the Earth's polar axis rotates about the normal to the plane of the Earth's orbit about the Sun at a rate which to a first approximation is equal to

$$\frac{3}{2}\left\{ \frac{n_M^2}{\varpi}\left(\frac{M_M}{M_E}\right) + \frac{n_\odot^2}{\varpi} \right\} H \cos\theta$$

where θ is the inclination of the Earth's axis to the normal to the ecliptic, H is the ratio $(C - A)/C$, sometimes called the *dynamical ellipticity* of the Earth (Chapter 3), n_M is the mean angular velocity of the Moon about the Earth, $n_\odot$ that of the Sun, ϖ the spin angular velocity of the Earth and M_M and M_E the masses of the Moon and Earth, respectively.

H can be found with high precision from astronomical observations of the precession and has the value 3279.30×10^{-6}. It follows at once that

$$\frac{C}{Ma^2} = \frac{J_2}{H} = 0.330\,79.$$

Now C/Ma^2 can be related to the variation of density within the Earth and in particular if the density is constant within a sphere

$$C/Ma^2 = 0.4.$$

The fact that C/Ma^2 is significantly less than 0.4 for the Earth demonstrates that the density increases considerably from the surface to the centre. Table 2.1 shows the values of C/Ma^2 for a sphere with a central core having a radius one half that of the surface, and with a range of ratios of the density of the core to that of the outer shell. The particular radius of the core is chosen because it corresponds, as will be seen below, to the actual state of the Earth. It is clear that a value of C/Ma^2 not very much less than 0.4 indicates an appreciable increase of density towards the centre and that the value of 0.33 corresponds to a density of the core about 3.4 times that of the shell. We shall find later that the outer planets have much lower values of C/Ma^2 and, correspondingly, a much greater increase of density towards the centre.

It may of course be quite wrong to suppose that a planet is divided into two parts each with a constant density. In general, densities will increase inwards as a result of compression under the gravitational attraction of the planet itself, and an alternative is to suppose that the density increases inwards solely because of the self-compression. Without other evidence it is not possible to decide between these or any other possible distributions of density, provided only that they give the correct mass and moment of inertia. For the Earth that other evidence comes from seismology.

An isotropic elastic solid can support two types of wave motion that propagate through the body of the solid; one, the P-wave, in which the displacement of the solid is along the direction of propagation, and the other, the S-wave, in which the displacement is perpendicular to the direction of propagation and can be regarded as a component of rotation about that direction. Denoting, as is usual, the velocity of P-waves by α and that of S-waves by β, the theory of elastic waves shows that

$$\alpha^2 = (K + \tfrac{4}{3}\mu)/\rho, \qquad \beta^2 = \mu/\rho,$$

where K is the bulk modulus, μ the shear modulus and ρ the density (see Bullen, 1975).

Clearly α is always greater than β; thus, if an earthquake generates both types of disturbance, the P-wave will arrive first at a recording station. The P and S notation comes from this circumstance; P stands for *primary*, the first signal to arrive, S for *secondary*.

Table 2.1. *Moment of inertia for a composite sphere, having a core of half the surface radius and a density ρ times the outer shell*

ρ	C/Ma^2
1	0.400
2	0.366
3	0.340
Earth	0.330 8
4	0.318
6	0.285
8	0.260
10	0.240
12	0.226
14	0.215

An earthquake gives rise to many more than two signals at a distant observatory because there are many paths involving reflexions and refractions within the Earth by which seismic waves may travel between two points on the surface. The first pair to arrive, however, travel through the outermost parts of the Earth, following paths that are concave outwards because the velocities α and β increase inwards. If Δ denotes the angular separation of the earthquake and the observatory and T the time of travel, then to a good approximation the first P and S signals arrive at times that satisfy

$$T = a\Delta - b\Delta^3.$$

Such signals may be traced out to a separation of just over $100°$. The signals then disappear or become very faint, but at separations of about $140°$ and greater one set of signals, which can be identified as the P-waves, reappear quite strongly and continue to the antipodes. The S-waves however do not reappear. Figure 2.2 is a somewhat idealized

Figure 2.2. Travel-time diagrams of P and S arrivals. (From Cook, 1973.)

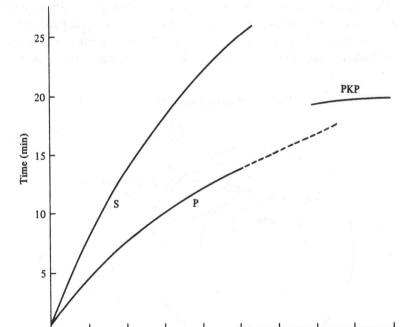

travel-time diagram of the first P and S arrivals over the whole surface of the Earth.

The region from about 104 ° to 140 ° where the P arrivals fail is known as the *shadow zone*. Its origin (first realized by Oldham, 1906) may be understood from Figure 2.3. Consider a sequence of ray paths emerging from an earthquake at E and arriving at observatories O_1, O_2, O_3, etc. Out to O_2 the times increase steadily. Then at O_3 the rays just come in contact with the boundary between the outer shell and an inner core. If the velocity in the core is less than that in the shell, the next ray which just enters the core will be refracted towards the radius vector and will subsequently emerge from the core further round its boundary. Thus there will be a gap, $O_3 O_2'$ within which no rays from E can reach the surface of the Earth. Just such is the behaviour shown by the first P-wave arrivals round the Earth and it shows that there is a major discontinuity within the Earth, inside which the longitudinal wave velocity is less than outside.

The decrease in velocity could arise from a combination of circumstances. Since α^2 is equal to $(K + \frac{4}{3}\mu)/\rho$, the density could be greater in the core or $(K + \frac{4}{3}\mu)$ could be less. Now we already know from the value of C/Ma^2 that the density increases inward towards the centre, although we cannot say that it increases by a jump. But there is a strong suggestion that the density does increase discontinuously at the surface corresponding to the shadow zone. However, the behaviour of the S-waves shows that, in

Figure 2.3. The core shadow zone. (After Cook, 1973.)

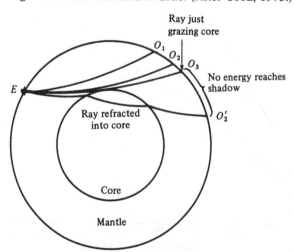

addition, $(K + \frac{4}{3}\mu)$ decreases. The S-waves do not reappear after the shadow zone as do the P-waves and it is supposed that the reason is that μ vanishes inside the boundary; in effect, the material of the core is liquid. Then, even though the bulk modulus is not reduced, $(K + \frac{4}{3}\mu)$ does fall.

The value of C/Ma^2 and the existence of the seismic shadow zone together show that the Earth has a central core, which is probably liquid and in which the density is on the average about three times as great as in the outer shell. Other evidence reinforces the conclusion. In addition to the forced motion of the axis of rotation of the Earth under the attraction of the Sun and the Moon, there is a free oscillation or nutation, the period of which depends on the elastic properties of the interior of the Earth. It has been shown that the observed period requires that the core should be liquid.

What is the radius of the core revealed by the shadow zone? If the velocities did not change with depth in the outer shell, it would be a simple matter to calculate the radius of the core (Figure 2.4). In practice, it is necessary to know how α and β change with radius in order to calculate the curvature of successive rays or, what is the same thing, the maximum depth reached by a ray emerging at an angular distance Δ from the source. That can be done if the variation of T with Δ is known out to the maximum value of Δ. Thus the core radius is now estimated to be about 3485 km.

Figure 2.4. Principle of calculation of radius of core. Δ is the angular radius to the shadow zone. The radius, r, of the core is given by $r = a \cos \frac{1}{2}\Delta$.

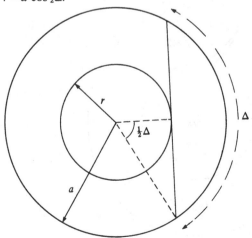

It might still be argued from the evidence so far presented that it is not necessary for the fall in velocity at the core to be accompanied by a jump in density. Other seismic evidence requires a jump in density as well. Seismic waves can reach the surface from a distant earthquake by reflexion at the boundary of the core (Figure 2.5). Two possible paths are shown as PcP and ScS; c denotes reflexion at the core boundary. Other possibilities, PcS and ScP, involve conversion from P-waves to S-waves or vice versa over reflexion. Now the reflexion coefficients for all four possible reflexions can be calculated from wave theory; they depend on the velocity and density, and the observed values show that the density must increase at the boundary of the core.

With rather simple evidence we have now arrived at a fairly clear picture of the Earth as composed of a dense liquid inner core and an outer less dense solid mantle. To go further, as we can for the Earth, we must look in detail at the transmission of seismic waves through the Earth and study the way in which the Earth vibrates elastically as a whole. That we shall consider in the next section, in order to establish a model of the Earth which shall be a reference for our study of the planets, as well as to enable us to derive relations between the pressure and density of material within the Earth. Before leaving the simple physical arguments advanced in this section, arguments which do not involve detailed numerical analysis but only a physical understanding of behaviour in general, two points need emphasizing. One is that results that depend on general

Figure 2.5. Reflexion of waves from core.

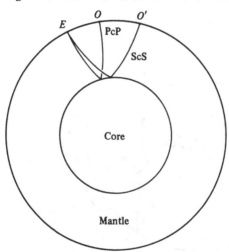

rather than particular arguments are all that for the most part are available in the study of the planets, and so the reasoning for the Earth has been set out at some length; the second point is that for the other planets we have to rely on even more elementary arguments, and that apart from the Moon there is no evidence about elastic waves within them and none about precession.

2.3 Construction of models of the Earth

The data used to construct models of the interior of the Earth are primarily the times of travel of P-waves and the periods of free oscillations of the Earth. Models, which comprise graphs or tables of the variations of density and the two elastic moduli, K and μ, with radius, must also give the correct values of the mass and C/Ma^2.

The first attempts to construct models of the interior of the Earth were based on the approach of Williamson and Adams (1923). Because α^2 is equal to $(K + \frac{4}{3}\mu)$ and β^2 to μ/ρ, it follows that

$$\alpha^2 - \tfrac{4}{3}\beta^2 = K/\rho.$$

$\alpha^2 - \tfrac{4}{3}\beta^2$ is commonly designated by Φ.

Now suppose that the change of volume of a solid under pressure, p, is solely due to elastic compression and not at all to change of crystal structure. Then, by the definition of bulk modulus,

$$\frac{d\rho}{dp} = \frac{\rho}{K} = \Phi^{-1}.$$

Subject then to the condition that only elastic compression is involved, $d\rho/dp$ is known wherever α and β are known in the Earth.

Next suppose that the Earth is in hydrostatic equilibrium, so that pressure depends on radius according to the equation

$$\frac{dp}{dr} = -g\rho,$$

where g is the attraction of gravity at the radius, r.

It follows, by combining the two equations, that

$$\frac{d\rho}{dr} = \frac{d\rho}{dp}\frac{dp}{dr} = -\frac{g\rho}{\Phi}.$$

Finally, g is given by $-GM_r/r^2$, where M_r is the mass within radius r.

If the Earth were uniform, and if the conditions of hydrostatic equilibrium were satisfied, then the equations could be integrated, starting from the outside of the Earth, where M_r is equal to the total mass of the Earth, and using the value of C/Ma^2 to determine the surface density, which is so far unspecified. In essence this was the procedure followed until the discovery of the free vibrations of the Earth as a whole, but it suffers from a number of limitations.

In the first place, reliable values of α and β are required throughout the Earth. In principle they are obtained from the changes of the travel times of P-waves and S-waves with distance through the solution of an integral equation. The equation can be solved provided α and β increase uniformly as the radius decreases, but it is not certain that they do so; β probably and α possibly have minima within the outer part of the shell and it is not then possible to obtain unique solutions. More significant is the fact that β is not so well known as α. The times of arrival of P-waves are well established because the signals start suddenly out of a quiet background, but they themselves generate unresolved noise in detectors so that the S signals arise out of a noisier background and the times are not so reliable.

More important perhaps than inadequacies in the data are failures of the assumptions on which the Adams–Williamson procedure is based. As one goes inward through the Earth, density may change with radius not only because pressure increases with radius, but also because composition changes with radius. Furthermore the increase of temperature with depth may affect the change of density, although, because coefficients of thermal expansion decrease as pressure increases, that is probably not a significant effect. Formally, the most troublesome feature is that there are discontinuous changes of density within the Earth which introduce arbitrary constants into the integration of the Adams–Williamson equations, constants which cannot be determined uniquely since there are no further global properties beyond mass and C/Ma^2 by which they might be fixed. Thus special hypotheses have to be invoked to enable the equations to be integrated, but then of course the results to which they lead are not unique.

The primary discontinuity in the Earth is that between the core and the shell, or *mantle*, as it is usually known. There are also discontinuities within the mantle, which is divided into the upper and lower mantle, and within the core, divided into outer and inner core. The crust, distinguished between oceanic and continental, lies over the upper mantle. The evidence for the upper mantle comes from the observation of sharp changes in the slopes of the travel-time curves of P-waves and S-waves at a distance of about 20°; the corresponding boundary is known as the 20° discontinuity and corresponds to rather sharp changes in the gradients of velocity, elastic moduli and density at a depth of some 600 km. As for the inner core, it was first observed by Miss I. Lehmann that some energy leaks into the shadow zone that cannot be accounted for by diffraction round the curved boundary of the core, and she suggested that within the liquid core there was an inner core in which the P velocity was greater than in the core. Because density must increase with pressure for stability, she suggested that the shear modulus was not zero and that the inner core was solid. With the setting up of large discriminating arrays of seismometers it has been possible to detect waves that have passed through the inner core as shear waves, so confirming the solidity of the inner core. In addition, some of the periods of free oscillations of the Earth show the inner core to be solid.

The first systematic attempts to integrate the Adams–Williamson equations were those of Bullen who used two criteria to obtain unique solutions. If some distribution of density is obtained in the mantle, the moment of inertia (C_m) and mass (M_m) of the mantle can be calculated. Given those of the Earth as a whole, those for the core follow:

$$M_c = M - M_m; \qquad C_c = C - C_m.$$

Hence the ratio $C_c/M_c a_c^2$ can be found.

An evident condition on C_m and M_m is that $C_c/M_c a_c^2$ should not exceed 0.4. That alone, however, is not sufficient, and to proceed further Bullen observed that the velocities α and β are related by simple algebraic formulae to K, μ and ρ, so that if α and β, or $d\alpha/dr$ and $d\beta/dr$ vary continuously with radius, so must ρ, $d\rho/dr$ and so on; equally, discontinuities in α, $d\alpha/dr$, would entail discontinuities in ρ, $d\rho/dr$, etc. Bullen also had to make some estimate of the density at the top of the upper mantle which of course is inaccessible to direct observation.

The outcome of Bullen's study was a set of models of his type A, of which some characteristic properties are shown in Table 2.2 and Figure 2.6. Bullen's next step came as a result of his observation that the bulk modulus appeared to undergo no change in passing from mantle to core despite the great increase in density. Indeed it seemed that to a good approximation K was a linear function of pressure throughout at least much of the lower mantle and the core. Bullen supposed that this was a general property of terrestrial matter at a sufficiently high pressure and used it as a condition to be satisfied by Earth models. Thus he produced his models of type B. Evidently since his incompressibility–pressure hypothesis arose from the properties of the models of type A, the difference between the models A and B can only be in detail and in such matters as the maximum central density of the inner core, which is rather uncertain.

Table 2.2. *An Earth model of type A* (Bullen, 1975)

Region	Depth (km)	ρ (kg/m³)	α (km/s)	β (km/s)	p (10^{11} Pa)	K (10^{11} Pa)
A	—	2840	6.30	3.55	—	0.65
	33	3320	7.75	4.35	0.009	1.15
B	413	3640	8.97	4.96	0.141	1.73
C	984	4550	11.42	6.35	0.379	3.49
D	2000	5110	12.79	6.92	0.87	5.10
	2898	5560	13.64	7.30	1.36	6.39
E	2898	9980	8.10	0	1.36	6.55
	4000	11 420	9.51	0	2.47	10.33
F	4980	12 170	10.44	0	3.20	13.26
	5120	12 250	9.40	—	3.28	—
G	5120	—	11.16	—	3.28	—
	6371	1251	11.31	—	3.61	—

The regions A–G are as labelled by Bullen (1975).

Figure 2.6. Properties of Bullen model A.

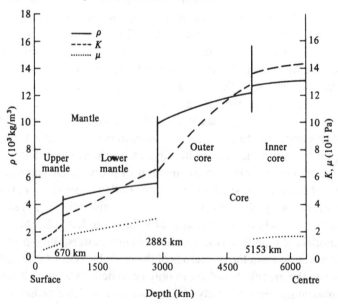

Matters of detail are not important in relating the structure of the planets to that of the Earth; it is important to concentrate on aspects of Bullen's models that might be relevant to other planets. In the first place we need to understand the nature of the major divisions of the Earth to see if similar divisions may occur in other planets; secondly we need to investigate the incompressibility–pressure hypothesis to see whether it may be applicable to conditions in other planets.

In the years since Bullen set up his models of type B, important new data have become available on which to base a discussion of those questions, namely the values of periods of free oscillations of the Earth. Following a great earthquake in Chile on 22 May 1960, it was found that some very sensitive detectors of long period had recorded trains of signals lasting for many hours which could be resolved into a number of components, each with a well-defined period from 50 min downwards. Almost at once the oscillations were identified as the free elastic oscillations of the Earth as a whole. Whereas the motion of a high frequency disturbance has a wavelength short compared with the radius of the Earth and propagates locally in the Earth almost as if there were not boundaries, at low frequencies the motion of the whole Earth is coherent and the frequency is determined by the elastic properties of the Earth and by the finite size of the Earth, because suitable boundary conditions (vanishing of stress) must be satisfied on the outer surface and on internal discontinuities. Since the free oscillations were first detected, they have been observed from subsequent large earthquakes and the theory has been developed in great detail.

Corresponding to the two types of body wave, there are two types of free oscillation: one, called spheroidal, in which there is a radial component of the displacement, and the other, called torsional, in which there is no radial component. The different modes of oscillation are distinguished by the letter S for spheroidal and T for torsional and by triplets of numbers (l, m, n), of which l denotes the number of nodal surfaces (spheres) within the Earth on which the motion vanishes, m denotes the number of nodal cones with axes on the polar axis and n the number of nodal meridional planes. Thus the disturbance is in general proportional to

$$f(r)P_m^n(\cos\theta) \begin{matrix} \cos \\ \sin \end{matrix} n\lambda$$

where θ is the co-latitude and λ the longitude.

For any realistic model of the Earth, the equations of motion must be solved numerically. However, it is possible to see from the form of the respective equations that the periods of the torsional modes depend on the distributions of ρ and μ within the Earth, as well as upon the radii of surfaces of discontinuity, whereas the periods of spheroidal modes depend in addition on the distribution of K.

The periods of over 1000 free modes of oscillation are now quite well established and constitute a body of data of great importance for the study of the interior of the Earth (Gilbert and Dziewonski, 1975). In principle it should be possible to set up models of the interior of the Earth which fit the observed periods and reproduce the observed mass and moment of inertia. The great advantage to be looked for in the use of free periods is that it should be possible to avoid the hypotheses that have to be introduced to obtain unique results from the Adams–Williamson procedure, and at the same time to dispense with the less accurate S-wave travel times. The problem of determining models from periods of free oscillation is the prime example, and one of the most complex, of the procedure known generally as 'inversion', whereby distributions of properties within the Earth are derived from quantities observed at the surface of the Earth. In general, the derived distributions are not unique, but it is often possible to establish ranges of uncertainties which depend both on the uncertainties of the data and the way in which the quantities actually observed cover the whole possible range. In practice, the inversion starts from an existing model and attempts to calculate improvements to it. Further, it is usual to include the travel times of P-waves in the data, for they are well established. Thus, most calculations using free oscillations have started from a model of Bullen's type, but it is important to note that the Bullen type of model is but a starting point and the calculation of the improvements from the free oscillations does not require adherence to any arbitrary constraints that may have been applied to obtain the initial model. There is one important restriction however. It is always supposed that the differences between the initial model and the improved model that is being sought are so small that all differences between the periods calculated for the two models are linear functions of the differences between the distributions of ρ, K and μ in the models. That limitation may restrict the range of models that can be reached starting from the chosen starting point, and so it is not clear that uncertainties calculated for that linearized procedure necessarily cover all possible models that would fit the data.

2.4   Models and composition of the interior of the Earth

It turns out that Bullen's models of type A yield periods of free oscillations in somewhat better agreement with the data than do the models of type B. For our present purpose, therefore, where we are not concerned with details of the structure of the Earth, it will suffice to set out the main zones of the Earth as they are in a model of type A, supplemented in some instances by data from free oscillations.

Roughly speaking, the Earth may be divided into five zones, the crust, the upper mantle, the lower mantle, the outer core and the inner core.

The crust is the result of tectonic activity over the lifetime of the Earth. Over the continents it is roughly 30 km thick, more under mountains, and comprises acidic and intermediate types of rock with sediments at the surface. The oceanic crust is not much more than 7 km thick, the lower 5 km being of basalt, the upper 2 km of oceanic sediments. Oceanic crust is formed at mid-oceanic ridges where new material comes to the surface; it is destroyed where it is driven against the boundary of a continent by the movement of tectonic plates and in the process mountains and new continental crust are formed. The crust as we see it on the Earth is almost certainly peculiar to the Earth, for it originates from tectonic activity that has no parallel on the Moon or the other planets.

The crust everywhere rests on material of higher density, about 3300 kg/m^3, and higher seismic velocities; the P-wave velocity is about 8 km/s. The boundary between crust and upper mantle is known as the *Mohorovičić discontinuity*. Below it, the upper mantle continues to a depth of almost 800 km. Throughout the first 400 km or so properties increase steadily inwards and then there is a sharp increase in gradients which manifests itself at the surface as the 20° discontinuity in the travel-time curves. Thereafter the gradients decrease to the rather low values they have in the lower mantle, which extends to the core and is characterized by steadily increasing velocities, density and elastic moduli. It is instructive to reduce the actual densities to values at zero pressure, which can be done because the bulk modulus is known. To a good approximation (Figure 2.7) the reduced density is constant in the upper part of the upper mantle to 400 km, increases in the transition zone between 400 and 700 km and thereafter remains constant to the top of the core. Within the core, the density and bulk modulus increase steadily, corresponding nearly to pure hydrostatic compression, until the boundary of the inner core, where the density and bulk modulus both increase discontinuously and the shear modulus, instead of being zero as in the outer core, becomes non-zero. The radius of the inner core appears to lie between 1215 and 1250 km.

How are the models, as distributions of density and elastic moduli, to be interpreted in terms of chemical composition? The first point to notice is that we know the pressure within the Earth, for, given the density as a function of radius, we can calculate the mass M_r out to a given radius, and hence the value of gravity at that radius, namely GM_r/r^2, and so, finally the pressure by integrating inwards from the surface, since

$$\frac{\mathrm{d}p}{\mathrm{d}r} = -g\rho.$$

Thus we have density, bulk modulus and pressure as functions of the radius, and so may express density and bulk modulus as functions of pressure. Now a great many observations of pressure–density relations for likely components of the Earth have been made (Chapter 4) and, from comparisons of those experimental data with properties of the Earth, the following identifications may be made with reasonable confidence.

The upper mantle consists mainly of silicates of iron and magnesium, having the olivine and pyroxene crystal structures in the outer 400 km.

Figure 2.7. Reduced density within the Earth: (*a*) actual density in the Earth; (*b*) density reduced to zero pressure. (From Cook, 1973.)

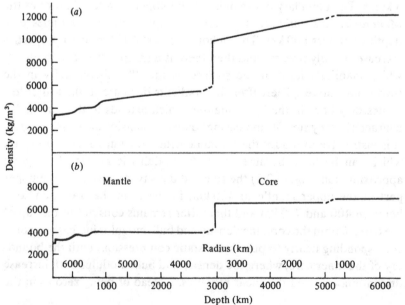

The increase in density in the lower part of the upper mantle between 400 and 1000 km is in part the consequence of a change in crystal structure from olivine or pyroxene to the more compact spinels. However, experiments indicate that change of crystal structure alone could only account for about half the increase of density and so it is supposed that the proportion of iron increases with depth in the lower mantle. Experiments also indicate that at sufficiently high pressures silicates transform into the constituent oxides; it is not clear where such a change occurs in the Earth but, while it is possible that the lower mantle is composed of iron-rich spinel throughout, it may be that the change to a mixture of iron, magnesium and silicon oxides takes place in the deepest part of the upper mantle.† It is not really possible to compare terrestrial data with experiment much more closely because the range of possible composition would make it difficult to distinguish between changes of composition and changes of crystal structure (see Chapter 4).

The composition of the core has long been debated. The pressure–density curve is well established and the question is how well can it be matched by experimental data? Iron has often been suggested as the material and the pressure–density curve for iron has been established from shock-wave studies (Chapter 4). The two curves lie nearly parallel, but that for the core lies somewhat below that for iron (Figure 2.8). To account for the difference some admixture of such materials as silicon, sulphur or oxygen (as oxide) is suggested. It may also be supposed that the iron contains about 10 per cent of nickel as usually accompanies it in the solar system; the density of nickel is, however, greater than that of iron so that if nickel is present rather more material of low density would be needed. An iron core and a silicate or oxide mantle are not the only possibilities for the major divisions of the Earth; for example, models dominated by iron and iron oxide have been suggested. It has also frequently been proposed that the core is a metallic form of the material of the mantle ionized under pressure. However (Chapter 4), as more information about such metallic transitions becomes available from theory and experiment it looks less and less likely that the corresponding change of density would be great enough.

The nature of the inner core is unclear. All we know with some assurance is that density, bulk modulus and shear modulus all increase at the boundary. A natural suggestion is that the molten material of the

---

† Dziewonski, Hales and Lapwood (1975) argue that the dependences of P-wave and S-wave velocities in the lower mantle are consistent only with an oxide composition.

outer core is solidified in the inner core because the rise of melting temperature as a consequence of elevated pressure exceeds the increase of actual temperature with decreasing radius. Unfortunately, we have no knowledge whether from theory or experiment about how the elastic properties of molten iron would behave on solidification at the pressure and temperatures of the interior of the Earth and so it is profitless to push further the question of the composition of the inner core. It is also of little value from the standpoint of the study of the other planets, for at the present time we have no prospect of obtaining any evidence for or against an inner core in any of them.

2.5   The incompressibility–pressure hypothesis within the Earth

The essence of Bullen's suggestion is that throughout the lower mantle and the core the bulk modulus, K, varies smoothly with pressure, p, irrespective of composition; it applies, as originally formulated, to the range 0.4–4×10^{11} Pa in pressure and to materials predominant in the Earth. The most striking evidence for the correctness of the hypothesis was the observation that, whereas the density changed from about 5600 to 10 000 kg/m^3 on crossing from the mantle into the core, any change in

Figure 2.8. Plot of the equation of state of the core compared with iron and similar materials.

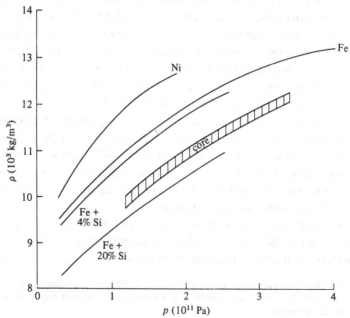

bulk modulus was within the uncertainty of the determination. Bullen's original proposal was based on P and S travel times interpreted by the Adams–Williamson procedure, but it is now possible to test the hypothesis against models based on P-wave travel times and periods of free oscillations interpreted by methods not restricted by the arbitrary assumptions required in the Adams–Williamson method. The models give both K and p as functions of radius, so it is straightforward to tabulate K as a function of p. The graph in Figure 2.9 shows the result of the comparison for two models.

It is clear from Figure 2.9 that Bullen's original proposition is still amply justified: any possible change of K from mantle to core is within the uncertainty of the determinations, namely, about 10^{10} Pa in K. Bearing in mind that any application of the hypothesis to another planet may not be correct in detail, the variation of K with p is well represented by the linear relation

$$K = 2.3 + 3.21p.$$

Figure 2.9. Dependence of bulk modulus on pressure.

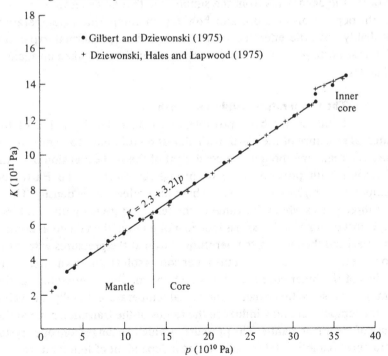

A quadratic relation has also been proposed, but it is clearly unnecessary.

In setting up the behaviour of terrestrial material as a reference for studying the interiors of planets, we may state Bullen's hypothesis in the form: the bulk modulus at terrestrial pressures of 0.4–4×10^{11} Pa is represented by

$$K = 2.3 + 3.21p$$

in units of 10^{11} Pa, to within $\pm 10^{10}$ Pa.

The limitations of the statement need emphasis, and indeed Bullen himself has emphasized them: the statement applies between pressures of 0.4×10^{11} and 4×10^{11} Pa and it applies to terrestrial material. The first restriction excludes, in its application to the Earth, the upper mantle and the inner core; the second might lead us to be cautious in extending the hypothesis to other planets. In the upper mantle, K still appears to change continuously with pressure, but does so much more rapidly than elsewhere within the Earth. As to the inner core, there is quite clear evidence that K does increase discontinuously on going from the outer to the inner core.

The behaviour of K in the upper mantle and inner core leads us to ask how far the departures from the simple rule that holds elsewhere in the Earth depend on pressure and how far on composition (temperature probably has little effect because the coefficient of thermal expansion decreases with pressure). The discussion of that issue is taken up again in Chapter 4.

2.6 The temperature within the Earth

It has so far been possible to conduct the discussion of the internal structure of the Earth with almost no reference to temperature, for, as already mentioned, the coefficient of thermal expansion of solids decreases with pressure so that throughout much of the Earth the temperature, high though it be, can have little effect on the density. There are three places where the value of the temperature is significant. First, and most obvious, if the core is indeed molten iron, then the temperature must exceed the melting temperature of iron at the pressures within the core. Secondly, it may be that the inner core is solid iron, in which case the radius of the inner core is that at which the melting temperature at the ambient pressure first exceeds the actual temperature. Finally, the value of the temperature may influence the radius of the transition zone at the base of the upper mantle. The pressure at which a change from one crystal structure to another takes place is not independent of temperature, for, associated with the change, as with melting or vaporization, there is a change of internal energy (or latent heat) and so the transition temperature and pressure are related by the Clausius–Clapeyron equation. For this reason it cannot be assumed that a similar transition will occur in other planets at the same pressure, even could it be supposed that the chemical compositions were the same.

The high temperature within the Earth is important for other reasons. When the mantle of the Earth is deformed relatively quickly, over times from about 1 s up to 1 y, it behaves as an almost perfectly elastic solid. Yet there is plain evidence that rocks at the surface have undergone plastic deformation by creep or flow, no doubt in very much longer times. Can it be that the mantle of the Earth would also flow if the temperature were high enough? The temperature must be high for two reasons. In the first place, the rate at which solids can creep under a steady stress is negligible at temperatures less than half the melting temperature but then increases towards the melting temperature. Secondly, the force which drives the movement may be the buoyancy force exerted on a body of material when the temperature gradient exceeds the adiabatic gradient. Thus, if movements in the mantle are to be invoked, for example, to account for the movements of tectonic plates, it is necessary for the temperature within the mantle to be close to the melting point of silicates or oxides.

Unfortunately, it is extremely difficult to obtain independent evidence of the distribution of temperature within the Earth. The rate at which heat flows out through the surface of the Earth has been measured in many places and the average value, much the same on the average for continents and oceans, is about 0.06 W/m^2. To calculate the temperature within the Earth it would be necessary to know how the effective thermal conductivity varies with depth, how sources of heat are distributed within the Earth, whether the temperature has settled to a steady value and, if not, what was the initial temperature throughout the Earth. Something is known of phonon conductivity (ordinary thermal conductivity) in silicates and something of the transport of heat by radiation, but for almost all the rest it is necessary to guess. A different approach is needed.

A starting point might be the temperature in the core. If the core is iron, its temperature must at least exceed the melting temperature at the local pressure. At the same time, convection in it will probably transport heat quite rapidly, so that the temperature will be close to the melting temperature throughout. If then we knew the melting temperature of iron as a function of pressure we would know the temperature throughout the outer core, while in the inner core it would be less than the melting temperature. It may be thought that such a model is on the whole a likely one for the core, but it cannot be proved to be correct. We do not know that the core is iron but, whatever it is, it is liquid and the temperature is probably close to the melting temperature. Even if the core is iron, we still do not know the temperature, for the melting temperature is not known at the core pressures. Empirical melting point formulae have been proposed which may apply to iron, but in the absence of experimental data we cannot be sure. Figure 2.10 does, however, show a plausible range of temperature within the core. On this basis, the temperature at the outer boundary of the core would lie between 3500 and 5000 K.

If such a value for the temperature of the outer boundary of the core is reasonable, we have values for the temperature of the base of the mantle and for the outer surface (290 K) and a simple calculation shows that, if the thermal conductivity of the mantle were uniform and if there were no heat sources within the mantle, the temperature gradient at the surface would be 0.5 deg/km. The actual value is close to 30 deg/km. Evidently the effective conductivity of the mantle is much greater than the conductivity of rocks at the surface, or there are sources of heat within the

Figure 2.10. Possible temperature within the Earth.

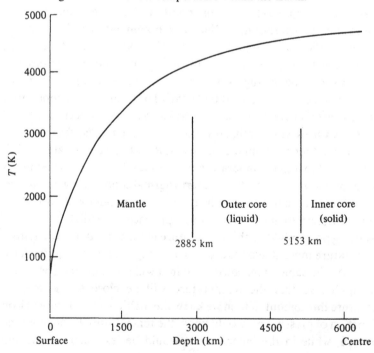

mantle, or both. We know, in fact, that there are sources of heat close to the surface in the radioactive uranium, potassium and thorium in rocks, mostly granite, and indeed most of the heat flowing out from the surface could be accounted for by the radioactivity in the continental, though not in the oceanic, crust. No doubt the upper mantle also contains sources of radioactive heat. On the other hand, it is known that olivines are semi-transparent to low frequency infra-red radiation and at moderate depths within the mantle radiation transport is probably more effective than phonon conduction. Can we set further limits? Is it possible to show that radiation transport is inadequate to move heat from the core? If it is, a persuasive argument can be made for convection in the mantle, for in that case the temperature in much of the mantle would approach the temperature of the boundary between the core and the mantle and in consequence the gradient in the outer parts would exceed the adiabatic gradient, so giving rise to buoyancy forces, while the high temperature would afford the possibility of creep. Thus some sort of convective motion might be expected in the parts of the mantle where both the temperature and the gradient are high; in the lower part of the mantle, on this picture, the gradient might be less than the adiabatic gradient. The argument is plausible but it cannot be taken much further partly because we do not know if radiative transport is inadequate, partly because we do not know if creep in a quasi-solid material would indeed give rise to convection. The possibility is there; it may provide a driving force for tectonic movements at the surface and it may need to be considered in other planets.

2.7   The electrical conductivity within the Earth

The magnetic field of the Earth is now usually supposed to arise from dynamo action in the liquid core; in consequence the conductivity of the core must be high enough to allow electrical currents to flow. It is generally supposed that if the core is mainly iron its conductivity, while not known for certain, will be within the right range, and one may suppose also that, if any other planet has a liquid core like that of the Earth, it will similarly have a conductivity that would permit of dynamo action.

The conductivity of iron is a function of temperature and pressure and it is because experiments have not been carried out at pressures and temperatures close enough to those of the core that we do not have close estimates of the conductivity. The electrical conductivity of the material of the mantle can perhaps be even less well estimated from experiment. Silicates at the temperature and pressure of the mantle develop conductivity of the sort shown by semi-conductors; that is to say, electrons can pass from filled valence bands or from impurity levels to bands empty at room temperature, in which the electrons are then mobile. The dependence of the conductivity so developed upon temperature is of the general form

$$\sigma = \sigma_0 \exp{(-E/kT)},$$

where σ_0 depends on the number density of electrons in the impurity or valence levels and E is an activation energy (gap energy). The conductivity depends on pressure through changes in the gap energy, E.

For olivine

$$\sigma_0 = 10 \text{ S/m} \qquad \text{intrinsic}$$
$$\sigma_0 = 10^{-2}\text{--}10^{-4} \text{ S/m} \quad \text{impurity}$$
$$E = 3 \text{ eV} \qquad \text{intrinsic}$$
$$E = 0.5\text{--}1.0 \text{ eV} \qquad \text{impurity}$$

(the conductivity of iron at room temperature is 10^7 S/m; of germanium and silicon, about 10^3 S/m).

Experiments have not been carried to a high enough temperature or pressure to explore the behaviour of silicates under conditions that obtain deep in the mantle; furthermore, the conductivity depends strongly on composition through the gap energy; thus it would be hopeless to try to predict in detail the conductivity throughout the mantle, but, on the other hand, it is possible to predict that it should increase greatly with temperature and, therefore, with depth.

It is then of interest to obtain some estimate of the conductivity within the mantle, for if a similar estimate were available for the Moon or a planet supposed to have a mantle of similar composition it might be possible, by comparison with the Earth, to arrive at some idea of the temperature in the Moon or planet. Disturbances of the magnetic field at the surface of the Earth provide the means for estimating the electrical conductivity. The surface field has its origin principally within the Earth, but a small variable part is generated by currents in the ionosphere and magnetosphere; the field of external origin can be separated from that of internal origin by analysis into spherical harmonics. Suppose that separation effected: then, if the mantle conducts electricity, a component of the external field will induce currents in the mantle which will then generate a component of the internal field, varying in time in the same way as the external component. The relation between the external component and the internal component induced by it depends on the variation of conductivity with depth. Once again we meet an inverse problem in geophysics. The direct problem and external component from the distribution of conductivity with depth can always be solved numerically; given a number of such relations determined experimentally for components with different rates of change, the inverse problem is to find distributions of conductivity with depth to reproduce the data. Again there is no unique solution, but Figure 2.11 shows the range within which the conductivity probably lies within the upper mantle.

2.8   Departures from the hydrostatic state

The Earth is very nearly in a hydrostatic state; that is to say, the stresses are almost everywhere due to normal pressures, and shear

Figure 2.11. Electrical conductivity in upper mantle. (From Cook, 1973.)

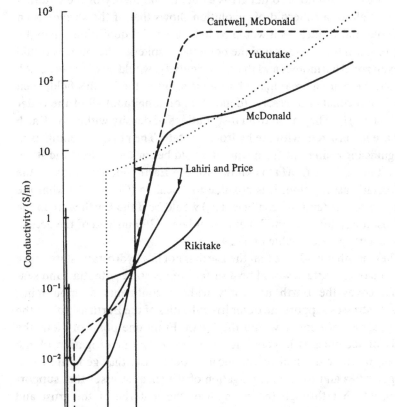

stresses are generally negligible. The hydrostatic assumption is made in the Adams–Williamson procedure, but not when deriving Earth models by the inversion of data from free oscillations, and the models that have been constructed in that way do not depend on assuming hydrostatic conditions in the Earth. However, when for the other planets far less information is available than for the Earth, hydrostatic conditions must be supposed in order to make much progress in interpretation. Thus, whereas the moment of inertia of the Earth can be found without having to make any assumption, nothing equivalent to precession can be observed for any of the other planets, and the moment has to be estimated from the gravitational coefficient J_2. To do so, it is necessary to assume hydrostatic conditions (Chapter 3). It is therefore of some importance to see how far the Earth departs from the hydrostatic state.

That the Earth does so depart is evident. The theory of the external gravity field of a spheroid of revolution shows that, if the surface is an equipotential surface, as it would be in hydrostatic equilibrium, then the only terms in the expansion of the potential in spherical harmonics would be even zonal harmonics and the coefficients J_{2n} would be of the order J_2^n. In fact, there are a great many harmonic terms in the Earth's field, even zonal, odd zonal and tesseral (dependent on longitude) all of the order $(10^{-3}-10^{-4})J_2$. They arise from distributions of density within the Earth that are inconsistent with the hydrostatic state. The polar flattening itself is significantly different from what it would be for the hydrostatic state. Since the value of C/Ma^2 is known for the Earth independently of the hydrostatic assumption, it is possible to calculate (Chapter 3) what the polar flattening (and the coefficient J_2) would be if the Earth were in the hydrostatic state; it is found that f would be $1/299$ instead of the actual, significantly greater, value of $1/298.25$.

There is other evidence that the Earth is not in a hydrostatic state: were it so the outer surface would be a surface of constant potential, and seas would cover the Earth uniformly without continents or mountains. Clearly stresses support the outer irregularities of the Earth as well as the irregularities of density within the Earth. From one point of view, the study of tectonics is largely encompassed by the investigation of the forces, whether thermal or chemical or otherwise, that generate those irregularities and by the investigation of the shear stresses that support them, whether through the strength of the material of the crust and mantle or through convective motions within the mantle. The question will be taken up again in discussing possible tectonic activity on the Moon or the terrestrial planets; so far as concerns the Earth two things may be said. First, it is evident that movements do take place within the Earth and lead to tectonic activity in the crust and uppermost parts of the mantle and to the development of oceans, continents and mountains. Secondly, these movements are probably generated by thermal forces: they could not occur in a hydrostatic Earth for there would be no forces to drive them. Thus departures from hydrostatic equilibrium represent both the sources of energy that drive tectonic movements and the irregularities produced by those movements.

We do not know how to separate the effects of the sources and consequences upon the gravitational potential. Any particular disturbance of gravity comes from an irregularity of density. This may be due to an abnormality of temperature, possibly driving a convective cell acting on a tectonic plate, or it may be due to some irregularity supported by the strength of the mantle or crustal material in shear. Near the surface one would suppose that strength in shear is more important in supporting departures from the hydrostatic state, but, at the higher temperatures at depth, strength may fail and irregularities may be supported dynamically by the motions of material of the mantle. Thus one might argue that irregularities of density at small depths are supported by shear strength, and those at greater depths are the consequences of convective currents. But we cannot distinguish between them in practice because of an inherent ambiguity in deriving a distribution of density from a potential: the density variations may be put at any depth less than a maximum set by the order of harmonic variation and in practice the limits set for the Earth are so wide as to be quite unhelpful in the present matter.

In studying the Moon and the inner planets, we shall again have to ask whether departures from the hydrostatic state are supported statically by shear strength or dynamically by internal motions.

2.9   The magnetic field of the Earth

The Earth has a magnetic field which to first approximation is that of a dipole the centre of which is slightly displaced from the centre of the Earth and which is inclined to the axis of rotation at an angle of some 10 ° so that the magnetic north pole lies not at the geographical pole but in Canada. The field is not, however, adequately represented by that simple description and many relatively large terms are needed in a spherical harmonic representation of the field. Table 2.3(a) lists some of the larger coefficients. Alternatively, the non-dipole part of the field may be represented by dipoles (or loops of current) in the surface of the core, and

Table 2.3(*a*). *Some harmonic coefficients of the geomagnetic field*

n	m	g_{ni}^m (T)	h_{ni}^m (T)
1	0	-3.04×10^{-5}	—
2	0	-0.16×10^{-5}	—
3	0	0.13×10^{-5}	—
4	0	0.10×10^{-5}	—
5	0	-0.02×10^{-5}	—
6	0	0.14×10^{-5}	—
1	1	-0.021×10^{-5}	0.58×10^{-5}
2	1	0.30×10^{-5}	-0.19×10^{-5}
3	1	-0.20×10^{-5}	-0.05×10^{-5}
4	1	0.08×10^{-5}	0.02×10^{-5}
2	2	0.15×10^{-5}	0.02×10^{-5}
3	2	0.13×10^{-5}	0.02×10^{-5}
4	2	0.06×10^{-5}	-0.03×10^{-5}
3	3	0.08×10^{-5}	—
4	3	-0.03×10^{-5}	—
6	3	-0.02×10^{-5}	—

Note: g_{ni}^m and h_{ni}^m are respectively the coefficients of $P_n^m (\cos \theta) \cos m\lambda$ and $P_n^m (\cos \theta) \sin m\lambda$ in the spherical harmonic expansion. The integral of any harmonic over the unit sphere is $(2n + 1)^{-1/2}$.

Table 2.3(*b*). *Representation of the geomagnetic field by radial dipoles*

	M/a^3 (T)	Co-latitude (deg)	Longitude (deg E)
Central dipole	-7.0×10^{-5}	23.6	208.3
Radial dipoles at 0.25a	1.0×10^{-5}	13.7	341.9
	1.1×10^{-5}	46.0	179.9
	-0.3×10^{-5}	54.9	40.1
	0.8×10^{-5}	77.4	241.7
	0.3×10^{-5}	91.3	120.8
	-0.7×10^{-5}	139.8	319.3
	-1.2×10^{-5}	141.1	43.0
	0.4×10^{-5}	102.9	180.1

a is the radius of the Earth.
The unit of magnetic moment is T m^3, so that M/a^3, the moment divided by a^3, is in tesla.

Table 2.3(*b*) gives the principal ones. The dipole field and the main non-dipole components may be shown to originate from currents within the Earth.

The main internal fields are not constant: they vary with a secular variation which may be represented by the growth and decay of loops of current in the surface of the core.

Beyond the relatively small secular variation, it is known from the study of the magnetization of rocks that the direction of the main dipole field has reversed on many occasions in geological history so that what is now the north pole was on many occasions the south pole. The frequency of reversals has varied in geological history; in the relatively recent past it has been every few million years, the change occurring relatively rapidly in some 10 000 years. At other times in the past, tens of millions of years or more have passed without change.

In addition to the fields of internal origin there are fields that vary in time and are produced by currents in the ionosphere and magnetosphere. They have already been mentioned in connexion with the estimation of the electrical conductivity of the Earth. The magnetosphere and ionosphere are both controlled by the main field of internal origin. The magnetosphere is that region around the Earth where the field is essentially that of the geocentric dipole; outside the magnetosphere the field transported by the solar wind is dominant and the magnetopause which bounds the magnetosphere is the region where the field lines of the Earth's field are deflected by the field and ions are transported by the solar wind. The magnetopause moves as the strength of the solar wind varies as a result of currents flowing in it and generating magnetic disturbances at the Earth. The position of the ionosphere around the Earth is determined by the depth to which ionizing radiation from the Sun can penetrate in the atmosphere. The actual structure and the variation from day to night are, however, influenced by the main field, for the ionized particles of the ionosphere tend to move along field lines.

The origin of the main field is now almost always sought in dynamo action in the core of the Earth (see Moffat, 1978, and Chapter 9). The general idea is that motions in the fluid electrically conducting core, taking place in a magnetic field, induce electrical currents which then generate a magnetic field. If the fluid motions are of the proper sort, and if the conductivity and viscosity of the core are suitable, the generated field will reinforce the original one and so a field will be maintained by electromagnetic induction in the core. The theoretical problem is to find motions which can maintain a field in this way, and a difficulty that has to be met is that it is known that self-induction in a fluid will not maintain a field if the motions have axial symmetry. What drives the motions in the core of the Earth and are they suitably asymmetrical? Those questions are not answered, although some plausible ideas of answers can be developed. It is usually supposed that the relatively high rate of spin of the Earth is important, but a spinning core by itself would not generate a field; some other components of motion must be present and have been sought in thermal convection within the core. It has, however, been argued that the core is stably stratified and does not convect, although the argument has been strongly criticized; yet, if the core is stable to convection, internal hydrodynamic waves may travel in it, the motions of which may generate a field. Apart from thermal convection it has been suggested that mechanical coupling between the mantle and core would generate internal motions that could induce a field.

There is no agreement at present on the details of the dynamo process nor on the source of energy by which it is driven. No doubt, in looking at fields of other planets, the obvious first idea would be to think in terms of thermal convection in a spinning core as responsible for the field, but there are other possibilities.

2.10 Conclusion

The study of the planets must start with the Earth. The available methods are there most fully worked out, and the ideas of constitution we shall use were developed from terrestrial studies. What themes should we bear in mind as we leave the security of the Earth for the vagueness of the planets, about which we are so ignorant?

One important point concerns the roles of the polar moment of inertia and a magnetic field relative to the internal constitution. The former shows us how strongly the density increases towards the centre, while the latter, if present, will strongly suggest, on the basis of a dynamo theory of its origin, that part of the planet is liquid and conducts electricity.

We may ask if the planet is in a hydrostatic state or not. Departures will show up as topographic irregularities at the surface and in the existence of significant coefficients in the spherical harmonic expansion of the potential beyond the second zonal coefficient, the one that corresponds to the polar flattening. Departures from the hydrostatic state may be relics of irregularities in the initial formation of a planet or they may be the result of a continuing process: we suppose the latter to be the case on the Earth, while, for the Moon, the former may be more probable. The Earth warns us not to take too simplistic a view: the obvious surface irregularities of the Earth, the mountains, oceans and continents, which are the results of tectonic procesees, affect the potential but slightly because of the maintenance of isostatic equilibrium, while large scale features of the non-hydrostatic potential appear to bear little relation to present tectonic processes.

If a planet is in hydrostatic equilibrium, the coefficient J_2 in the spherical harmonic expansion of the potential is determined by the polar moment of inertia, and the latter may be inferred from the former, given the spin angular velocity (Chapter 3). If hydrostatic equilibrium fails, the moment of inertia derived in this way will be in error and that must be considered for the terrestrial planets.

Temperature has a relatively small effect on the density at a given pressure because the coefficient of thermal expansion in general decreases very greatly at high pressures. However, the pressure at which a phase change occurs, whether it be melting or a change of crystal structure, is dependent on pressure through the Clausius–Clapeyron relation, so that it is wrong to assume that phase changes, if they exist, will occur in other planets at the same pressure as within the Earth.

Important though temperature is in this way, we have seen how difficult it is to arrive at any firm idea of its variation through the Earth; how much less then can these estimates be made for other planets?

The chemical constituents of the Earth have been identified as metal silicates and oxides in the mantle and iron in the core. On that basis, the gross average composition of the Earth may be compared with that of chondritic meteorites, often taken to be typical of the composition of the heavier elements in the solar system, i.e. excluding the more fugitive gases (see Table 1.3). Although in gross composition the Earth appears to fit its chondritic pattern it is clear that it is unusual among the planets, for the substantial differences between the bulk densities (Chapter 6) show that neither the terrestrial planets nor still less the major outer planets can have a common composition.

The Earth shows upon its surface the results of tectonic activity. We relate that activity to forces that probably derive from the heat flowing out of the Earth, and tectonic activity may be taken as evidence of an internal temperature high enough to permit movement of solids by creep or otherwise. They also show that a planet is not in hydrostatic equilibrium. Evidence for surface structure and tectonic activity on the inner planets is now accumulating from photographs obtained by space craft and must be borne in mind in discussing the internal states.

The distribution of density within the Earth is found from seismic data, no assumption being made about composition. Because seismic data are not available for the other planets, a model of the internal structure can only be obtained by taking an assumed equation of state and using it to construct a variation of density with radius that fits the known size, mass and moment of inertia; the procedure will be discussed further in Chapter 4. How then can we obtain equations of state, especially for the inner planets? How far do they depend on composition, can anything useful be said about equations of state if the composition is not known? We have seen that in the Earth a very simple rule relates bulk modulus and pressure to high accuracy, and we shall have to consider if that can be extended to the inner planets.

3

Methods for the determination of the dynamical properties of planets

3.1   Introduction

Leaving aside the special case of the Moon, the properties of planets that can at present be determined are certain gross quantities descriptive of a planet as a whole; these are the size, the spin angular velocity, the mass and mean density, the moments of inertia and the coefficients in a spherical harmonic expansion of the gravitational potential, together with some features of the magnetic field and possibly electromagnetic induction in the planet. The Moon alone is open to the study of the variation of properties with depth by seismology. The investigation of the internal state of a planet depends on what can be inferred from the measured gross properties, and fails unless those properties can be measured with precision. Given only integral properties, a wide range of internal distributions of density is consistent with the data, but the more precisely the integral properties are known the more restricted the range of possible distributions.

The various dynamical properties of a planet are not independent, for all are determined by three factors: the spin, the chemical composition and the temperature. Suppose the spin acceleration at the surface at the equator (where it has its greatest value) to be small compared with the acceleration of the self-gravitational attraction. Then composition and temperature together determine the equation of state, the latter mainly by its control of the occurrence of any polymorphic phase changes. Given an equation of state, in general different in different zones, the radius is determined for a given mass, and vice versa, and so is the variation of density with radius, and thus the moment of inertia. The moment of inertia and the spin determine the polar flattening (section 3.7). The composition, spin and temperature accordingly fix the moment of inertia, the size and the polar flattening if the mass is given. The converse is not, however, true, and the equation of state cannot be found uniquely from

51

mass, size, spin, flattening and moment of inertia, the properties which can be measured or estimated for most planets. There is an important limitation which severely restricts discussion of Venus and Mercury: if the spin is negligible the flattening may be indetectable, and the moment of inertia cannot thus be determined.

Consider a planet in hydrostatic equilibrium spinning with angular velocity ϖ. Then, as is shown in section 3.7, the value of the polar flattening f is given by

$$\frac{5}{2}\frac{m}{f} = 1 + \frac{25}{4}\left(1 - \frac{3}{2}\frac{C}{Ma^2}\right)^2,$$

where m is the ratio of centrifugal to gravitational acceleration at the equator, equal to $\varpi^2 a^3/GM$, M is the mass of the planet, a is the equatorial radius, and C is the polar moment of inertia.

The value of J_2 is related to the polar flattening. Let the potential of the gravitational attraction of the mass of the planet be denoted by V and let it be expressed in spherical harmonics. Because the planet is in hydrostatic equilibrium and is spinning, it will be symmetrical about the polar (spin) axis and so the potential (and all the other quantities) depend only on the radial distance, r, and the co-latitude, θ, measured from the north pole, and are independent of longitude. Further, if the centre of co-ordinates is taken to be the centre of mass, there is no first harmonic in the expansion of the potential, which may then be written as

$$V = -\frac{GM}{r}\left[1 - \left(\frac{r_0}{r}\right)^2 J_2 P_2(\cos\theta) + \cdots\right].$$

r_0 is for the moment an unspecified scale factor.

To this must be added the potential from which the spin acceleration may be derived. The radial and meridional tangential components of the acceleration are $r\varpi^2 \sin^2\theta$ and $r\varpi^2 \sin\theta\cos\theta$. They are equal to

$$\left(-\frac{\partial}{\partial r}, -\frac{1}{r}\frac{\partial}{\partial\theta}\right)(\tfrac{1}{2}r^2\varpi^2 \sin^2\theta),$$

showing that the spin acceleration may be regarded as deriving from a potential

$$-\tfrac{1}{2}r^2\varpi^2 \sin^2\theta.$$

Thus the total potential V, or the *geopotential* as it is called, is equal to

$$V - \tfrac{1}{2}r^2\varpi^2 \sin^2\theta.$$

Now the surface of a rotating body in hydrostatic equilibrium is one in which the potential V is constant.

Let that surface have the form

$$r_s = a(1 - f \cos^2 \theta)$$

or

$$r_s = a[1 - \tfrac{1}{3}f - \tfrac{2}{3}fP_2(\cos \theta) + \cdots],$$

that is, it is a spheroid of revolution and the polar flattening, the relative difference between equatorial and polar radii, is f.

On substituting for r_s in the expression for V, and remembering that $\sin^2 \theta = \tfrac{2}{3}[1 - P_2 \cos \theta)]$, we find that the part independent of θ is

$$-\frac{GM}{a}(1 + \tfrac{1}{3}f) - \tfrac{1}{3}a^2\varpi^2,$$

while the coefficient of $P_2(\cos \theta)$ is

$$-\tfrac{2}{3}f\frac{GM}{a} + \frac{GMr_0^2}{a^3}J_2 + \tfrac{1}{3}a^2\varpi^2.$$

If the potential is to be a constant, independent of θ, this coefficient must be zero.

It is usual to choose the arbitrary scale factor r_0 to be a. As before, write m for $a^3\varpi^2/GM$, the ratio of spin to gravitational acceleration at the equator. Then

$$f = \tfrac{3}{2}J_2 + \tfrac{1}{2}m.$$

The foregoing result applies to the physical surface of a planet in hydrostatic equilibrium, but is more general. It applies to any surface on which the potential is constant; thus, on the Earth, the form of the sea level surface is so related to J_2, and even though the land surface is irregular, and not an equipotential surface, it is still possible in principle to define a surface (the *geoid*) which continues the sea level equipotential surface below the land. When a planet is not in hydrostatic equilibrium, the geometrical flattening of the physical surface does not necessarily agree with the flattening calculated from J_2 and m. Thus, evidence about the internal state of the planet may be obtained from a comparison of the geometrical with the dynamical flattening derived from J_2 and m; those for the Moon and Mars in particular are discrepant.

3.2 Distance and size

The sizes of the major planets have still to be found by multiplying the angular diameter as found from telescopic measurements by the known distance from the Earth. It is therefore worth recalling that the relative distances of the planets follow from Kepler's law

$$n^2 a_p^3 = GM_\odot,$$

where n is the mean orbital angular velocity, a_p the semi-major axis of the orbit, and $M_\odot$ the mass of the Sun; that the ratios of the angular velocities are very accurately known and so therefore are the ratios of semi-major axes; and, finally, that the scale of the whole system is found from radar measurements of distances from the Earth to the inner planets. The scale is expressed by the accepted value of the semi-major axis of the Earth's orbit, the astronomical unit, which by international agreement is taken to be $1.496\,00 \times 10^8$ km.

Telescopic measurements of angular diameters may seem straight-forward, and so they are for Jupiter and Saturn, but the angular diameters of Uranus, Neptune and especially Pluto, are small and, as may be seen from Table 3.1, are none too well known. (It should be recalled that a relative error ε in the linear dimensions generates a relative error 3ε in the volume and mean density.)

Telescopic observations will also give the geometrical flattening, again subject to appreciable errors with small planets. A relative error of ε in each of the polar and equatorial diameters generates a relative error of 2ε in the flattening.

The form of the planet observed telescopically is that of its projection on a plane perpendicular to its radius vector from the Earth, and so its

Table 3.1. *Some planetary properties*

Planet	Distance from Sun (AU)[a]	Equatorial Radius (km)	Equatorial Diameter[b] (arc sec)	Orientation of spin axis to orbital plane	Spin angular velocity (rad/s)	Polar flattening f
Mercury	0.387	2439	5.4	10°	1.22×10^{-6}	0
Venus	0.723	6052	30.5	3°	-2.99×10^{-7}	0
Earth	1	6378	—	23°27'	7.29×10^{-5}	0.003
Mars	1.524	3397	8.9	23°59'	7.09×10^{-5}	0.005
Jupiter	5.203	71 200	23.4	3°04'	1.77×10^{-4}	0.06
Saturn	9.539	60 000	9.8	26°44'	1.71×10^{-4}	0.10
Uranus	19.18	25 650	1.8	82°5'	8×10^{-5}	0.02
Neptune	30.06	24 800	1.1	28°48'	8×10^{-5}	0.02
Pluto	39.44	1500	0.2	unknown	1×10^{-5}	unknown

[a] 1 AU $= 1.496 \times 10^8$ km.
[b] The equatorial diameter in angular measure is the greatest value as seen from the Earth.

true polar flattening can only be calculated if the inclination of the polar axis to the plane is known. For most planets there is little difficulty because the spin axes are all roughly parallel and perpendicular to the ecliptic (Table 3.1), but Uranus is an exception, for its spin axis lies almost in the plane of the ecliptic.

The size of Saturn, as of many natural satellites, has been obtained from pictures taken from space craft.

A powerful method for the determination of the size and flattening of a planet makes use of occultations; the most accurate results are obtained from occultations of a space probe or artificial satellite, but occultations of stars also provide valuable data for planets inaccessible to space probes or unprovided with artificial satellites. Suppose that the position of an occulted object is known, relative to the planet, as a function of time. The simplest case is that of an artificial satellite in orbit about the planet and tracked by measurement of the Doppler shift of radio transmissions that it radiates. The size of the orbit can readily be found if the plane of the orbit contains the Earth. Let the maximum and minimum velocities along the line of sight then be v_1 and v_2, so that the range of tangential velocities is $(v_1 - v_2)$. In a circular orbit the tangential velocity would be constant and equal to $\frac{1}{2}(v_1 - v_2)$.

The period, T, is the interval between successive instants at which the minimum velocity recurs. Hence the mean angular velocity, n, is $2\pi/T$, and the radius of a circular orbit is

$$\tfrac{1}{2}(v_1 - v_2)/n$$

or

$$(1/4\pi)(v_1 - v_2)T.$$

Artificial satellites often have nearly circular orbits but, if they deviate, the eccentricity and time of pericentre can be derived from a more detailed study of the line-of-sight velocity as a function of time.

Doppler tracking thus provides the scale of the orbit. Now, since the plane of the orbit of the satellite contains the centre of mass of the planet, the satellite, if its orbit also contains the Earth, must pass behind the planet as seen from the Earth and so be occulted. The diameter of the circle in which the planet intersects the plane of the orbit (necessarily a great circle of the planet) may then be found from the times at which the satellite disappears behind the planet, cutting off its radio signals, and at which it reappears and the signals are heard again.

The case of a space probe is similar, save that only one disappearance and reappearance may take place instead of a sequence. The trajectories of space probes lie for the most part in the plane of the Earth's orbit about the Sun so that the occultation of a space probe by a planet will in the simplest case give the diameter of a section by the ecliptic. Doppler tracking of a space probe gives its line-of-sight velocity, but the complete trajectory may be found by integrating the equations of motion of the space probe in the gravitational field of the Sun and the planets, while ensuring that the solutions fit the measured line-of-sight velocity. Thus the tangential velocity of the space probe may be found and so the ecliptic diameter from the times of disappearance and reappearance.

Occultations of stars are slightly different in that the planet may be thought of as moving across a fixed field of stars instead of remaining stationary, whilst a space probe or satellite goes behind it. Times of disappearance and reappearance of stars are measured photoelectrically.

Stars, in general, and space probes, in some instances, do not pass along a diameter of the projected disc of the planet, but if the distance of the path from the centre of the disc is known, a diameter may be found.

Given a number of occultations, it may be possible to determine a geometrical flattening as well as an equatorial diameter. Details will not be pursued further, since the aim of this chapter, here and subsequently, is to set out principles and not to be a detailed treatise, but examples will appear in the discussion of individual planets.

3.3   Spin

It would seem a straightforward matter to determine the period of rotation of a planet from observations of the passage of distinctive features across the disc, but, in fact, the spin of only one planet – Mars – can be found unambiguously in that way. The innermost planets – Mercury and Venus – present difficulties because each of them rotates very slowly, while the solid surface of Venus is obscured by a thick atmosphere. The spins of each have been derived from the Doppler shifts of radar returns, but for some time the results were in doubt because the effects to be observed are very small.

The spin period obtained for Mercury from radar observations (Goldstein, 1971) has been confirmed by photographs taken from Mariner 9 as it passed the planet (Klaasen, 1975) as well as by photography from telescopes on the Earth (Murray, Dollfus and Smith, 1972).

The period of rotation of Mercury (58.6 d) is two-thirds of its period about the Sun. Venus rotates with a period of 243.09 d (Shapiro, 1967) such that it rotates four times with respect to the Earth between every inferior conjunction with the Earth.

These resonant rotations of Mercury and Venus are remarkable because the values of J_2 for each, to which torques coupling them to the Sun or Earth would be proportional, are known to be less than 10^{-5}.

There is no difficulty in observing the motions of surface markings on Jupiter and Saturn, but the surfaces are not solid, nor no doubt are the interior, and so the significance of the observed surface spin must be open to some doubt. Uranus and Neptune are too distant for reliable observations to be made of surface markings, but it is possible to obtain the spins from the Doppler shifts of spectrum lines in light reflected from the surface. Pluto is too distant for the spin to be determined.

3.4 Masses

The masses of planets are found from the gravitational attraction they exert upon other bodies. The actual measurements are of accelerations, a, and distance, d; since the acceleration of a test body is given by

$$a = -GM/d^2,$$

it is the product GM that is determined and not M itself in kilogrammes. The point has already been made in Chapter 2. In consequence, the absolute masses of the planets are known only with the very low precision (a few parts in 1000) with which G is known, whereas the relative masses are for the most part known much more precisely, as may be seen from the subsequent discussion of the separate planets.

If a planet has satellites, whether natural or artificial, it is straightforward to obtain a well-determined mass. By Kepler's law

$$GM = n^2 a^3,$$

where n is the mean orbital angular velocity of the satellite and a the semi-major axis of its orbit. Both n and a may be obtained from telescopic observation and in that way the masses of Mars, Jupiter, Saturn and Uranus have long been determined, for those planets all have two or more natural satellites.

If the satellite is artificial and emits radio transmissions, the Doppler shifts may be used directly to determine GM. As has already been seen, the tangential velocity in a circular orbit in a plane containing the Earth is $\frac{1}{2}(v_1 - v_2)$, where v_1 and v_2 are the greatest and least velocities in the line of sight. But the tangential velocity is an, while n is $2\pi/T$, where T is the period found from the recurrence interval of maximum or minimum velocity. Accordingly,

$$GM = \frac{T}{16\pi}(v_1 - v_2)^3.$$

Artificial satellites have been used to obtain the masses of the Moon and Mars but not so far of other planets.

While satellites afford the possibility of observations continued over a long time, and so may give the most accurate results, masses may also be derived from the effects on the motions of other planets or of space probes. Consider, as the simplest case, a space probe travelling towards a planet along the line joining the Earth to the planet. When the space probe is at a distance x from the planet, its velocity is $\dot{x}$, and this is measured by the Doppler shift of radio transmissions from the space probe and is known as a function of time; x is not known.

The acceleration, $\ddot{x}$, is given by

$$\ddot{x} = -\frac{GM}{x^2},$$

the integral of which is

$$\dot{x} = (GM)^{1/2}/x^{1/2}.$$

Hence $x^{3/2} = \frac{3}{2}(GM)^{1/2}(t - t_0)$, where t_0 is a constant of integration.

It follows that

$$\dot{x} = (GM)^{1/3}[3(t - t_0)/2]^{-1/3},$$

from which, knowing $\dot{x}$ as a function of time, GM may be found.

In practice, space probes do not travel along straight lines between Earth and planet, so that solving the equations of motion and fitting the observations to theory is somewhat more complex than in the idealized case.

The masses of Venus and Mercury and of Io, a satellite of Jupiter, have now been derived from observation of space craft.

Planets, to a first approximation, move in elliptical orbits about the Sun, but each attracts the others and all orbits are perturbed to a greater or lesser degree by those attractions; observations of the perturbations should enable the relative masses of all planets to be estimated. Thus indeed were the masses of Venus and Mercury estimated until space probes approached them and so were Neptune and Pluto discovered.

The mass of the Moon is well determined from artificial satellites, but in addition a method peculiar to the Moon gives an accurate value. The Moon and the Earth each move in orbits about their common centre of mass, the diameter of the Earth's orbit being m/M times that of the Moon's, where m is the mass of the Moon and M that of the Earth. The ratio is about $1/81$. Let d be the diameter of the Moon's orbit and D the distance of the Earth and Moon from the Sun. Then the orbit of the Earth subtends an angle of md/MD at the Sun; at the fixed stars the angle subtended is negligible, and so, as the Moon and Earth go round about each other once a month, the position of the Sun relative to the stars appears to vary through the angle md/MD. The angle is very small, about 6 arc sec, and the value of m/M found from it was not very accurate. With the advent of space probes more precise measurements can be made. Consider a space probe moving away from the Earth; the velocity of the Earth in its orbit tangential to the line of sight of the space probe will vary between $+n_m md/M$ and $-n_m md/M$, where n_m is the mean angular velocity of the Moon and the Earth in their respective orbits about their common centre. Thus the velocity of the space probe derived from Doppler shifts of radio transmission received 'from it on Earth will fluctuate with a monthly period and amplitude of $n_m md/M$. The ratio m/M found from such fluctuations agrees well with the value found from artificial satellites of the Moon.

3.5 The gravity fields of planets

In the free space outside a planet, the gravitational potential, V, satisfies Laplace's equation

$$\nabla^2 V = 0.$$

Because planets are nearly spherical, it is convenient as in Chapter 2 to write and solve Laplace's equation in spherical polar co-ordinates (r, θ, λ), where r is the radial distance from the centre of mass, θ is the co-latitude measured from the north pole and λ is the longitude measured from a convenient arbitrary meridian. In those co-ordinates, the solutions of Laplace's equation are spherical harmonics, and the potential may be written as a series:

$$V = -\frac{GM}{r}\left[1 - \sum_{n=2}^{\infty} \left(\frac{a}{r}\right)^n J_n P_n(\cos \theta)\right.$$

$$\left. + \sum_{n=2}^{\infty} \sum_{m=1}^{m=n} \left(\frac{a}{r}\right)^n (C_{nm} \cos m\lambda + S_{nm} \sin m\lambda) P_n^m (\cos \theta)\right].$$

In this expression, a is a scale factor, usually the equatorial radius of the planet, $P_n(\cos \theta)$ is a Legendre function and $P_n^m(\cos \theta)$ is an associated Legendre function (Whittaker and Watson, 1940). These functions contain arbitrary numerical factors and a convention is needed to standardize them. It is usual in the present context to define the associated Legendre functions so that the integral of the square of a surface harmonic over a unit sphere is 4π, that is to say

$$\int [P_n^m (\cos \theta)]^2 \begin{array}{c} \cos^2 \\ \sin^2 \end{array} m\lambda \ dS = 4\pi.$$

The Legendre functions are sometimes standardized in the same way, but another choice, for which

$$\int_{-1}^{+1} [P_n(\cos \theta)]^2 \ d \cos \theta = 2/(2n+1)$$

is also commonly used.

The J_n, C_{nm} and S_{nm} are numerical coefficients which characterize the field of an individual planet.

The expansion so far as the term in $P_2(\cos \theta)$ has already been introduced in connexion with the polar flattening of a planet. Let us now see more generally how the coefficients J_n, and so on, are related to the internal distribution of matter in the planet, and then how they may be determined from the orbits of satellites and otherwise. It should first be pointed out that the value of J_2 for the Earth is about 10^{-3} and of all other coefficients about 10^{-6} or less. The value of J_2 for most other planets seems to exceed that of all other coefficients by a large factor, but some coefficients for the Moon, probably for Mercury and possibly for Venus, are of comparable magnitude, indicating that the shapes of those planets and their gravity fields are not determined predominantly by hydrostatic pressures, but that non-hydrostatic stresses are of comparable importance.

Suppose that the density at a point inside a planet with co-ordinates $(r_1, \theta_1, \lambda_1)$ is ρ. The potential at an external point with co-ordinates (r, θ, λ) is equal to the integral of contributions from all elementary volumes of the planet, namely

$$G \int_T \frac{dm}{D},$$

where dm is the mass of an elementary volume $d\tau$, that is, $\rho d\tau$, and D is the distance from that volume to the external point; T is the volume of the planet.

Now

$$D^2 = r_1^2 + r^2 - 2r_1 r \cos \chi,$$

where χ is the angle between the radius vectors r_1 and r.

By the cosine rule,

$$\cos \chi = \cos \theta_1 \cos \theta + \sin \theta_1 \sin \theta \cos(\lambda_1 - \lambda).$$

D^{-1} may be expanded in powers of r_1/r:

$$D^{-1} = \frac{1}{r}\left[1 + \sum_{n=1}^{\infty} \left(\frac{r_1}{r}\right)^n P_n(\cos \chi)\right],$$

and this expansion may be converted into one in terms of (θ_1, λ) and (θ, λ) by means of the expansion theorem for spherical harmonics:

$$P_n(\cos \chi) = P_n(\cos \theta_1)P_n(\cos \theta)$$

$$+ \sum_{m=1}^{\infty} P_n^m(\cos \theta_1)P_n^m(\cos \theta)\cos[m(\lambda_1 - \lambda)].$$

Consider now a typical term of the integral. It will be

$$\frac{G}{r}\int_T d\tau \left(\frac{r_1}{r}\right)^n P_n^m(\cos \theta_1)P_n^m(\cos \theta)$$

$$\times (\cos m\lambda_1 \cos m\lambda + \sin m\lambda_1 \sin m\lambda),$$

that is, with a equal to the equatorial radius,

$$\frac{GM}{r}\left(\frac{a}{r}\right)^n P_n^m(\cos \theta)\left[\cos m\lambda \int_T \rho\frac{d\tau}{M}\left(\frac{r_1}{a}\right)^n P_n^m(\cos \theta_1)\cos m\lambda_1\right.$$

$$\left. + \sin m\lambda \int_T \rho\frac{d\tau}{M}\left(\frac{r_1}{a}\right)^n P_n^m(\cos \theta_1)\sin m\lambda_1\right].$$

This is of the form already given for a typical term of the potential if we write

$$C_{nm} = \int_T \frac{d\tau}{M}\left(\frac{r_1}{a}\right)^n P_n^m(\cos \theta_1)\cos m\lambda_1$$

$$S_{nm} = \int_T \frac{d\tau}{M}\left(\frac{r_1}{a}\right)^n P_n^m(\cos \theta_1)\sin m\lambda_1.$$

Furthermore, for the terms independent of λ,

$$J_n = \int_T \rho\frac{d\tau}{M}\left(\frac{r_1}{a}\right)^n P_n(\cos \theta).$$

The coefficients J_n, C_{nm}, S_{nm} are those identified as the dimensionless *multipole moments* of the density of the planet. The value of J_2, the *quadrupole moment*, has already been seen to be $(C - A)/Ma^2$.

This formulation shows that it is not possible to obtain the distribution of density within a planet from the variation of potential outside it; the most that can be done is to obtain some of the multipole moments. Any density distribution that gives these moments will be consistent with the observed external potential. On the other hand, if the density is given as a function of radius and angle, the coefficients of the harmonic terms may be calculated. In particular, as is set out below, a fairly detailed theory has been developed for calculating the potential and form of a planet in hydrostatic equilibrium, that is to say, one for which surfaces of constant density are also surfaces of constant potential and constant pressure.

The methods for the determination of the potential all depend on the fact that the multipole moments of a planet are small, say less than 0.1. It is, therefore, a good first approximation to say that a satellite in orbit about a planet moves in a Keplerian ellipse, that is that it moves as it would about a point mass in a strictly periodic elliptical orbit which lies in a fixed plane and of which the semi-major axis maintains the same direction. The effects of the higher harmonic terms in the potential appear as changes in the plane and orientation of the orbit, in its size and eccentricity, and in its period, and because those changes are relatively small they can be calculated by a perturbation theory based on the Kepler orbit. For that, the Hamiltonian formulation of the equations of motion is most convenient.

First, however, it is necessary to define the geometry of an elliptical orbit. It is convenient to refer the plane of the orbit to the equatorial plane of the planet (Figure 2.1). Let the two planes intersect in the line NN', the line of nodes, where N is the intersection when the satellite is moving upwards from south to north. The direction of NN' must be related to some direction fixed in space and it is usual to take the direction of the first point of Aries (Υ). The angle that NN' makes with that fixed direction is the *longitude of the node*, Ω. The angle between the two planes is the *inclination*, i.

Let the semi-major axis of the elliptical orbit be denoted by AP; P is the pericentre where the satellite is closest to the planet and A is the apocentre where the satellite is most distant. The angle that NP subtends at the centre of mass of the planet (which is one of the foci of the ellipse) is called the *longitude of pericentre*, ω; Ω, i and ω determine the orientation of the orbit in space.

The other necessary quantities determine the size and eccentricity of the ellipse and the speed of the satellite in it. They are a, the semi-major axis, e, the eccentricity and n, the mean angular velocity of the satellite in its orbit, or *mean motion*.

Finally, it is necessary to identify a particular satellite, for there might be a number in the same orbit, and that is done by specifying the time, τ, at which it passes through pericentre.

No difficulty arises in defining the elements of an invariant orbit about a mass point, but what is to be done when the orbit is no longer a constant ellipse and indeed not even periodic? Evidently, if six elements, Ω, i, ω, e,

a and n define the orbit, then, given those elements, the position and velocity vectors, r and v, of a satellite at any point upon it may be calculated. Conversely, if position and velocity vectors are given, six numbers in all, the corresponding six orbital elements can be defined. The orbit they specify is known as the *osculating ellipse* corresponding to r and v. It is inherent in this way of specifying the osculating ellipse that it may vary in the course of one revolution of the satellite in its orbit as well as from orbit to orbit. It may therefore sometimes be necessary to specify a position to which the osculating elements refer, and it is usual to take the ascending node. One object of satellite theory is to calculate the evolution of the osculating elements as a function of time for a given potential and orbit.

Now let us see how to formulate the equations of motion in Hamilton's form, the general expression of which is

$$\dot{p}_k = \frac{\partial \mathcal{H}}{\partial q_k}, \qquad \dot{q}_k = -\frac{\partial \mathcal{H}}{\partial p_k}.$$

$\mathcal{H}$ is Hamilton's function, equal to the sum of the kinetic energy T and potential energy V.

The p_k and q_k are momenta and co-ordinates, in number corresponding to the number of degrees of freedom of the system. For satellite theory we require three momenta and three co-ordinates. The momenta and co-ordinates are said to be canonical if the equations of motion can be written in Hamilton's form. It turns out that the equations of motion for elliptical motion about a mass point can be put in the special form such that $\mathcal{H}$ is independent of all the q_k, and of all but one of the p_k; that is to say

$$\dot{p}_k = 0, \qquad p_k = \text{constant for all } k$$

and

$$\dot{q}_k = 0, \qquad q_k = \text{constant for all } k$$

except one, for which $\dot{q}_n = \beta$, $q_n = \alpha + \beta t$.

The p_k and q_k are then said to be *canonical constants*. Not all dynamical systems admit of canonical constants, but the orbital problem does, as might be seen from the fact that all six orbital elements are constants.

There is some choice in how the canonical constants are specified; one choice is the Delauney elements of which the momenta are

$$L = (GMa)^{1/2}$$
$$G = L\eta, \qquad \text{where } \eta^2 = 1 - e^2$$
$$H = G \cos i,$$

and the fixed co-ordinates are ω and Ω, while the variable co-ordinate is M, equal to $n(t - \tau)$, the mean anomaly.

The Hamiltonian for the Kepler orbit is $T - GM/r$. Now consider what happens if an additional term is introduced into the potential so that it is no longer possible to find a set of canonical constants.

Let $\mathcal{H}_0$ be the Hamiltonian for the Kepler orbit and $\mathcal{H}$ that with the additional terms, V' say. Then if p'_k and q'_k are the perturbed momenta and co-ordinates

$$\dot{p}'_k = \frac{\partial(\mathcal{H}_0 + V')}{\partial q'_k}, \qquad \dot{q}'_k = -\frac{\partial(\mathcal{H}_0 + V')}{\partial p'_k}.$$

But $\partial\mathcal{H}_0/\partial q_k$ and $\partial\mathcal{H}_0/\partial p_k$ are all zero except one that is constant. So

$$\dot{p}'_k = \frac{\partial V'}{\partial q_k}, \qquad \dot{q}'_k = -\frac{\partial V'}{\partial p_k},$$

ignoring the differences between q'_k, p'_k and q_k, p_k respectively in the differentials.

If V' can be written as a function of the p_k and q_k, the equations for p'_k and q'_k may be integrated, usually straightforwardly.

If higher accuracy is required, the difference between the p'_k, q'_k and p_k, q_k must be allowed for in the differentials.

Suppose, for instance, we wish to find the speed of the node corresponding to the second zonal harmonic in the potential. Then we take one of the $\dot{q}'_k$ to be $\dot{\Omega}$ so that the appropriate equation of motion is

$$\dot{\Omega} = -\frac{\partial V}{\partial H},$$

H having been substituted for the momentum corresponding to Ω. Now

$$V' = \frac{GM}{a}\left(\frac{a}{r}\right)^3 J_2 P_2(\cos\theta),$$

where r is the amplitude of the radius vector of the satellite.

We have to write r in terms of the orbital elements, but it is independent of i and so its differential with respect to H is zero. On the other hand, $\cos\theta$ depends on the inclination and therefore on H, and on the position of the satellite in the orbit. Expressing $P_2(\cos\theta)$ as a function of H, and performing the differentiation with respect to H, we obtain the instantaneous value of $\dot{\Omega}$ for an arbitrary position of the satellite in its orbit usually expressed as the longitude of the satellite measured round the orbit from the ascending node. Integration round the orbit from one crossing of the ascending node to the next gives an expression for the change of the longitude of the node, $\Delta\Omega$, in one nodal period.

The algebra is always heavy, even for the simplest problems, because the potential V' is naturally written as a function of spherical polar co-ordinates, and its expression in terms of the canonical constants (p_k and q_k), which involve the elements of the orbit, is complex and takes the form of a Fourier series in which the arguments are linear combinations of the mean longitude of the satellite, the longitude of the node and the longitude of pericentre, and the coefficients are polynomials involving powers of the eccentricity and of the sine of the inclination of the orbit to the equator (Kaula, 1966).

Many special methods are available for integrating the equations of motion in the Hamiltonian or equivalent form, and there is an extensive literature (see Cook, 1963). Where high accuracy is needed (as it is for the Earth, but not as yet for planets) the differences between the perturbed constants, p'_k and q'_k, and the canonical constants, p_k and q_k, must be allowed for in the differentials.

A summary of the principal results can be set out according to the symmetry of spherical harmonics, which are primarily divided into those independent of longitude (zonal harmonics) and those that depend on longitude (tesseral and sectorial harmonics). Zonal harmonics may be further subdivided into even harmonics, which are polynomials in even powers of $\cos \theta$ and are symmetrical about the equator, and odd harmonics, polynomials in odd powers of $\cos \theta$ and anti-symmetrical about the equator.

Some results for the first few even and odd zonal harmonics are given in Tables 3.2 and 3.3, which give respectively the changes of longitude of

Table 3.2. *Secular and long periodic changes of node and perigee: even zonal harmonics*

n	Ω_n	ω_n
2	$-\frac{3}{2}\cos i$	$3(1-\frac{5}{4}S)$
4	$15\cos i(1-\frac{7}{4}S)(1+\frac{3}{2}e^2)$ $-\frac{45}{16}\cos i(1-\frac{7}{3}S)e^2\cos 2\omega$	$-\frac{15}{32}(16-2S+49S^2)$ $+(18-63S+\frac{189}{4}S^2)e^2$ $+\frac{15}{32}(6-35S+\frac{63}{5}S^2)e^2\cos 2\omega$
6	$-\frac{105}{16}\cos i(1-\frac{9}{2}S+\frac{33}{8}S^2)(1+5e^2+\frac{15}{8}e^4)$ $-\frac{525}{32}\cos i(1-6S+\frac{99}{16}S^2)(1+\frac{1}{2}e^2)$ $\times e^2\cos 2\omega+O(e^4)\cos 4\omega$	$\frac{525}{6}[\frac{8}{5}(1-8S+\frac{129}{8}S^2-\frac{297}{32}S^3)$ $+(2-6S+\frac{33}{8}S^2)S\cos 2\omega$ $+O(e^2)]$

$\Delta\Omega = 2\pi J_n(a/p)^n\Omega_n$, $\Delta\omega = 2\pi J_n(a/p)^n\omega_n$.
a is the equatorial radius.
p is the semi-latus rectum.
$S = \sin^2 i$.

node and perigee in one nodal period (the time between successive passages through the ascending node) in the forms

$$\Delta\Omega = 2\pi J_n\left(\frac{a}{p}\right)^n \Omega_n,$$

and

$$\Delta\omega = 2\pi J_n\left(\frac{a}{p}\right)^n \omega_n,$$

where J_n is the coefficient of the nth zonal harmonic in the potential, a is the equatorial radius of the Earth, p is the semi-latus rectum of the orbit, and Ω_n and ω_n are functions of the eccentricity, e, the longitude of perigee, ω, and the inclination, i.

The even zonal harmonics generate perturbations of node and perigee which include terms independent of ω (secular variations) and terms of arguments that are even multiples of ω, namely of order

$$e^{2n} \frac{\cos}{\sin} 2n\omega;$$

the odd zonal harmonics, on the other hand, generate perturbations with arguments that are odd multiples of ω and of order

$$e^{2n-1} \frac{\cos}{\sin} (2n-1)\omega.$$

Let us concentrate first on the even harmonics, as is reasonable for the Earth and most other planets for which the flattening part of the potential, the second zonal harmonic, dominates. It will be seen from Tables 3.2 and 3.3 that the second zonal harmonic generates steady motions of the node and perigee amounting in each nodal period to

$$\Delta\Omega = 2\pi J_2\left(\frac{a}{p}\right)^2 (-\tfrac{3}{2}\cos i)$$

Table 3.3. *Long-periodic changes of node and perigee: odd harmonics*

n	Ω_n	ω_n
3	$\tfrac{3}{2}(1-\tfrac{15}{4}e)\cot i \sin\omega$	$(3/2e)\sin i[(1-\tfrac{5}{4}S)-\text{cosec}^2 i +(\tfrac{5}{4}S-\tfrac{9}{4})e^2]\sin\omega$
5	$\tfrac{15}{4}\cot i (1-\tfrac{21}{5}+\tfrac{105}{8}S^2)(1+\tfrac{3}{4}S^2)e\sin\omega$ $-\tfrac{105}{32}\cos i(1-\tfrac{15}{8}S)e^3\sin 3\omega$	$(105/16e)\,\text{cosec}\,i[(-\tfrac{4}{7}+2S-\tfrac{3}{2}S^2)S +(\tfrac{4}{7}-\tfrac{87}{4}S+\tfrac{67}{2}S^2-\tfrac{357}{16}S^3)e^2 -(1-\tfrac{9}{8}S)S^2e^2\cos\omega]\sin\omega$

Notation as in Table 3.2.

and

$$\Delta\omega = 2\pi J_2\left(\frac{a}{p}\right)^2 [3(1 - \tfrac{5}{4}\sin^2 i)]$$

respectively. The net steady motions of node and perigee will comprise those parts together with contributions (usually much smaller) from the higher even harmonics. The success of the application of early artificial satellites about the Earth to the study of the flattening of the Earth was a consequence of the fact that steady, as distinct from periodic, changes can be found with high accuracy by continuing observations over a sufficiently long period; similarly the flattening of planets with artificial or natural satellites can be found from the steady changes of longitude of the node (perigee is more difficult to observe since most satellites have orbits of low eccentricity).

When the second zonal harmonic dominates the potential, the steady motion of perigee is essentially determined by that harmonic. Thus $\sin \omega$ in the expressions for the perturbation generated by the third zonal harmonic is a known periodic function of time (typically with a period of 50–100 d for Earth satellites) and it is a straightforward matter to find the amplitude of the perturbation of that period in say Ω or $\dot{\omega}$, the eccentricity of inclination.

It will be seen that all even harmonics contribute to the steady evolution of Ω and ω and that all odd harmonics contribute to the term with the period of ω (the perturbation of long period) in the evolution of the elements Ω, ω and e and i as well. In consequence, the steady motion of the node of, say, a single satellite is proportional to a linear combination of all even zonal harmonic coefficients and, if the coefficients are to be found separately, more than one satellite must circle the planet. In fact, it is impossible in principle to make a complete separation because there is an infinite set of coefficients and only a limited number of satellites, and there is as yet no completely satisfactory way of dealing with the problem, which is especially acute for the Earth, where the details of the potential are of great interest (Cook, 1978). However, if the second zonal harmonic dominates, it will be possible to estimate it fairly well from only one satellite and probably quite well from two or more, although the higher harmonics may not be well determined.

The perturbations corresponding to tesseral and sectorial harmonics contain no steady terms or terms of long period, but are in general proportional to the angle $m(\varpi t - \Omega)$, where m is the degree of the harmonic, ϖ is the spin angular velocity of the planet and Ω is the longitude of the node of the orbit. The product ϖt enters because it is a measure of the rotation of the planet, and the potential at a point fixed in space depends on the orientation of the planet about its axis of spin. These perturbations of short period (fractions of a day) are much more difficult to determine than those of long period and also require that the geocentric co-ordinates of the observatories be known. Thus, in general, it is not yet possible to determine them for satellites about planets.

Let us now review the problem of finding the harmonic terms in the potential so far as is required for the study of the internal structures of the planets. Our primary aim is to find J_2 because of its connexion with the moment of inertia. If a planet is spinning fast, as are Mars and the outer planets, the flattening dominates non-hydrostatic terms in the potential, so that J_2 is much greater than the general J_n, and then it will be possible to make quite a good estimate of the value of J_2 from just two satellites, or perhaps from just one. At the same time, some idea of the order of magnitude of coefficients of higher harmonics will be obtained but the actual values will probably be rather uncertain. Coefficients of higher harmonics are wanted for two purposes. First, if the planet is in hydrostatic equilibrium, the odd zonal coefficients are zero, but the even ones are of order $(J_2)^{n/2}$; the factors depend on the details of the constitution, and so, if J_2 is large enough, J_4 is not insignificant, and the actual value of J_4 will bear on the internal constitution. The terrestrial value of J_2 (10^{-3}) is too small for this argument to be used, because the expected value of J_4 is about 10^{-6}, but the Jovian value of J_2 (10^{-1}) is much greater, so that J_4 is also greater and can be used in studies of the constitution (Chapter 8). The other reason for wanting to know the higher harmonics is so that it may be possible to estimate the degree of departure from hydrostatic equilibrium. For that an order of magnitude suffices, necessarily so since, for a complete description, the tesseral and sectorial harmonics are required and as yet they cannot be found for any of the planets except Mars. In summary, the most important aim of an analysis of the potential is to obtain an estimate of J_2, but values of the higher zonal coefficients are useful if they can be derived, essentially because they may be used to check the assumptions made when using J_2 in the study of the internal structure.

The element of the orbit of a satellite which is most readily obtained, whether by telescopic observation of natural satellites or by radio tracking of artificial satellites, is the longitude of the node. Fortunately, almost all planetary satellites move in nearly equatorial orbits so that the inclination is nearly zero and $\cos i$, to which the motion of the node is proportional, takes almost its maximum value of 1.

Table 3.4 gives estimates of the steady motions of the nodes of some planetary satellites.

In the foregoing discussion no reference has been made to bodies other than the planet or a single satellite. This assumption is often unrealistic. Of the planets with natural satellites, all save the Earth have more than one and in at least two cases, Jupiter and Uranus, the satellites are massive enough for there to be significant interactions between them. Furthermore, the effect of the Sun is often important. It dominates the motion of the Moon and has to be considered for most other satellites.

The behaviour of satellites about the Moon is rather complex. Just as the motion of the Moon about the Earth is strongly influenced by the Sun, so is the motion of an artificial satellite of the Moon strongly influenced by the Earth. At the same time the second zonal harmonic does not dominate the gravitational potential of the Moon as it does that of any planet; the sectorial harmonics of second order are as large. Orbits about the Moon therefore evolve in a more complex way than do orbits about the planets and need special discussion; analytical treatments have proved inadequate for the most part.

One of the outer planets, Uranus, has the unusual feature that its spin axis lies nearly in the plane of its orbit instead of roughly perpendicular to it as for other planets. The satellites of Uranus, however, move very nearly in the equatorial plane of the planet. Thus the solar attraction on a satellite is perpendicular to the plane of the orbit instead of in the plane of the orbit.

More details of special features, when they are relevant to the determination of the gravitational fields of particular planets, will be given in the chapters in which those planets are discussed. Because of the special nature of the Moon, the determination of its potential is discussed separately in Chapter 5.

Table 3.4. *Motions of the nodes of some planetary satellites*

Primary	Mars	Jupiter	Saturn
Satellite	Phobos	Ganymede	Tethys
Mean motion of satellite (deg/d)	1125	50	191
Distance of satellite from primary (km)	9000	1.07×10^6	2.95×10^5
J_2 of primary	2×10^{-3}	1.5×10^{-2}	1.7×10^{-2}
Mean motion of node of satellite (deg/d)	0.48	0.005	0.20

The innermost planets, Venus and Mercury, have no satellites, but space probes have passed close to them and values of harmonic coefficients have been estimated from tracking the vehicles close to the planets. It is possible to do that because the potential departs from the $1/r$ form.

For simplicity, consider a space probe moving towards a planet in its equatorial plane. The potential within which it moves is then

$$V = -\frac{GM}{r}\left[1+\frac{1}{2}\left(\frac{a}{r}\right)^2 J_2 + \cdots\right],$$

(the odd zonal harmonics vanish in the equatorial plane). Thus

$$\ddot{r} = -\frac{GM}{r^2}\left[1+\frac{3}{2}\left(\frac{a}{r}\right)^2 J_2 + \cdots\right],$$

so that

$$\dot{r}^2 = \frac{GM}{r}\left[1+\frac{1}{2}\left(\frac{a}{r}\right)^2 J_2 + \cdots\right].$$

Now we know that if J_2 is zero the solution is

$$r = (\tfrac{3}{2})^{2/3}(GM)^{1/3}(t-t_0)^{2/3};$$

let this be written as

$$r_0 = K(t-t_0)^{2/3}.$$

When J_2 is not zero, put $r = r_0 + r_1$, where r_1 is small compared with r_0. Then

$$\dot{r}_1 = \frac{3}{4(t-t_0)}\left[-r_1+\frac{a^2 J_2}{2K(t-t_0)^{2/3}}+\cdots\right],$$

whence

$$r_1 = -\frac{3a^2 J_2}{K}(t-t_0)^{-5/3}.$$

Just as for the determination of the mass of a planet by tracking a space probe in its neighbourhood, the actual situation is more complicated than the simplified model chosen for illustration, because the space craft will not be moving directly along the straight line from the Earth to the planet nor will its path lie in the equatorial plane of the planet; a numerical solution of the equations of the motion must therefore be made, and the mass and J_2, and possibly higher harmonic coefficients, adjusted until the numerical solution fits the tracking data.

The theoretical methods used for the study of orbits of satellites and space probes in the gravity fields of planets, as well as the analysis of the data, are both strongly influenced by the fact that the fields of all planets are to a very good approximation those of mass points and that departures from the field of a mass point are for most planets dominated by the second zonal harmonic. It is a consequence of these facts that all theories, both of artificial satellites and of space probes, can be constructed by perturbation methods in which the departures from motions in the field of a mass point are treated as small effects. It is a second consequence that the value of J_2 can usually be estimated from the data without having to pay serious attention to the determination of higher harmonics. It has been pointed out above that this is a fortunate circumstance because the values of higher harmonics are probably only determined so far as orders of magnitude. The Moon has been seen to be an exception.

Although values of J_2 for Mars, Jupiter, Saturn and Neptune have been known for some time from the analysis of telescopic observation of satellites of those planets, our present knowledge of the gravity fields of the planets depends to a great extent on observations of space probes sent towards a number of planets, especially those that lack natural satellites, for, not only have those artificial objects enabled fields to be estimated for planets that lack natural satellites, but also, the high precision of radio Doppler tracking, which greatly exceeds that of telescopic observation, has yielded more detail about two planets (Mars and Jupiter) for which values of J_2 (and J_4 for Jupiter) had already been estimated from telescopic observations. Now that the Pioneer 11 space craft has reached the neighbourhood of Saturn details in the field of that planet have been confirmed.

So far in this section we have supposed that the only perturbing forces acting upon a satellite or space probe are the gravitational attraction of the parent planet or possibly of the Sun or other satellite. It is well known that other forces have to be taken into account when studying the motion of artificial satellites about the Earth and that corrections must be applied for them when estimating harmonic coefficients from the observed motions. Close satellites are subject to the resistance of the atmosphere and to the pressure of solar radiation, as well as to the gravitational attraction of the Sun and the Moon, while general-relativistic effects must be taken into account for the most exact work. Luni-solar attraction and radiation pressure are more significant relative to the effects of the Earth's gravitational field the further the satellite is away from the Earth. Similar extraneous forces act on the satellites and space probes of the

Moon and other planets. Thus, because the Earth is more massive than the Moon, the effect of terrestrial attraction upon a lunar satellite is relatively greater than that of lunar attraction on a terrestrial satellite. Again the gravitational attraction of the Sun and the repulsion of solar radiation pressure are both greater for planets closer to the Sun than is the Earth, and less for those further away. Atmospheric resistance is important for Venus and in some circumstances for Jupiter, but it has no significant effect upon long-established natural satellites.

Atmospheric drag causes the orbit of a satellite to contract, but to first order has no effect on the motion of the node, so that it will not cause serious errors in the estimation of zonal harmonics; the solar attraction and repulsion, however, must be calculated carefully if errors are not to be made in the estimation of zonal harmonics of the potential. The attraction of natural upon artificial satellites is generally negligible apart from that of the Moon upon terrestrial satellites (and the Earth upon lunar satellites).

3.6   Precession and libration

It was shown in Chapter 2 that the moments of inertia of the Earth cannot be obtained from the value of the harmonic coefficient, J_2, by itself, but that it is possible to do so by combining J_2 with the the the dynamical ellipticity, H, obtained from the precession of the Earth under the attraction of the Sun and the Moon. The relevant perturbations of satellite orbits about the Earth and the luni-solar torques which generate the precession are in consequence of Macullagh's expression for the potential, both proportional to the difference of moments of inertia $[C - \frac{1}{2}(A + B)]$, or $(C - A)$ in most cases. But the perturbation of an artificial satellite depends on the ratio of the perturbing torque to the angular momentum of the satellite in its orbit, which is determined by the mass of the planet and the radius of the orbit, so that the perturbations are proportional to $(C - A)/Ma^2$, or J_2. The gyroscopic precession of the Earth's axis of spin is proportional to the ratio of the gravitational torque, again proportional to $(C - A)$, to spin angular momentum of the Earth, or $C\varpi$, where ϖ is the spin angular velocity of the Earth; in consequence the luni-solar precession is proportional to $(C - A)/C$.

Precession as a consequence of a solar torque is negligible for most planets (torques due to natural satellites other than the Moon would be yet smaller); nonetheless an outline of the theory of precession will be set out as the basis for a brief discussion of the possibility of observing the precession and using it to determine C/Ma^2.

A convenient way of deriving the precession is from the Lagrangian form of the equations of motion of the Earth. If L is the Lagrangian, equal to the difference of the kinetic energy T and the potential energy V, the equations in Lagrangian form are

$$\frac{d}{dt}\frac{\partial L}{\partial \dot{\xi_i}} = \frac{\partial L}{\partial \xi_i},$$

where the ξ_i are a set of co-ordinates sufficient to specify the state of the system; in this case they are the Eulerian angles which specify the position of the spinning planet, namely the inclination, θ, of the equator to the ecliptic, the longitude, Ω, of the direction of the intersection of the equator and ecliptic measured from a suitable origin, ordinarily the first point of Aries, and χ, the longitude of some arbitrary meridian in the planet measured from the ascending node (Figure 3.1).

Assume that the planet is symmetrical about the axis of spin, so that χ may be ignored, and take the spin angular velocity to be ϖ. It is then easy to see that the square of the angular velocity about an axis co-incident with the intersection of equator and ecliptic is

$$\dot{\theta}^2 + \dot{\Omega}^2 \sin^2 \theta,$$

while that about the spin axis is

$$(\varpi + \dot{\Omega} \cos \theta)^2.$$

Figure 3.1. Geometry of precession.

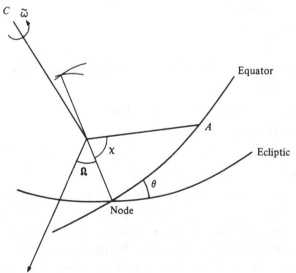

Then, because all directions in the equator are equivalent, the kinetic energy is given by

$$2T = A(\dot{\theta}^2 + \dot{\Omega}^2 \sin^2 \theta) + C(\varpi + \dot{\Omega} \cos \theta)^2.$$

Let $M_\odot$ be the mass of the Sun, r the distance of the planet from the Sun, a the equatorial radius of the planet and ϕ the co-latitude of the Sun measured from the north pole of the planet. The second harmonic in the potential energy is then given by

$$2V = \frac{GM_\odot}{r^3}(C - A)\tfrac{1}{2}(3 \cos^2 \phi - 1)$$

$$= \frac{GM_\odot}{r}\left(\frac{a}{r}\right)^2 \frac{C - A}{a^2}\tfrac{1}{2}(3 \cos^2 \phi - 1).$$

For all planets, except possibly Mercury and Venus, ϖ is much greater than $\dot{\Omega}$ and $\dot{\theta}$ and so $\dot{\Omega}^2$ and $\dot{\theta}^2$ may be ignored in comparison with $\varpi\dot{\Omega}$ and $\varpi\dot{\theta}$.

If λ is the longitude of the Sun measured from the first point of Aries,

$$\cos \phi = \sin \theta \sin (\lambda - \Omega).$$

The Lagrangian equations of motion then read

$$-C\varpi \sin \theta \, \dot{\theta} = \frac{3}{2} \frac{GM_\odot}{r}\left(\frac{a}{r}\right)^2 \frac{C - A}{a^2} \sin [2(\lambda - \Omega)] \sin^2 \theta;$$

$$C\varpi \sin \theta \, \dot{\Omega} = -3\frac{GM_\odot}{r}\left(\frac{a}{r}\right)^2 \frac{C - A}{a^2} \sin^2 (\lambda - \Omega) \sin \theta \cos \theta.$$

Let k denote the factor

$$\frac{GM_\odot}{r}\left(\frac{a}{r}\right)^2 \frac{(C - A)}{C\varpi a^2}.$$

Now n, the mean motion of the planet about the Sun, is given by

$$n^2 = GM_\odot/r^3$$

and so k is equal to $n(n/\varpi)(C - A)/C$. Then

$$\dot{\theta} = -\tfrac{3}{2}k \sin [2(\lambda - \Omega)] \sin \theta$$

and

$$\dot{\Omega} = -\tfrac{3}{2}k\{1 - \cos [2(\lambda - \Omega)]\} \cos \theta.$$

The steady motion of the node upon the ecliptic, $-\tfrac{3}{2}k \cos \theta$, is the solar precession, while the periodic terms in $\dot{\theta}$ and $\dot{\Omega}$ constitute the nutation, a rotation of the spin axis about its mean position at the speed $2n$. Such an elementary theory is inadequate for the Earth, especially when modern sensitive methods of observation are used, particularly laser range measurements to the cube-corner reflectors on the Moon, but it will be quite adequate for a discussion of planetary precession.

Table 3.5 contains estimates of $[(C - A)/C]$, k, and the rate of precession under the attraction of the Sun for Mars, Jupiter and Saturn. The values for all other planets are less; for Mercury and Venus because J_2 is very small and for the outermost planets because they are so distant from the Sun.

We may ask then whether there is any hope of observing precession. That of Mars might be observable if radio devices were placed on the surface to measure the distance from a Martian satellite, the latter constituting an inertial reference. There seems to be no hope for Jupiter and Saturn since they do not have solid surfaces to which measurements might be made. We must conclude that only for Mars is there some future possibility of using the ratio J_2/H to obtain the polar moment of inertia. It will be seen in the next section how the moment of inertia may be *calculated* from J_2 or the flattening if a hydrostatic state is assumed. Now that is almost certainly a good assumption for Jupiter, Saturn, Uranus and Neptune, but it may well not be so good for Mars, Venus or Mercury; because the precessions of Venus and Mercury are negligible, Mars remains the one planet for which measurements of precession may be possible and would certainly be valuable.

Just as the Earth is subject to the torques exerted upon its equatorial bulge by the Sun and the Moon, so the Moon is subject to torques exerted by the Earth. The consequent motions it undergoes, the *physical librations*, are somewhat different from precession and nutation and because they are a special case are discussed in Chapter 5 and Appendix 3. As seen from the Moon, the Earth remains almost in the same direction relative to the principal axis of inertia of the Moon, but, because the Moon's orbit is slightly eccentric and is inclined to the equator of the Moon, as defined by the A and B axes of inertia, the direction of the Earth oscillates about the C and B axes and so exerts torques

Table 3.5. *Solar precession of the planets*

Planet	H	k (rad/s)	Precessional period (centuries)
Mars	5.3×10^{-3}	8.4×10^{-13}	1800
Jupiter	0.0557	9.1×10^{-14}	14 400
Saturn	0.0807	2.2×10^{-14}	67 000

$H = (C - A)/C$.

upon the Moon, which in consequence rocks slightly around its principal axes. The main effects are the slight inclination of the C axis of the Moon to the pole of the ecliptic and a small oscillatory rotation about the C axis. The librational motion of the Moon differs from the precession and nutation of the Earth for two reasons: the Earth is almost fixed in relation to the Moon and does not rotate steadily about it as the Moon does about the Earth, and the spin angular momentum of the Moon is relatively much less than that of the Earth.

The librational movements of the Moon have in the past been determined from telescopic observations, but are now found with high accuracy from laser ranging to cube-corner reflectors on the Moon (Cook, 1976).

3.7   The potential of a planet in hydrostatic equilibrium

Because it is not yet possible to observe the solar precession of any planet but the Earth, the moments of inertia of the other planets cannot be estimated directly from observed dynamical behaviour, and so, if any idea of them is to be obtained, recourse must be had to theory. No doubt many hypotheses could be constructed about the relation between density and radius in a planet that would give a relation between the observable J_2 and the moment of inertia, but, in the absence of any direct evidence, the only hypothesis that can be justified is that the planet is in hydrostatic equilibrium. From that assumption there follows, as is shown below, an almost exact relation between the surface ellipticity or J_2, and the moments of inertia.

Such a relation could only be applied with confidence to the study of the interior structures of the planets if there were reasonable assurance that the planet was indeed in hydrostatic equilibrium. It is clearly not applicable to the Moon (it is not needed either) which, as we shall see in Chapter 5, departs grossly from hydrostatic equilibrium. It is almost applicable to the Earth, for which, as was shown in Chapter 2, the moment of inertia calculated from J_2 on the hydrostatic hypothesis is close to but not identical with the value given by the ratio J_2/H. The value of J_2 cannot so far be determined for Venus and Mercury, so that moments of inertia cannot be found from it; the field of possible application of the hydrostatic hypothesis is therefore restricted to Mars and the outer planets. More detailed discussions will be given later in the relevant chapters but, briefly, it may be said that the hypothesis fails to some extent on Mars, where the surface topography and values of higher harmonics in the potential show that that planet cannot be in hydrostatic equilibrium, irregularities of density being supported near the surface at least by the strength of the solid material. As to the outer planets, they are no doubt fluid and so at first sight in hydrostatic equilibrium, but fluid currents, as well as material strength, can support departures from hydrostatic equilibrium and some may perhaps occur. Therefore, in applying the hydrostatic theory to particular planets, the effect of departures from the hydrostatic hypothesis must be considered.

A significant parameter in this connexion is the central pressure of the planet. If it is very much larger than the strength of solid planetary material, we may suppose that the distribution of density and thus the moment of inertia is not greatly affected by shearing stresses, while, in a fluid planet, we may compare dynamic stresses with the central pressure. We return to these questions at the end of the section and meanwhile take the point of view that in the absence of more detailed information the hydrostatic hypothesis is the only one that can be adopted, remembering always that we should be alert for any fact that may confirm or deny it.

The equation of hydrostatic equilibrium is

$$\nabla p = -\rho \nabla U,$$

where p is the pressure, ρ the density and U the potential, here the sum of the gravitational potential and that of the spin acceleration.

It follows that

$$\nabla p \wedge \nabla U = -\rho \nabla U \wedge \nabla U,$$

which is zero, so that ∇p is parallel to ∇U and so, if U is constant on some surface, then p is also.

Further,

$$\nabla \wedge \nabla p = -\nabla \rho \wedge \nabla U - \rho \nabla \wedge \nabla U.$$

But $\nabla \wedge \nabla p$ and $\nabla \wedge \nabla U$ are identically zero and, therefore, $\nabla \rho \wedge \nabla U$ is zero, and, consequently, ρ is constant on an equipotential surface.

Now the spin acceleration at any point (r, θ, λ) is a function of r and θ but not of λ; it follows that p, ρ and U are functions of r and θ but not of λ: the planet is symmetrical about the axis of spin†.

We therefore take equipotential surfaces within the planet to have the form

$$r = a[1 + \sum e_n P_n(\cos \theta)],$$

where we suppose, in a theory of first order, that squares of e_n may be neglected and the odd coefficients are zero because the planet must be symmetrical about the equator; a is the mean radius of the surface and

---

† It is, however, possible to have triaxial ellipsoids in hydrostatic equilibrium.

the e_n are functions of a, as are ρ, p and U; if ρ is as supposed given as a function of r then the gravitational part of U, V say, may be calculated on the surface, as may the spin part U_s, say. Also, V is the sum of two parts, the potential of matter inside the surface and that of matter outside it.

Now, since the surface is an equipotential, the potential upon it cannot depend on θ but must be a function of a only. Thus the coefficients of spherical harmonic terms in an expansion of the potential upon the surface a must all be zero, leading to a differential equation for each e_n. A transformation of a variable closely related to the moment of inertia leads to a first order differential equation by means of which the outer surface value of e_2 (at $a = a_0$, the mean radius of the outer surface) is related to the moment of inertia.

To calculate the potential at a point (r, θ) within the planet, consider a spheroidal shell of mean radius a and thickness da; the potential at (r, θ) outside the shell is

$$dV_e = \tfrac{4}{3}\pi G\rho \frac{\partial}{\partial a}\left[\frac{a^3}{r} + \sum \frac{3}{2n+1}\frac{a^{n+3}}{r^{n+1}}e_nP_n(\cos\theta)\right]da,$$

while that at (r, θ) inside the shell is

$$dV_i = \tfrac{4}{3}\pi G\rho \frac{\partial}{\partial a}\left[\tfrac{3}{2}a^2 + \sum \frac{3}{2n+1}\frac{r^n}{a^{n-2}}e_nP_n(\cos\theta)\right]da.$$

The total potential is the integral of the external potential of all shells within a together with the integral outside a. Thus

$$V = \frac{4\pi}{3}G\int_0^a \rho\frac{\partial}{\partial a}\left[\frac{a^3}{r} + \sum \frac{3}{2n+1}\frac{a^{n+3}}{r^{n+1}}e_nP_n(\cos\theta)\right]da$$
$$+ \frac{4\pi}{3}G\int_a^{a_0} \rho\frac{\partial}{\partial a}\left[\tfrac{3}{2}a^2 + \sum \frac{3}{2n+1}\frac{r^n}{a^{n-2}}e_nP_n(\cos\theta)\right]da.$$

To V must be added the potential of the spin acceleration, namely

$$U_s = -\tfrac{1}{2}r^2\varpi^2\sin^2\theta = -\tfrac{1}{3}r^2\varpi^2 - \tfrac{1}{3}r^2\varpi^2 P_2(\cos\theta).$$

The mean density, $\bar{\rho}_0$, of the whole planet is

$$\frac{3}{\pi a_0^3}\int_0^{a_0} \rho a^2\,da,$$

and that within the surface of mean radius a is

$$\bar{\rho} = \frac{3}{\pi a^3}\int_0^a \rho a^2\,da.$$

In the expression for V, $1/r$ is replaced by

$$\frac{1}{a}\Big[1-\sum e_n P_n(\cos\theta)\Big],$$

but elsewhere, where powers of r are multiplied by quantities of order e_n, it is replaced by a^{-1}.

The total potential U is then

$$-\frac{4\pi G}{3}\Big\{\frac{1-\sum e_n P_n(\cos\theta)}{a}\int_0^a 3\rho a^2\,\mathrm{d}a$$

$$+\sum\frac{3P_n(\cos\theta)}{2n+1}\Big[\frac{1}{a^{n+1}}\int_0^a \rho\,\mathrm{d}(e_n a^{n+3})+a^n\int_a^{a_0}\rho\,\mathrm{d}\Big(\frac{e_n}{a^{n-2}}\Big)\Big]\Big\}$$

$$-\tfrac{1}{3}a^2\varpi^2-\tfrac{1}{3}a^2\varpi^2 P_2(\cos\theta),$$

and that is to be a function of a only, so the coefficients of all the P_n must vanish; that is to say

$$-\frac{e_n}{a}\int_0^a \rho a^2\,\mathrm{d}n$$

$$+\frac{1}{2n+1}\Big[\frac{1}{a^{n+1}}\int_0^a \rho\,\mathrm{d}(e_n a^{n+3})+a^n\int_a^{a_0}\rho\,\mathrm{d}\Big(\frac{e_n}{a^{n-2}}\Big)\Big]$$

must vanish, except for $n=2$, for which it will be $a^2\varpi^2/8\pi G$. On multiplying by a^{n+1} and differentiating, it is found that

$$-\Big(a^n\frac{\mathrm{d}e_n}{\mathrm{d}r}+na^{n-1}e_n\Big)\int_0^a \rho a^2\,\mathrm{d}a+a^{2n}\int_a^{a_0}\rho\frac{\mathrm{d}}{\mathrm{d}a}\Big(\frac{e_n}{a^{n-2}}\Big)\,\mathrm{d}a$$

is equal to 0 or $-5a^4\varpi^2/8\pi G$.

Now divide by a^{2n} and again differentiate, so reducing the right-hand side to zero in all cases. Then

$$\Big(\frac{\mathrm{d}^2 e_n}{\mathrm{d}a^2}-\frac{n(n+1)}{a^2}e_n\Big)\int_0^a \rho a^2\,\mathrm{d}a+2\rho a^2\Big(\frac{\mathrm{d}e_n}{\mathrm{d}a}+\frac{e_n}{a}\Big)=0$$

or

$$\bar\rho\Big(\frac{\mathrm{d}^2 e_n}{\mathrm{d}a^2}-\frac{n(n+1)}{a^2}e_n\Big)+\frac{6\rho}{a}\Big(\frac{\mathrm{d}e_n}{\mathrm{d}a}+\frac{e_n}{a}\Big)=0.$$

This differential equation was first obtained by Clairaut in 1743.

By studying the behaviour of a Taylor series expansion in the neighbourhood of $a=0$, it may be shown that e_n increases with a. It must then increase all the way to the surface, for suppose on the contrary there is an

a at which e_n ceases to increase and so $de_n/da = 0$. Then

$$\frac{d^2 e_n}{da^2} = \left[n(n+1) - \frac{6\rho}{\bar{\rho}} \right] \frac{e_n}{a^2}.$$

Now for $n \geqslant 2$, $n(n+1) \geqslant 6$ and ρ is less than $\bar{\rho}$, so $d^2 e_n/da^2$ has the sign of e_n, and $|e_n|$ would again increase.

Further study of the original condition on the coefficients of $P_n(\cos \theta)$ shows that all e_n except e_2 must vanish, and that e_2 must be positive at the surface of the planet (and so throughout) and thus, as Newton (1687) originally showed, a spinning planet in hydrostatic equilibrium is oblate (Jeffreys, 1970, p. 186).

Putting $a = a_0$, and e_2 at a_0 equal to e_{02},

$$-a_0^3 \bar{\rho}_0 \left(a_0^2 \frac{de_{02}}{da_0} + 2a_0 e_{02} \right) = -\frac{5}{8\pi} \frac{\varpi^2 a_0^4}{G}.$$

Now the quantity m introduced in the theory of the external field is

$$\varpi^2 a_0^3 / GM = \varpi^2 / \tfrac{4}{3} \pi G \bar{\rho}_0,$$

and so

$$a \left(\frac{de_{02}}{da} \right)_0 + 2a_0 e_{02} = \tfrac{5}{2} m.$$

We now make a transformation introduced by Radau (1885) which reduces Clairaut's differential equation to one of first order and also enables the surface value of e_2 to be related to the polar flattening.

Let η be defined to be $d\ln e/d\ln a$, or

$$\eta = \frac{a}{e} \frac{de}{da},$$

where e has been written for e_2 since all other e_n are zero. Then

$$\frac{de}{da} = \eta \frac{e}{a}, \qquad \frac{d^2 e}{da^2} = \left(\frac{1}{a} \frac{d\eta}{da} + \frac{\eta^2 - \eta}{a^2} \right) e.$$

On substituting in Clairaut's equation,

$$a \frac{d\eta}{da} + \eta^2 - \eta - 6 + \frac{6\rho}{\bar{\rho}} (\eta + 1) = 0$$

and observing that

$$\frac{\rho}{\bar{\rho}} = 1 + \tfrac{1}{3} \frac{a}{\bar{\rho}} \frac{d\bar{\rho}}{da},$$

it follows that

$$a \frac{d\eta}{da} + \eta^2 + 5\eta + 2 \frac{a}{\bar{\rho}} \frac{d\bar{\rho}}{da} (1 + \eta) = 0.$$

The following identity is now used to eliminate $d\eta/da$:

$$\frac{(d/da)[\bar{\rho}a^5(1+\eta)^{1/2}]}{\bar{\rho}a^5(1+\eta)^{1/2}} = \frac{1}{\bar{\rho}}\frac{d\bar{\rho}}{da} + \frac{5}{a} + \frac{1}{2(1+\eta)}\frac{d\eta}{da}.$$

Thus

$$\frac{d}{da}[\bar{\rho}a^5(1+\eta)^{1/2}] = 5\bar{\rho}a^4\psi(\eta),$$

where

$$\psi(\eta) = (1 + \tfrac{1}{2}\eta - \tfrac{1}{10}\eta^2)/(1+\eta)^{1/2}.$$

This is Radau's equation and it owes its importance to the behaviour of $\psi(\eta)$, for

$$\frac{1}{\psi}\frac{d\psi}{d\eta} = \frac{1}{20}\frac{\eta(1-3\eta)}{(1+\eta)(1+\tfrac{1}{2}\eta-\tfrac{1}{10}\eta^2)},$$

which shows that ψ attains a minimum where $\eta = 0$ and a maximum when $\eta = \tfrac{1}{3}$. When $\eta = 0$, $\psi = 1$, when $\eta = \tfrac{1}{3}$, $\psi = 1.00074$. Thereafter ψ decreases.

On the Earth the surface value of η is 0.57, for which $\psi = 0.99961$. The largest value of η on any planet is about 1.4, for which $\psi = 0.974$.

To a good accuracy, therefore, for all planets

$$\frac{d}{da}[\bar{\rho}a^5(1+\eta)^{1/2}] = 5\bar{\rho}a^4.$$

The moment of inertia is intimately related to this equation, for

$$C = \tfrac{8}{3}\pi \int_0^{a_0} \rho a^4 \, da$$

$$= \tfrac{8}{9}\pi \int_0^{a_0} \left(3a^4\bar{\rho} + a^5\frac{d\bar{\rho}}{da}\right) da.$$

On integrating the second term by parts

$$C = \tfrac{8}{9}\left(\bar{\rho}a_0^5 - 2\int_0^{a_0} a^4\bar{\rho} \, da\right).$$

But, from the reduced form of Radau's equation,

$$\int_0^{a_0} \bar{\rho}a^4 \, da = \tfrac{1}{5}\bar{\rho}a_0^5(1+\eta_0)^{1/2}$$

and so

$$C = \tfrac{8}{9}\pi\bar{\rho}a_0^5[1 - \tfrac{2}{5}(1+\eta_0)^{1/2}]$$

or

$$\frac{C}{Ma_0^2} = \tfrac{2}{3}[1 - \tfrac{2}{5}(1+\eta_0)^{1/2}].$$

Let us now revert to our earlier notation in which f denotes the polar flattening, the same as the surface value of e. Then, η_0 is equal to $\frac{5}{2}(m/f) - 2$.

Equivalent expressions for C/Ma^2 in terms of the flattening f and J_2 are therefore

$$\frac{C}{Ma^2} = \frac{2}{3}\left[1 - \frac{2}{5}\left(\frac{5}{2}\frac{m}{f} - 1\right)^{1/2}\right]$$

and

$$\frac{C}{Ma^2} = \frac{2}{3}\left[1 - \frac{2}{5}\left(\frac{4m - 3J_2}{m + 3J_2}\right)^{1/2}\right].$$

We must now consider the limitations of the first order theory and in particular whether it will be adequate for application to the major planets which spin very fast and for which m is much larger than for the Earth.

First notice that the formula for C/Ma^2 may be re-arranged to give the following expression for J_2 in terms of C/Ma^2 and m:

$$\frac{J_2}{m} = \frac{4 - \frac{25}{4}\left(1 - \frac{3}{2}\frac{C}{Ma^2}\right)^2}{3 + \frac{75}{4}\left(1 - \frac{3}{2}\frac{C}{Ma^2}\right)^2}.$$

This shows how the spin and internal distribution of density determine J_2 and the flattening, and how J_2 increases with spin, m, for a given value of C/Ma^2. It is in consequence of this behaviour that J_2 is relatively very much larger for the major than for the terrestrial planets, because m is very much greater. The following list shows how J_2/m varies with C/Ma^2.

C/Ma^2	J_2/m
0.4	0.5
0.3	0.2
0.2	0.08

Yet another way of expressing the result of hydrostatic theory is to give the value of f/m:

$$\frac{f}{m} = \left[\frac{2}{5} + \frac{5}{2}\left(1 - \frac{3}{2}\frac{C}{Ma^2}\right)^2\right]^{-1}.$$

The following figures show the relation:

C/Ma^2	f/m
0.4	1.2
0.3	0.87
0.2	0.61

Evidently the terrestrial value of C/Ma^2, about 0.33, gives values of J_2/m and f/m close to the observed values, which are approximately 0.33 and 1 respectively.

In fact, the foregoing theory is not adequate for the major planets and more can be derived by working to the second or third order in the ellipticity. Clairaut's equation is an equation for the ellipticity or, what comes to the same thing, the coefficient of P_2 (cos θ) in the expression for the radius vector of an equipotential surface. By retaining further terms in the various expansions, equations may be obtained for the coefficients of the higher harmonics, $P_4(\cos \theta)$ and $P_6(\cos \theta)$. It then becomes preferable to use integral rather than differential equations for the coefficients. The details are given in Chapter 8; the principles are the same as in the first order work. An expression is written down for the radius vector of a surface of constant density and potential, and the po.ential is calculated. The condition is then applied that the radius vector is to be such that the potential shall be a constant; on substituting the general expression for the radius vector into the general expression for the potential and applying the condition that the potential is to be a function of radius only, a series of conditions is obtained in the form of integral equations for the coefficients of the harmonics in the expression for the radius vector. Thus, given the density as a function of radius, the coefficients $J_2, J_4, J_6 \ldots$ in the external potential may be calculated on the hydrostatic hypothesis, providing the basis on which to check the supposed density function against the observed J_2, J_4, J_6. Values of J_2, J_4 and J_6 are now available for Jupiter and J_2 and J_4 for Saturn.

The theory by Darwin (1899) and by Zharkhov and Trubitsyn (1970) supposes that the density is given as a function of radius, but the theories of the outer planets start from an equation of state that gives the density as a function of pressure. Thus, to the integral equations of hydrostatic theory must be added the equation of state

$$\rho = \rho(p)$$

and the hydrostatic equation

$$\nabla p = -\rho \nabla U.$$

In the latter equation, U, the potential, is the sum of the potential of self-gravitation of the planet and the potential of spin acceleration.

Darwin's relation between C/Ma^2 and J_2 for a hydrostatic planet takes the simple form it does because the function

$$\psi(\eta) = (1 + \tfrac{1}{2}\eta - \tfrac{1}{10}\eta^2)(1 + \eta)^{-1/2}$$

is taken to be 1, a value to which, as has been seen, it is close for all values of η encountered in practice. One should, in fact, allow for the departure of $\psi(\eta)$ from 1 for the major planets, but then the relation between C and J_2 would depend on the variation of ρ with radius, or would involve J_4. Rather than proceed to this elaboration, it seems best to keep the simple form of Darwin's relation, use the value for C/Ma^2 so obtained as a guide to the structure of the planet, and then use comparisons between the calculated and observed values of J_2, J_4, $J_6 \ldots$ in order to check the validity of any postulated density function or equation of state.

Darwin's simple relation between C/Ma^2 and J_2 is thus inadequate for the major planets because terms of order J_2^2 are important. It fails for the terrestrial planets because they are not in hydrostatic equilibrium and so we now consider whether it is possible to estimate the error committed by using Darwin's formula when there are departures from hydrostatic equilibrium as shown by the presence of significant odd zonal and tesseral and sectorial harmonics in the gravitational potential.

Consider a spinning planet and take, as usual, spherical polar co-ordinates (r, θ, λ) with the origin at the centre of mass and the co-latitude measured from the north pole of the axis of spin.

Let the density, ρ, be a function of (r, θ, λ). The polar moment of inertia, C, is given by

$$C = \tfrac{2}{3} \int_{T} \rho r^4 [1 + P_2(\cos \theta)] \, d(\cos \theta) \, d\lambda \, dr,$$

where the integral extends throughout the volume, T, of the planet.

Now suppose that at any radius the density departs from the value, ρ_H, corresponding to hydrostatic equilibrium by the amount ρ', so that

$$\rho = \rho_H + \rho'.$$

ρ_H does vary with angular position at a given radius, but it is constant on surfaces of constant potential, and so if such a surface is specified by say its polar radius, then ρ_H is a function of that radius only.

ρ' may then be written in a series of spherical harmonics as

$$\sum_{n=1}^{\infty} x_n(r) Y_{nm}(\theta, \lambda).$$

Now write

$$C = C_H + C'$$

where C_H is the polar moment of inertia for the density ρ_H and C' that for ρ'. Then

$$C' = \tfrac{2}{3} \int_T r^4 \sum x_n(r) Y_{nm}(\theta, \lambda)[1 + P_2(\cos\theta)]\, d(\cos\theta)\, d\lambda\, dr.$$

It immediately follows from the orthogonality property of spherical harmonics that the only part of C' that gives a non-zero integral is

$$x_2(r)P_2(\cos\theta),$$

and that gives

$$C' = \frac{8\pi}{15} \int_0^a x_2(r)r^4\, dr,$$

a being the surface radius of the planet.

Now the moment of inertia, A, about an axis in the equatorial plane, is

$$\int_T r^4 (\cos^2\theta + \sin^2\theta \sin^2\lambda)\, d(\cos\theta)\, d\lambda\, dr.$$

If we take the average value of A for all axes in the equatorial plane, that is, if we assume that ρ' does not depend on azimuth, then

$$A = \pi \int \rho r^4 (2\cos^2\theta + \sin^2\theta)\, d(\cos\theta)\, dr$$

or

$$\frac{2\pi}{3} \int \rho r^4 [2 + P_2(\cos\theta)]\, d(\cos\theta)\, dr.$$

Thus, replacing ρ by ρ' as before. we have

$$C' - A' = \frac{2\pi}{3} \int \rho' r^4 P_2(\cos\theta)\, d(\cos\theta)\, dr$$

or, with $\rho' = x_2(r)P_2(\cos\theta)$,

$$C' - A' = \frac{4\pi}{15} \int_0^a x_2(r)r^4\, dr$$

$$= \tfrac{1}{2}C'.$$

Now the value of J_2 is

$$\frac{2\pi}{3Ma^2} \int \rho r^4 P_2(\cos\theta)\, d(\cos\theta)\, dr$$

and if we write, with an obvious notation,

$$J_2 = J_{2H} + J_2',$$

it follows that

$$J_2' = \frac{4\pi}{15Ma^2}\int x_2(r)r^4\,dr$$
$$= (C' - A')/Ma^2$$
$$= \tfrac{1}{2}C'/Ma^2.$$

This result provides the relation between the departures of C and J_2 from the values they would have for a density distribution in hydrostatic equilibrium. It is not, however, the result that relates the value of C calculated on the hydrostatic hypothesis from the *actual* value of J_2, to the true value of J_2. In that calculation we use Darwin's result in the form

$$\frac{C}{Ma^2} = \frac{2}{3}\left[1 - \frac{2}{5}\left(\frac{4m - 3J_2}{m + 3J_2}\right)^{1/2}\right],$$

but we use an erroneous value for J_2, namely the actual one, and not one corresponding to a hydrostatic distribution of density; that is we use J_2 when we should use J_{2H} subsequently adding the correction $2Ma^2J_2'$ to the calculated value, C_H, of C.

To calculate the error committed, differentiate Darwin's formula:

$$\frac{\delta C}{Ma^2} = \frac{2\delta J_2/m}{(4 - 3J_2/m)^{1/2}(1 + 3J_2/m)^{3/2}},$$

where δJ_2 is J_2'.

Write this expression as

$$\delta C = k\delta J_2$$

where

$$k = \frac{2Ma^2/m}{(4 - 3J_2/m)^{1/2}(1 + 3J_2/m)^{3/2}}.$$

If J_2 is $\tfrac{1}{3}m$, as it roughly is for the Earth,

$$k = Ma^2/2.4m.$$

This result gives a discrepancy δC that is much larger than C', and the calculated value of C is

$$C_H + Ma^2J_2'/2.4m$$

instead of

$$C_H + 2Ma^2J_2'.$$

Thus the error committed by using Darwin's formula is

$$\left(\frac{1}{2.4m} - 2\right)Ma^2J_2'.$$

Suppose, for example, that J_2' is 10^{-6} and m is about 3.3×10^{-3}. The error in C calculated on the hydrostatic hypothesis is then $1.2 \times 10^{-4} Ma^2$, whereas the true value of C departs from the hydrostatic value by only $2 \times 10^{-6} Ma^2$; C itself lies between $0.4 Ma^2$ and $0.2 Ma^2$ for all planets. The application of these ideas to Mars will be considered in Chapter 6.

4

Equations of state of terrestrial materials

4.1 Introduction

The dynamical properties of a planet depend on the way in which the density varies with radius, and seismological properties depend also on the way in which the elastic moduli and elastic dissipation vary with radius. The data we have for the Earth are sufficiently complete that the variations of density and elastic moduli with radius can be derived from them and we are then presented with the problem of inferring the mineralogical and chemical composition consistent with them. When, as for the other planets, seismic data are lacking, we must proceed in a different way and derive the variation of density from a postulated composition, asking if it leads to the observed mass and moment of inertia. In either case, we must know how the density depends on pressure, temperature and composition, for all these vary with radius, and when we discuss the Earth and the Moon, for which we have seismic data, we must also examine the dependence of the elastic moduli upon the three variables. Some idea of the problems that arise, of the theoretical principles, of possible experimental methods, and of the systematics of equations of state of minerals has already been given in Chapter 1, and it is the aim of this chapter to give a more extensive and systematic account.

Our aim is to discuss the variation of density and other properties with radius. If a planet is in hydrostatic equilibrium, then all harmonic components of the gravitational potential except the even zonal ones will be zero, and the coefficient J_{2n}, of the term proportional to $P_{2n}(\cos \theta)$, will be of order $(J_2)^n$. We know that many other harmonic components of lower symmetry occur in the potential of the Earth, the Moon and Mars so that the distributions of density in those bodies do not correspond to the hydrostatic state. We do not know for certain how the corresponding stress differences are supported, but it is reasonable to suppose that in the outer parts of a planet, where the temperature is low and the strength correspondingly high, the stress differences are supported by the strength of the material, whereas in the deeper parts the temperature may be high enough for the material to creep steadily, when it is likely that the stress differences are supported by the movement of the material as in convection. However, it is not at present necessary to settle that issue in order to discuss the major variations of density with radius, for the non-hydrostatic harmonic coefficients in the potential are of order 10^{-4} or less and it is clear that, except in the Moon, the hydrostatic pressure deep within any planet greatly exceeds the strength of minerals, and it is possible to consider as a first, probably rather good, approximation that the density does indeed depend only on radius. Strictly the argument applies only to the small terrestrial planets, for which the surface centrifugal acceleration is small compared with the gravitational acceleration; when the surface centrifugal acceleration is not small, as for the major planets, the surfaces of constant density are not surfaces of nearly constant radius. In this chapter, we shall ignore all lateral variations of density, both those consistent and those inconsistent with the hydrostatic state, the former because we are concerned with the terrestrial planets and the latter because they are small and, in any case, we know little about their effects on equations of state.

Suppose then that the density is taken to be a function of radius alone:

$$\rho = \rho(r).$$

Suppose also that the variation with radius is a consequence of the variation of pressure, temperature and composition with radius:

$$\rho = \rho(p(r), T(r), C(r)).$$

Here C stands for an empirical function which gives the density as a function of mineral composition, crystal structure and chemical composition.

Thus, to find ρ as a function of radius, we need to know how pressure, temperature and composition vary with radius. The change of pressure with radius is a straightforward matter. Suppose that

$$\rho = \rho(p).$$

At any radius, r,

$$\frac{dp}{dr} = -g\rho.$$

Furthermore the value of gravity at radius r is equal to the attraction of the mass within radius r concentrated at the centre:

$$g = \frac{G}{r^2} \int_0^r \rho(r') \, d\tau,$$

where $d\tau$ is an element of volume at radius r'.

If ρ is given as a function of p, the foregoing equations may be solved (numerically in general) to give ρ as a function of r. Let M be the mass within radius r. Then

$$\rho = \frac{1}{4\pi r^2} \frac{dM}{dr},$$

and $g = -GM/r^2$.

Now if

$$\frac{dp}{dr} = -g\rho,$$

then

$$\frac{dp}{d\rho} \frac{d\rho}{dr} = -g\rho,$$

or

$$\frac{d\rho}{dr} = -g\rho \frac{d\rho}{dp}.$$

In terms of M,

$$\frac{d\rho}{dr} = -\frac{2}{4\pi r^3} \frac{dM}{dr} + \frac{1}{4\pi r^2} \frac{d^2M}{dr^2}.$$

Thus the equation of hydrostatic equilibrium reads

$$-\frac{2}{r} \frac{dM}{dr} + \frac{d^2M}{dr^2} = \frac{GM}{r^2} \frac{dM}{dr} \frac{d\rho}{dp}.$$

Since $d\rho/dp$ is supposed to be known, this equation may be solved for $M(r)$ and then $\rho(r)$ may be found by differentiation and $g(r)$ and $p(r)$ by substitution.

It has been tacitly assumed that density is a continuous function of pressure, but that is in general not the case. The equation for M would then have to be integrated piecewise, allowing for discontinuous increases of pressure with density. Such discontinuities do indeed occur and correspond to changes of phase, whether from solid to liquid or vice versa or from a less compact to a more compact crystal structure.

Evidently it is crucial to the study of the interiors of planets to investigate theoretically and experimentally the behaviour of the densities of single phases as functions of pressure and also the changes of phase that may occur at sufficiently high pressure; and those are the principal topics of this chapter. Density is also a function of mineralogical and chemical composition and consequently in order to estimate densities within planets it is necessary to have some idea both of the likely compositions of the planets and also of the dependence of density on composition. The evidence bearing on the likely composition will be considered when the structures of the Moon and the terrestrial planets are discussed in Chapters 4 and 5, respectively, while in this chapter synoptic relations between density and such parameters as mean atomic weight will be considered.

The density of a simple substance in general depends on temperature as well as pressure so that we may write

$$\frac{d\rho}{dr} = \frac{\partial \rho}{\partial p}\frac{dp}{dr} + \frac{\partial \rho}{\partial T}\frac{dT}{dr}.$$

$\partial \rho / \partial p$ is a function of temperature and $\partial \rho / \partial T$ a function of pressure.

If K is the bulk modulus,

$$\frac{\partial \rho}{\partial p} = \frac{\rho}{K}$$

so that

$$\frac{\partial}{\partial T}\left(\frac{\partial \rho}{\partial p}\right) = \frac{1}{K}\frac{\partial \rho}{\partial T} - \frac{\rho}{K^2}\frac{\partial K}{\partial T}.$$

Let α be the volume coefficient of thermal expansion:

$$\alpha = -\frac{1}{\rho}\frac{\partial \rho}{\partial T},$$

then

$$\frac{1}{K}\frac{\partial \rho}{\partial T} = -\frac{\alpha \rho}{K}.$$

Also

$$\frac{\partial}{\partial T}\left(\frac{\partial \rho}{\partial p}\right) = \frac{\partial}{\partial p}\left(\frac{\partial \rho}{\partial T}\right) = \frac{\partial}{\partial p}(-\alpha\rho) = -\rho\frac{\partial \alpha}{\partial p} - \frac{\alpha\rho}{K}.$$

Thus, comparing the two expressions for $\partial^2\rho/\partial T\partial p$, it follows that

$$\rho\frac{\partial \alpha}{\partial p} = \frac{\rho}{K^2}\frac{\partial K}{\partial T}$$

or

$$\frac{\partial K}{\partial T} = K^2 \frac{\partial \alpha}{\partial p}$$

and

$$\frac{\partial^2 \rho}{\partial T \, \partial p} = -\frac{\rho}{K}\left(\alpha + K\frac{\partial \alpha}{\partial p}\right).$$

If $\partial\alpha/\partial p$ is small, K is nearly independent of temperature and then ρ as a function of pressure is also nearly independent of temperature. Wildt (1963) and Ramsey (1963) have argued that those are indeed the conditions within the planets.

Values of α at high pressures do not appear to have been measured, but there are a few values of $\partial K/\partial T$ available. The experimental value of the adiabatic compressibility of sodium chloride is -1.09×10^7 Pa/deg. The bulk modulus is 2.5×10^{10} Pa, $K_S^{-1} \, \partial K_S/\partial T$ is therefore -4.4×10^{-4}/deg (K_S is the isoentropic compressibility) and $\partial\alpha/\partial p$ is -1.8×10^{-14}/deg Pa. Since α is about 1.2×10^{-4}/deg it is clear that if $\partial\alpha/\partial p$ is independent of pressure, α will vanish at about 7×10^9 Pa (70 kbar), a pressure attained high in the mantle of the Earth. In fact, because of the factor $1/K^2$, which varies as p^{-2} at high pressures, $\partial\alpha/\partial p$ may be expected to tend to zero at high pressure; nonetheless it is clear that α must become quite small at rather moderate pressures.

As another example, consider olivine, for which $\partial K_S/\partial T$ is about -1.5×10^7 Pa/deg (Kumazawa and Anderson, 1969) while K_S is about 1.3×10^{11} Pa at zero pressure. Thus $K_S^{-1} \, \partial K_S/\partial T$ is -1.2×10^{-4}/deg and $\partial\alpha/\partial p$ is -0.9×10^{-15}/deg Pa. With a linear dependence on pressure, α, which is 2.4×10^{-5}/deg at low pressure, would vanish at 2.6×10^{10} Pa (260 kbar).

The low pressure values of $\alpha^{-1} \, \partial\alpha/\partial p$ are 1.5×10^{-10}/Pa for salt and 0.6×10^{-10}/Pa for olivine, values which are not appreciably different despite the larger differences of K and α.

If the temperature and pressure vary with position in a column of a compressible fluid, then the fluid will be mechanically stable if the pressure and temperature are so related that the work done on an element of fluid when its pressure changes exceeds the change of internal energy. If the change of internal energy is greater, net energy becomes available as kinetic energy of convection. The gradient at which the work done and the change of internal energy just balance is the adiabatic gradient. Convection can transport heat rapidly and it is usually supposed that, provided the supply of heat is sufficient, the temperature gradient in a fluid will attain the adiabatic gradient but will not greatly exceed it. Thus, it is supposed that the gradient in the core of the Earth is close to the adiabatic gradient and that convective motions are sustained. Much higher gradients could be maintained in solids if the stresses produced are less than the strength of the material, but at sufficiently high temperatures most solids creep at a steady rate by diffusion or by movements of dislocations, and so it has been argued, particularly by Tozer (1967), that in the presence of a temperature gradient steady state creep will occur and the gradient in solids also will be close to the adiabatic gradient.

For an adiabatic change the entropy, S, is constant. Now

$$dS = \left(\frac{\partial S}{\partial T}\right)_p dT + \left(\frac{\partial S}{\partial p}\right)_T dp.$$

$(\partial S/\partial T)_p$ is C_p/T, where C_p is the specific heat at constant pressure, while, by one of Maxwell's thermodynamic relations,

$$\left(\frac{\partial S}{\partial p}\right)_T = -\left(\frac{\partial V}{\partial T}\right)_p = -\alpha V.$$

Thus, along an adiabat,

$$\frac{dT}{dp} = \frac{\alpha VT}{C_p}.$$

Under hydrostatic conditions,

$$\frac{dp}{dr} = -g\rho$$

and thus

$$\frac{dT}{dr} = \frac{dT}{dp}\frac{dp}{dr} = -\frac{g\alpha T}{C_p}$$

since ρ is the reciprocal of the volume V for unit mass.

There is considerable difficulty in estimating dT/dr within a planet. Values often quoted sometimes seem to allow inadequately for the decrease of α with pressure; on the basis of this argument alone it is unlikely that α would exceed 10^{-6} in the core of the Earth.

The specific heat, C_V, at constant volume, and at a sufficiently high temperature, is 25 kJ/deg kg mole, while the difference of C_p from C_V is given by (Wilson, 1966, p. 39)

$$C_p - C_V = \frac{T}{VK_T}\left(\frac{\partial V}{\partial T}\right)_p^2$$

Thus $C_p - C_V = (VT/K_T)\alpha^2$ (K_T is the isothermal bulk modulus).

V and K_T are well enough known at high pressure, so the problem again reduces to estimating α at high pressure.

In order to estimate the properties of materials in the interiors of the planets we need to know the variations of density, elastic moduli and coefficient of expansion with pressure and temperature; we need to know the changes of crystal structure and composition that minerals suffer at high pressure; and we need to know when materials melt. So far as the density of single minerals are concerned, it has been argued above, following Wildt (1963) and Ramsey (1963), that the thermal expansion can be ignored. Thus, a major part of this chapter is concerned with isothermal equations of state, although effects of temperature are not ignored. In subsequent sections changes of crystal structure and composition and the problem of melting are considered.

4.2 Theoretical equations of state

The components of planets are generally thought to be mainly of two sorts: minerals which are ionic crystals,† and metals. The outermost parts of the major planets are molecules with covalent bonding. Only in the simplest possible case – the metallic form of hydrogen (see Chapter 6) – is it possible to calculate the density at high pressures from first principles; in all other cases quasi-empirical methods must be used which in general are not trustworthy at the pressures deep in planets.

In principle, the properties of any solid or liquid could be calculated if it were possible to solve Schrödinger's equation for the complete assembly of nuclei and atoms comprising the material. We would naturally expect such a solution to show us how most of the electrons form cores round the nuclei, so that structures very like free atoms persist in condensed phases; we would expect to see how the internal energy depends on the linear scale of the system and thus to calculate the volume at a given pressure; and we would expect to identify certain motions with thermal motions and so identify the thermal properties. No such programme has yet been achieved for even the simplest condensed material – metallic hydrogen – although the isothermal properties do seem to be well understood. Since it is impossible to solve Schrödinger's equation in complete generality, some approximations must be made. It should be noted that we are here confronted with a rather different problem from that which Fröhlich (1973) has considered. Fröhlich discussed the way in which the macroscopic features of a problem could be extracted from the general quantum

† The silica tetrahedra of a mineral like olivine are themselves covalently bonded, but are held together by ionic bonds to metallic ions.

mechanical solution and, for example, showed how to derive the Navier–Stokes equation for the motion of a fluid. We could equally derive the equations of elasticity for a solid, but that does not help us to calculate the elastic constants by quantum mechanical methods because Fröhlich's averages are chosen to avoid our ignorance of the quantum mechanical solutions. We have to make approximations which retain some of the information that may be extracted from Schrödinger's equation.

One approximation allows the thermal properties to be treated separately from the isothermal behaviour. It is akin to the Born–Oppenheimer approximation used in the quantum theory of molecules and depends on the fact that the masses of nuclei are very much greater than those of electrons. Consequently, the electron velocities are much greater than nuclear velocities and it is possible to regard the total wave-function as a product of electronic and nuclear wave-functions. The eigenenergies of the solid are then the sums of eigenenergies of the electrons and nuclei separately. To the former correspond structures of the atomic cores and the forces that hold the cores in position in a crystal, while to the latter correspond the mechanical vibrations (phonons) of the crystal lattice which are excited when the material is heated. A somewhat similar division can be made for a liquid, complicated though it is by the fact that the atomic cores do not occupy fixed sites as in a crystalline solid. Thus, much as we can write the energy of a molecule as a sum of electronic, vibrational and rotational terms, so we may write the energy of a solid or liquid as the sum of electronic and vibrational terms (plus rotational and translational parts for a liquid); the electronic part corresponds to the internal energy at zero temperature and depends on configuration and pressure, while the vibrational part gives the thermal energy. Just as in molecules, so no doubt in solids and liquids, this separation is not exact, and perturbations of the order of the ratio of the electronic mass to the nuclear mass no doubt occur, but neither experiment nor theory are exact enough for them to matter. All minerals of importance are ionic crystals; that is to say, they consist of positive and negative ions held together in a lattice by Coulomb forces. There are no free electrons, and the pure crystals at low pressures are insulators. The potential of an ion is $\pm ze/r$, where z is the degree of ionization, e is the electronic charge and r the radial distance from the nucleus; close to the ion, however, the potential becomes repulsive and is often taken to be proportional to $\exp(-r/\rho)$, where ρ is a constant (Kittel, 1968) or to $1/r^n$. The repulsive form of the potential close to the nucleus is a consequence of the Pauli exclusion principle.

If the repulsive potential were known, then it should in principle be possible to calculate by classical methods the energy of ions arranged in a specified lattice. In principle, a fundamental quantum mechanical calculation would yield both the lattice structure and the repulsive potential, the latter following from the energy levels of each ion as perturbed by its neighbours. In practice, these two properties are taken as empirical facts.

It is often assumed in lattice calculations that the potential is of the form $-a/r^m + b/r^n$ (m will be 1 for Coulomb forces, n is much larger and a value such as 6 may be taken).

Given such a potential, the energy may be calculated as a function of r. If the external pressure is zero, the value of r for which the lattice energy is a minimum gives the zero pressure density, while, if the pressure is not zero, the density follows from the value of r for which the sum of lattice energy and external work is a minimum. Differentiation with respect to volume yields the pressure at zero temperature and further differentiation with respect to pressure gives the bulk modulus and its pressure derivative $\partial K/\partial p$. The last is important in studying planetary interiors and so some results of lattice calculations are given.

Fürth (1944) showed that $\partial K/\partial p$ at zero pressure is $\frac{1}{3}(m + n + 6)$, while Ramsey (1950) showed that at high pressure $\partial K/\partial p$ depends only on the repulsive potential and for a power law is $\frac{1}{3}(n + 6)$. Thus, if n is 6, $\partial K/\partial p$ would be 4, a value close to that found for a number of minerals. Anderson (1968) has obtained somewhat different results:

$$\left(\frac{\partial K_T}{\partial p}\right)_T = \tfrac{1}{3}(n+3) + \frac{\tfrac{1}{3}(m+3)(n-m)}{(n+3)(V_0/V)^{n-\frac{1}{3}m} - (m+3)},$$

where V_0 is the specific volume at zero pressure.

At zero pressure, $V = V_0$ and $(\partial K_T/\partial p)_T$ is $\frac{1}{3}(m + n + 6)$ as Fürth found, but, at high pressure, where V approaches zero, $(\partial K_T/\partial p)_T$ becomes $\frac{1}{3}(n + 3)$.

Lattice calculations thus predict that $(\partial K_T/\partial p)_T$ should decrease to a limit at high pressures, the limit being determined by the repulsive part of the potential.

It should be appreciated that the foregoing results are based on simplifications. Details of the lattice structure are ignored and a simple power law is assumed for the repulsive potential, whereas more complicated forms, different for different ions, might be appropriate. Calculations in which exact lattice sums were evaluated have been carried out for cubic structures by Anderson and Liebermann (1970), who show that if $m = 1$ (Coulomb attraction)

$$(\partial K_T/\partial p)_T = \tfrac{1}{3}(n + 7),$$

in agreement with the general expressions given above.

While some results of lattice calculations are of value in interpreting experimental results, they are not of great value for predicting behaviour. The difficulty is that potentials which might be appropriate for low pressures do not necessarily describe behaviour at high pressures.

Given the interionic potentials, other properties of a crystal may be calculated; in particular, its elastic moduli, its coefficient of thermal expansion and other thermal properties. The thermal properties are found from a knowledge of the spectrum of the modes of vibration of the lattice which are determined by the interionic potential and the mass of the lattice. Metals are distinguished from insulators by the gas of free electrons which permeates the lattice of positive ions and which makes the metal a good conductor of electricity. In the most elementary treatment of a metal the electronic wave-functions are supposed to be the single plane waves of a free electron, but that description cannot be correct in general although the properties of some metals are indeed close to those predicted by a free- or nearly-free-electron model. The free-electron plane wave model is adequate for hydrogen, where the potential of a single proton is that of a point charge, and for helium, but with more complex atoms, having a large un-ionized core of electrons, the potential within the core corresponding to bound states of the electrons is such that plane wave states outside the core cannot be matched to bound states within it. Yet the plane wave model works quite well and it is reasonable to suppose that many electronic, optical and mechanical properties of metals could be calculated if it were possible to find a potential which had the same effect outside the atomic core as the actual potential. Such potentials have been constructed and are known as pseudopotentials (Heine, 1970). The main requirement for the pseudopotential is that it leads to the correct eigenvalues of the conduction electrons, leading, that is, to the observed band structure. There is now an extensive literature concerned with pseudopotentials and, in particular, the way in which they may be constructed from experimental data has been reviewed by Cohen and Heine (1970).

If a pseudopotential can be constructed for a metal, then as discussed by Heine and Weaire (1970) the binding energy may be calculated as a function of the linear scale of the metal. Heine and Weaire illustrate the principle by the following simple model.

The energy of a free electron as a function of wave-number, k, *is*

$$E(k) = V_0 + \tfrac{1}{2}k^2,$$

where V_0 is the mean pseudopotential.

Consider a bare ion and an electron gas of density z electrons per ion. Take the pseudopotential of the ion to be

$$r < R_M : v_+(r) = -A_0$$
$$r > R_M : v_+(r) = -z/r,$$

where R_M is a constant.

To this add the potential v_e for a uniform gas of electrons contained in a spherical cell of radius R_A.

Within R_A, $r < R_A$, and

$$v_e = \frac{3}{2}\frac{z}{R_A}\left[1 - \frac{1}{3}\left(\frac{r}{R_A}\right)^2\right]$$

and, outside R_A, $r > R_A$, and

$$v_e = z/r.$$

Outside R_A, $v_e - v_+$ is zero. Within R_A the mean potential is

$$-\frac{0.3z}{R_A} + \frac{1.5zR_M^2}{R_A^3} - A_0\left(\frac{R_M}{R_A}\right)^3.$$

On summing the energy for all the occupied states of the electrons, the total is $zV_0 + \tfrac{3}{5}zE_F$ per atom, where E_F is the Fermi energy $(\hbar^2/2m)3\pi^2 N_e^{2/3}$, N_e being the electron concentration.

However, in this way the self-energy of the electron gas is counted twice and, therefore, the energy of a uniform sphere of negative charge z, namely $0.6z^2/R_A$, must be subtracted.

The exchange and correlation energies of the electrons in the gas (Chapter 7) must also be added. It is convenient to introduce the radius r_s of a sphere containing 1 electron, namely

$$r_s = z^{-1/3}R_A.$$

The exchange energy may be written as $0.458z/r_s$ and the correlation energy as $z(0.0575 + 0.0155 \ln r_s)$, and the total energy of the metal becomes

$$V_0 = z\left(\frac{1.105}{r_s^2}\right) - \frac{0.458z}{r_s} + z(0.0575 + 0.0155 \ln r_s)$$
$$-0.9\frac{z^2}{R_A} + \frac{3}{2}\frac{z^2}{R_A}\left(\frac{R_M}{R_A}\right)^2 - A_0 z\left(\frac{R_M}{R_A}\right)^3.$$

It will be noted that the expression contains a variable R_A (or r_s) which determines the atomic volume and two parameters, A_0 and R_M, by which the pseudopotential is described. The equilibrium volume for a given A_0 and R_M is found by minimizing the energy with respect to R_A (and r_s); to obtain the equilibrium volume at a finite pressure, it is necessary to minimize the sum of the electronic energy and the mechanical work done in compression.

For illustration, a simple pseudopotential has been adopted; in practice, more complex potentials would be necessary to reproduce the actual behaviour. There is, however, one metal, hydrogen, where the actual potential is just $-1/r$ and accurate calculations can be carried out, as will be seen in Chapter 7.

While it is clear how the zero-temperature equation of state could be calculated if a pseudopotential were given, it is unfortunately the fact that very few calculations have been done (see Heine and Weaire, 1970) and none for iron, the metal of greatest interest in the study of planets.

Of course, if the wave-functions for the bound states of an ion could be calculated with sufficient detail and precision, there would be no need to fit a pseudopotential. Some calculations have been done, in particular for aluminium and iron (Berggren and Fröman, 1969); those for aluminium yield an equation of state in good agreement with that from shock-wave experiments at high pressure, but those for iron, with its more complex electronic structure, deviate considerably from experiment.

So far, whether for ionic crystals or metals, the discussion has been concerned with the calculation of the energy (and hence pressure) as a function of volume at zero temperature. Because it seems, as was shown in the previous section, that the coefficient of thermal expansion decreases rapidly at high pressure, the study of thermal effects is of much less importance than the estimation of the isothermal dependence of density on pressure. Some consideration must, however, be given to the thermal energy, if only because it is a significant factor in the reduction of shock-wave observations to isothermal equations of state.

It was seen earlier that the Born–Oppenheimer approximation leads to the conclusion that the energy of a crystal (the argument applies equally to ionic crystals and to metals) divides into a part that depends only on pressure and a part that depends also on temperature. Correspondingly, the free energy, F, equal to $E - TS$, may be written as $\phi + \sum F_k$, where ϕ is the energy of the static lattice, while F_k involves products of displacements of nuclei from their equilibrium positions taken k at a time.

If F_k is zero for all k greater than 2 the vibrational energy will be a function of the squares and products of the displacements and the nuclei will execute harmonic oscillations. If, however, $F_3, F_4 \ldots$ are not zero the oscillations are not harmonic.

The free energy of any system is

$$F = -kT \ln Z,$$

where Z is the partition function of the system.

Treating the lattice vibrations as a set of independent harmonic oscillators,

$$Z = \prod_i Z(\omega_i),$$

where Z is the partition function of a mode of frequency ω_i.

Since the energy levels of the oscillator i are

$$E_n = (n + \tfrac{1}{2})\hbar\omega_i,$$

the partition function is

$$Z(\omega_i) = \sum_{n=0}^{\infty} \exp\left(-E_n/kT\right)$$

$$= \frac{\exp\left(-\tfrac{1}{2}\hbar\omega_i/kT\right)}{[1 - \exp\left(-\hbar\omega_i/kT\right)]}$$

and so

$$F = \phi_0 + kT \sum \left[(\tfrac{1}{2}\hbar\omega_i/kT) + \ln\left\{1 - \exp\left(\hbar\omega_i/kT\right)\right\}\right].$$

This expression accounts for the harmonic part of the vibrational energy.

Note that, in general, ω_i is a function of the wave-vector and of the polarization of a plane wave.

Now the pressure is given by

$$p = -\frac{\partial F}{\partial V},$$

so that

$$p = -\frac{d\phi_0}{dV} - \frac{1}{V}\sum \frac{d\ln\omega_i}{d\ln V}\hbar\omega_i\{\tfrac{1}{2} + [\exp\left(-\hbar\omega_i/kT\right) - 1]^{-1}\},$$

in which the thermal effects are in the summation through the vibrations of frequency ω_i.

Write

$$p = -\frac{d\phi_0}{dV} + \frac{1}{V}\gamma_i E_i$$

where

$$\gamma_i = -\frac{d\ln\omega_i}{d\ln V}$$

and

$$E_i = \hbar\omega \left[\frac{1}{2} + \frac{1}{\exp\left(-\hbar\omega_i/kT\right) - 1} \right].$$

If the γ_i are all equal, the thermal term becomes $(\gamma/V)E_v$, where E_v is the total thermal energy (including the zero point energies of the vibrations). In that case

$$p = -\frac{\mathrm{d}\phi_0}{\mathrm{d}V} + \frac{\gamma E_v}{V}.$$

Hence, if p_1 and p_2 are the pressures and E_1 and E_2 the energies of two states of different temperature but the same volume

$$p_1 - p_2 = \frac{\gamma}{V}(E_1 - E_2).$$

This is the Mie–Grüneisen equation of state which, as will be seen, plays a large part in the reduction of data from shock-wave experiments. As here derived it depends on the assumption that $\mathrm{d}\ln\omega_i/\mathrm{d}\ln V$ is the same for all modes of vibration. Neither theoretically nor experimentally does there seem to be much justification for that assumption. On the other hand, at a sufficiently high temperature the energy of each oscillator becomes kT (for two directions of polarization) and then

$$p_1 - p_2 = \frac{3NRT}{V}\frac{\sum \gamma_i}{3N} = \frac{3NRT\gamma}{V},$$

where N is the number of ions in the crystal and γ, equal to $\sum\gamma_i/3N$, is an average value of γ_i. It follows from the Mie–Grüneisen equation that

$$\gamma = V\left(\frac{\partial p}{\partial E}\right)_V.$$

Now

$$V\left(\frac{\partial p}{\partial E}\right)_V = \frac{V}{C_V}\left(\frac{\partial p}{\partial T}\right)_V$$

$$= -\frac{V}{C_V}\left(\frac{\partial p}{\partial V}\right)_T\left(\frac{\partial V}{\partial T}\right)_p.$$

which gives

$$\gamma = -\frac{V}{C_p}\left(\frac{\partial p}{\partial V}\right)_S\left(\frac{\partial V}{\partial T}\right)_p.$$

Now

$$V\left(\frac{\partial p}{\partial V}\right)_S = -K_S$$

and

$$\frac{\partial V}{\partial T} = V\alpha,$$

so that

$$\gamma = \frac{V\alpha K_S}{C_p}.$$

This expression is known as Grüneisen's ratio.

In view of the fact that it is not in general possible to calculate isothermal pressure–density equations of state from first principles, or from almost first principles, using either the idea of an ionic potential or of a metallic pseudopotential, it is desirable to have empirical equations which might be used to express experimental data in a compact form, and to interpolate among experimental data, and even perhaps to extrapolate beyond the range of experiment. The ideas behind the construction of such equations are extensions of those ideas of strain and strain energy functions which hold for infinitesimal strain to circumstances in which the strains are far from infinitesimal. Most treatments depend on the work of Murnaghan (1944, 1951) as developed by Birch (1947, 1952).

Suppose that x_i are the co-ordinates of a point after deformation and y_i those before, and let a strain tensor be defined as

$$e_{jk} = \frac{1}{2}\left(\frac{\partial y_i}{\partial x_j}\frac{\partial y_i}{\partial x_k} - \delta_{jk}\right),$$

where δ_{jk} is the Kronecker delta. Then, if the strain is taken to be hydrostatic, as is generally assumed in constructing equations of state,

$$x_i = (1+\alpha)y_i$$

and so

$$\frac{\partial x_i}{\partial y_j} = (1+\alpha)\delta_{ij}$$

and $e_{jk} = \frac{1}{2}[(1+\alpha)^{-2}-1]\delta_{jk} = \varepsilon\delta_{jk}$, say.

The corresponding changes in volume and density are

$$\frac{V_0}{V} = \frac{\rho}{\rho_0} = (1+2\varepsilon)^{3/2}.$$

While this definition of strain, e_{jk}, seems a natural extension of infinitesimal strain, Knopoff (1963) has pointed out that it is not unique and that there are other tensors of rank 2 that reduce to the infinitesimal strain form, and he has in particular suggested the form

$$\eta_{ij} = e_{ij} + \alpha_2 e_{ik}e_{kj}$$

Birch supposed that the strain energy, E, could be expressed as a power series in the strain, ε;

$$E = \sum_{n=2}^{\infty} a_n \varepsilon^n,$$

where the coefficients a_n are in general functions of temperature.

It may then be shown that the pressure $-\partial E/\partial V$ may be written as a function of ρ/ρ_0 as follows:

$$p = \tfrac{3}{2} K_0 \left(\frac{\rho}{\rho_0}\right)^{5/3} \sum_{n=2}^{\infty} a_n \left[\left(\frac{\rho}{\rho_0}\right)^{2/3} - 1\right]^{n-1}.$$

K_0 is the bulk modulus at zero pressure and α_n is equal to $na_n/2a_2$. For example α_2 is 1 and α_3 is $3a_3/2a_2$.

If the strain energy contains just the quadratic term $a_2\varepsilon^2$, it follows that

$$p = \tfrac{3}{2} K_0 \left[\left(\frac{\rho}{\rho_0}\right)^{7/3} - \left(\frac{\rho}{\rho_0}\right)^{5/3}\right].$$

This is a form of the Birch–Murnaghan equation, widely used for interpolations of experimental data, especially in the highly compressible alkali metals.

Since an important result of seismic studies is the function Φ, equal to $(\alpha^2 - \tfrac{4}{3}\beta^2)$ or K/ρ, it is desirable to see how the values of K predicted by theory or observed in experiment depend on pressure. The Birch–Murnaghan equation leads to

$$K = \tfrac{7}{3} p \left[1 + \frac{3}{7}\left(\frac{2}{3}\right)^{5/7}\left(\frac{K_0}{\rho}\right)^{2/7}\right],$$

so that $\partial K/\partial p$ tends to $\tfrac{7}{3}$ at high pressures and the somewhat higher value of 2.6 at low pressures. The values of $\partial K/\partial p$ found for many planetary materials are considerably greater, so that the Birch–Murnaghan equation does not represent their behaviour well. However, many results from shock-wave experiments can be represented quite well if the strain energy is supposed to contain a cubic term and α_3 is taken to be between $-\tfrac{1}{2}$ and $+1$ (Ahrens, Anderson and Ringwood, 1969; Takeuchi and Kanamori, 1966). Knopoff's expression for the strain, taken with a quadratic strain energy, leads to the equation

$$p = \frac{3K_0}{2(1+\alpha_2)}\left[\left(\frac{\rho}{\rho_0}\right)^{\frac{1}{3}(7+4\alpha_3)} - \left(\frac{\rho}{\rho_0}\right)^{\frac{1}{3}(5+2\alpha_3)}\right],$$

which of course reduces to the Birch–Murnaghan form when α_3 is zero.

The value of $\partial K/\partial p$ at high pressure is $\frac{1}{3}(7+4\alpha_3)$ and is larger at low pressures. In one respect, all these finite strain equations reproduce the behaviour of real materials, for it is found that $\partial K/\partial p$ is generally less at high than at low pressures.

In view of the fact that $\partial K/\partial p$ is often not very dependent on pressure, it might be sufficient to take

$$K = K_0 + K_0'p,$$

where $K_0' = \partial K/\partial p$ is supposed to be constant.

It then easily follows that

$$p = \frac{K_0}{K_0'}\left[\left(\frac{\rho}{\rho_0}\right)^{K_0'} - 1\right],$$

an equation due to Murnaghan (1944), which is almost equivalent to the Birch–Murnaghan equation.

Thomsen and Anderson (1971) have argued that it is not entirely consistent to obtain the total strain at a given pressure and temperature by first finding the isothermal strain from an equation of the Birch–Murnaghan type and then finding the thermal strain from the Mie–Grüneisen equation. They point out that formally all empirical equations can be regarded as Taylor series expansions about an initial point determined by the initial temperature and pressure and that the free energy should be therefore written in terms of the total strain, isothermal plus isobaric, measured from that initial condition.

Let the density at $p = 0$, $T = T_0$ be denoted by ρ_{00} and write x for the ratio ρ/ρ_{00}, where ρ is the density at (p, T). The strain, ε, is then $\frac{1}{2}(x^{2/3} - 1)$.

Anderson and Thomsen take the free energy, F, to be a double Taylor series in powers of strain ε and temperature deviation, t (equal to $T - T_0$).

Let differentiation with respect to t be denoted by the suffix t and that with respect to ε by the suffix ε. Let is be understood that all differential coefficients, such as $F_{\varepsilon t}$, are to be evaluated at $t = 0$, $\varepsilon = 0$.

Then the free energy at ρ, T, is

$$\begin{aligned}
F(\rho, T) &= F_{00} + F_t t + \tfrac{1}{2}F_{tt}t^2 + \cdots + g(t)\\
&\quad + (F_\varepsilon + F_{\varepsilon t}t + \tfrac{1}{2}F_{\varepsilon tt}t^2 + \cdots)\varepsilon\\
&\quad + \tfrac{1}{2}(F_{\varepsilon\varepsilon} + F_{\varepsilon\varepsilon t}t + \cdots)\varepsilon^2 + \cdots
\end{aligned}$$

In this expression, $g(t)$ takes account of anharmonic terms in the lattice vibrations and has the form:

$$g(t) = -3N_A k[(t + \theta)\ln(1 + t/\theta) - t],$$

where N_A is Avogadro's number.

It is found by experiment that zero order and first order coefficients are zero:

$$F_{00} = F_t = F_\varepsilon = 0.$$

Further,

$$F_{\varepsilon\varepsilon} = 9K_{00}/\rho_{00}$$
$$F_{\varepsilon\varepsilon\varepsilon} = -27K_{00}(K'_{00} - 4)/\rho_{00}$$
$$F_{\varepsilon\varepsilon\varepsilon\varepsilon} = -27K_{00}(K''_{00} + K'_{00} + \tfrac{1}{9})/\rho_{00}$$
$$F_{\varepsilon\varepsilon t} = -3\alpha_{00}K_{00}/\rho_{00}.$$

K'_{00} and K''_{00} denote the first and second derivatives with respect to pressure of K_{00}, evaluated at $p = 0$, $t = 0$.

The equation of state found from that free energy is somewhat similar to the Birch–Murnaghan equation with additional terms to allow for a cubic term in the strain energy and for thermal expansion:

$$p(\rho, T) = \tfrac{3}{2}K_{00}(x^{7/3} - x^{5/3})[1 - \xi(\theta)(x^{2/3} - 1)] - \frac{1}{qV_{00}}x^{5/3}F_{\varepsilon\varepsilon\varepsilon\varepsilon}\varepsilon^3$$
$$+ t\alpha_{00}K_{00}x^{5/3}[1 + 3\varepsilon(\delta_{00} - K'_{00} + \tfrac{5}{3})].$$

δ_{00} is the combination $(1/\alpha K)(\partial K/\partial T)_p$ evaluated at $p = 0$, $t = 0$,

The detailed compositions of the terrestrial planets are not known and it would be very helpful if some general rules for the behaviour of material at high pressures could be set out. Two rules in particular would be useful; one giving the density and bulk modulus at zero pressure as some functions of an average composition, and one giving the dependence of bulk modulus on pressure. So far as theory takes us, the zero-pressure values appear to be arbitrary; they enter explicitly in the empirical equations derived from the theory of finite strain, and they would require detailed calculations in lattice or pseudopotential theory. However, as will be seen later in this chapter, some general empirical rules appear to follow from experiment. On the other hand, theory suggests values of $\partial K/\partial p$ of the order of 4, decreasing as pressure increases, and the question arises whether it is possible to predict any asymptotic value to which $\partial K/\partial p$ would tend for all materials at high pressures.

An answer to that question is offered by the Thomas–Fermi–Dirac statistical theory of the atom, supposed to be valid when the density is so high that all eigenstates are mixed. The internal energy is then that of Z independent electrons (Z is the atomic number) in the field of the nucleus. Each nucleus is thought of as occupying its own spherical cell of volume equal to the atomic volume, with the cloud of Z free electrons around it.

In the original Thomas–Fermi theory, the potential was taken to satisfy Poisson's equation subject to the boundary conditions that it and its derivative should vanish at the boundary of the elementary sphere and that it should vary as $1/r$ at the centre. When the electrons satisfy Fermi–Dirac statistics, an exchange term has to be included in Poisson's equation. A great deal has been written about the Thomas–Fermi–Dirac equation and its properties (see, for example, Feynman, Metropolis and Teller, 1949; Teller, 1962; Boschi and Caputo, 1969; Gilvarry, 1954; Gilvarry and Peebles, 1954; March, 1955; and especially Gilvarry, 1969), but since the approximations inherent in the Thomas–Fermi procedure clearly fail at pressures within planetary interiors (for ionic crystals and metals still exist as such at those pressures) it suffices for the present purpose to consider only the high pressure asymptotic form of the equation, to which it is supposed that all other equations of state will tend.

Gilvarry (1969) shows that the pressure exerted by a material of atomic number Z at sufficiently high pressure and zero temperature is

$$p = \frac{h^2}{5m} \left(\frac{3}{8\pi}\right)^{2/3} \left(\frac{Z}{4V}\right)^{5/3} \left[1 - \frac{2me^2}{h^2}(4ZV)^{1/3}\right].$$

Here m is the mass and e the charge of the electron and V the atomic volume. The first term is the pressure exerted by the electrons in the volume V, the second is the effect of electron–electron and electron–nucleon interactions.

It follows that the bulk modulus $-V \, \partial p/\partial V$ of a Thomas–Fermi–Dirac substance at zero temperature is

$$\frac{5a}{3V^{5/3}} - \frac{4}{3}\frac{b}{V^{4/3}}$$

where

$$a = \frac{h^2}{5m} \left(\frac{3}{8\pi}\right)^{2/3} \left(\frac{Z}{4}\right)^{5/3}$$

and

$$b = 2\frac{me^2}{h^2}(4Z)^{1/3}.$$

The gradient of the bulk modulus is

$$\frac{\partial K}{\partial p} = \frac{5}{3}\left\{1 + \frac{4}{25}\frac{b}{a}\left(\frac{a}{p}\right)^{1/5} - \frac{4}{125}\left(\frac{b}{a}\right)^2\left(\frac{a}{p}\right)^{2/5} + \cdots\right\},$$

which approaches the value of $\frac{5}{3}$ at high pressures, the more rapidly the greater Z.

Attention may be called to the following features of the Thomas–Fermi–Dirac equation. In the first place, the relation between pressure and volume at zero pressure depends only on Z. That suggests that the equations of state of real materials at lower pressures might also depend primarily on atomic number. The bulk modulus, likewise, is a function only of atomic volume and Z, while, as has been said, $\partial K/\partial p$ approaches $\frac{5}{3}$. It will be seen later how far these properties are reliable guides to the behaviour of terrestrial materials at high pressures.

4.3   Experimental determinations of equations of state

At low pressures and moderate temperatures it is nowadays a straightforward matter to determine the density of a substance as a function of pressure and temperature by direct compression under hydrostatic pressure. Pressure can be measured by a pressure balance, temperature by thermocouples or resistance thermometers and density by change in the volume of the specimen or by X-ray diffraction measurements of lattice spacing. All these techniques fail or become very difficult at pressures and temperatures encountered at even moderate depths in the Earth. Liquids cannot be used to transmit pressures in hydrostatic systems at pressures greater than some 5×10^9 Pa because all suitable ones solidify. Thus direct measurement of pressure and hydrostatic compression both become impossible. Some way of producing high pressures without hydraulic amplification is needed, and some way of estimating the pressure that does not rely on direct hydrostatic measurement. High temperatures also become more difficult to attain because the strengths of materials, and thus the ability to confine systems at high pressure, decrease materially at temperatures of 1000 K or so.

The problem of producing high pressures has been tackled by designing presses in which the area of faces in contact with the specimen is much less than the area to which hydrostatic pressure is applied. There is then a problem of preventing the specimen being squeezed out from between the jaws of the press which has been overcome in presses of tetrahedral or spherical form. The first device to overcome the extrusion problem was, however, the 'belt' apparatus (Hall, 1960) in which the pistons of tungsten carbide compressed a specimen with a die of tungsten carbide surrounded by a steel ring or belt (Figure 4.1). It was with this apparatus that diamonds were first made artificially (Bundy, 1963). The tetrahedral press was invented by Hall (1958) and comprises four anvils of tungsten carbide in tetrahedral form squeezed together in a hydraulic press (Figure 4.2). Pressures of the order of 10^{10} Pa have been attained in such presses, sufficient to produce diamonds artificially. It was the tetrahedral press which enabled a very great deal of high pressure physics and chemistry to be undertaken, giving as it does a high pressure gain and overcoming the problem of containing the specimen. A somewhat similar design is that of Kawai (1971) who cuts a solid of revolution into tapering sections; either a sphere or a cylinder may be used (Figure 4.3). If the area of the internal faces is a and that of the external faces is A, a pressure p applied to the latter will cause a pressure Ap/a to be exerted by the internal faces on the specimen. As in the Hall press, the specimen is confined in a small volume in the centre. If the diameter of the sample is 2 mm and that of the outside surface 250 mm, the ratio of areas is 15 000,

Figure 4.1. The 'belt' high pressure apparatus. (After Hall, 1960.)

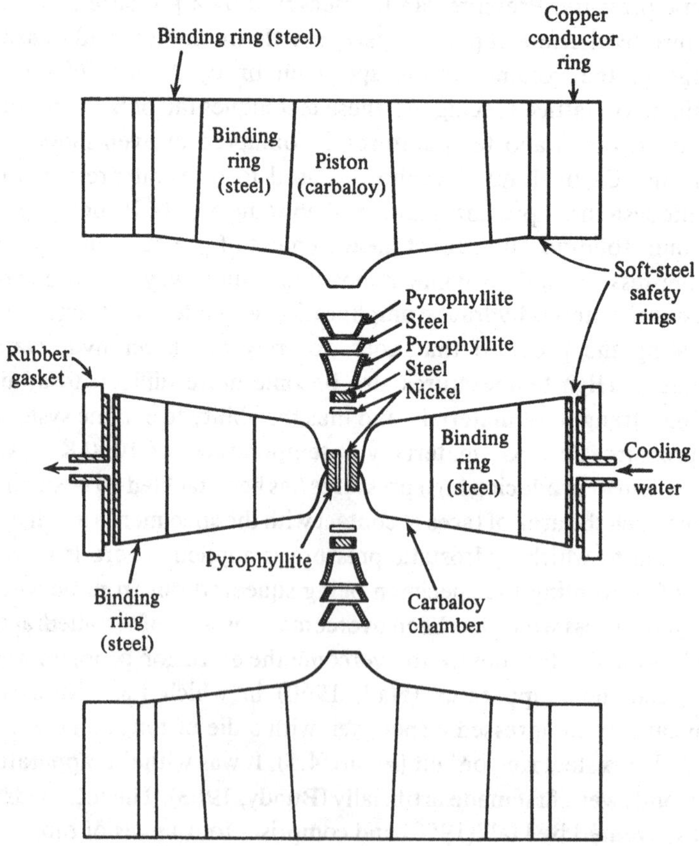

Figure 4.2. The tetrahedral press. (After Hall, 1960.)

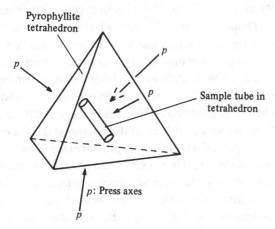

which should be the pressure gain. Kawai estimates that he can attain pressures of 8×10^{10} Pa. In a further development (Suito, 1972) an inner cubical press of eight tungsten carbide anvils is contained within an outer press of steel spherical segments. Anvils of the type devised by Bridgman have also been used extensively up to about 1.5×10^{10} Pa by Ringwood and his collaborators (Ringwood and Major, 1968).

The Hall and Kawai presses use tungsten carbide as the material of the anvils, and the maximum pressure that can be used is limited by the crushing strength of tungsten carbide. Diamond has a greater strength, and with the advent of artificial diamond it has become possible to

Figure 4.3. Spherical press. (After Kawai, 1971.)

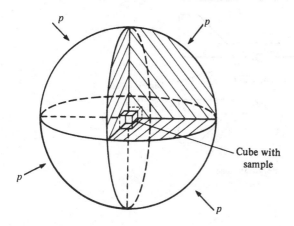

construct high pressure presses of it either as anvils (Figure 4.4, Merrill and Bassett, 1974) or in the form of a diamond piston working in a diamond cylinder. Diamond has a further advantage over tungsten carbide in that it is transparent to light and X-rays. This means that X-ray diffraction can be used to determine changes of lattice spacing and of crystal structure (Munro, 1967), while intense laser radiation can be used to heat specimens (see Liu, 1975*a*, *b* for description of a diamond anvil used with a YAG laser). A further advantage is that the temperature of the specimen may be measured with an optical pyrometer, so avoiding problems of thermocouple calibration at high pressure (see Kawai, 1971).

The determination of the stress in the specimen is a severe problem in all non-hydrostatic apparatus, that is, in all equipment used at the highest pressures. There are three difficulties. In the first place, the stress is almost certainly not hydrostatic; for example, because the material tends to be extruded from between anvils. Secondly, it is not possible to calculate the pressure from the ratio of areas of inner and outer faces of a Hall or Kawai type of press, or from the geometry of a piston and cylinder apparatus, because friction between components, which increases with increasing pressure, reduces the pressure on the specimen. Thirdly, direct static measurements of pressure cannot be made absolutely with a pressure balance above some 5×10^9 Pa because of the freezing of liquids. For all these reasons it is desirable at lower pressures, and essential at higher pressures, to establish a pressure scale based, like a practical temperature scale, on fixed points, such as phase transitions at known

Figure 4.4. Diamond anvil press. (After Merrill and Bassett, 1974.)

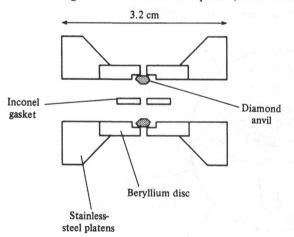

pressures. The types of such transition are phase transitions in bismuth, detected by variations of electrical resistance.

Even so, fixed points can only be established if a mechanical measurement of pressure can be made, and, above the range of the pressure balance, estimates of pressures of fixed points must be unsure. Thus, much recent work on phase transitions at high pressures has relied on a calculated pressure–density relation to establish pressure, in particular the relation for sodium chloride (Weaver, Takahashi and Bassett, 1971). In essence a suitable equation of state is used to interpolate between the properties measured statically at low pressures and those found dynamically at high pressures from shock-wave studies.

Let the energy of the solid be written as

$$E(V, T) = \phi(V) + E_v(V, T)$$

where ϕ is the static lattice energy and E_v the vibrational energy. Then since

$$p = -\left(\frac{\partial E}{\partial V}\right)_T + T\left(\frac{\partial p}{\partial T}\right)_V,$$

$$p = -\frac{d\phi}{dV} + p_v$$

and

$$K_T = V\frac{d^2\phi}{dV^2} + K_v.$$

p_v and K_v are the contributions from the lattice vibrations.

A central force model for the interionic potential is used:

$$\phi = \frac{\alpha_r e^2}{r} - \frac{c}{r^6} - \frac{d}{r^8} + b\,e^{-r/\rho},$$

the repulsive core being represented by $b\,e^{-r/\rho}$.

Then it is found that

$$p = -p_M x^{-4/3} - p_6 x^{-3} - p_8 x^{-11/3} + p_R x^{-2/3} \exp\left(-r_0 x^{1/3}/\rho\right) + p^*$$

and

$$K_T = -\tfrac{4}{3}p_M x^{-4/3} - 3p_6 x^{-3} - \tfrac{11}{3}p_8 x^{-11/3}$$
$$+ \tfrac{1}{3}p_R x^{-2/3}(2 + r_0 x/\rho) \exp\left(-r_0 x^{1/3}/\rho\right) + K^*$$

In these expressions, x is the volume ratio V/V_0 and p_M, p_6, p_8, p_R, r_0 and ρ are parameters fitted to the low pressure equation of state.

Various models were taken to obtain p^* and K^*, which are functions that match the low pressure to the high pressure equation of state. They are of the Mie–Grüneisen and Hildebrand type already discussed, but different formulations were chosen.

It was found that of the models studied, four agreed to within 4×10^8 Pa at 2.8×10^{10} Pa and so it was estimated that the best calculated values should give a pressure–temperature–density relation correct to about 1 per cent.

The experimental data used to determine numerical values for sodium chloride included

Madelung constant, α_r: 11.85×10^{-5} deg
Coefficient of volume expansion, α_0: $1.747\,56 \times 10^{-4}$/deg
Adiabatic bulk modulus, K_S: 2.5×10^{10} Pa
$-(\partial K_S/\partial T)_p$: 0.109×10^8 Pa/deg
Isothermal bulk modulus, K_T: 2.374×10^{10} Pa
$\partial K_T/\partial p$: 5.35

As will be seen below, K_S and K_T at zero pressure are much less than for many minerals while $\partial K_T/\partial p$ is greater.

Static methods, with which pressures of over 3×10^{10} Pa at temperatures of 3000 °C have been attained, have in recent years been used mainly for studies of phase changes, a subject which will be discussed in a later section. Even pressures of 3×10^{10} Pa are attained at relatively moderate depths in the Earth, scarcely into the lower mantle, and at those depths the changes of density are dominated by changes of phase and composition. Equations of state at greater pressures and depths, where hydrostatic compression is relatively more significant, must be studied with shock waves. It has been seen also that the pressure scale established through the equation of state of sodium chloride depends on shock-wave results; we turn therefore to the shock-wave method for determining the behaviour of minerals and metals at high pressures.

If a slab of material is given a strong blow on one side, intense enough that the resulting strain cannot be considered as infinitesimal, a pressure wave travels through the slab at a speed greater than the speed of sound. The pressure rises in a thin layer of material to the value set up by the blow and, behind the thin layer (the shock front), the material as a whole is set into motion. When the shock reaches the far side of the slab it is reflected as a rarefaction wave, so satisfying the condition that the pressure on that side should be zero (or atmospheric pressure which is near enough to zero). The whole slab is then moving at the speed of the material behind the shock.

The value of shock-wave studies lies in the fact that the equations of conservation of mass, momentum and energy, together with the measured velocities of the shock wave and the bulk material, enable the pressure, density and internal energy of the shocked material to be calculated.

Each shock of given strength thus gives one point specified by p and ρ in the shock. If material is given a series of shocks of different strengths, then a curve of density against pressure can be built up. It is neither an isotherm nor an adiabat, for the internal energy and the temperatures vary from point to point along the curve. It is known as the Hugoniot equation of state, and a major problem is the deduction of an isothermal equation of state from the Hugoniot equation.

Consider a shock front (Figure 4.5) propagating at a speed u_s into material of density ρ_0 and pressure p_0 at rest. The pressure rises in the front to p and the bulk material behind moves with a velocity u and attains a density ρ. Then the rate at which mass enters unit area of the shock is $\rho_0 u_s$, while the rate at which mass leaves the unit area of the shock is $\rho(u_s - u)$.

Conservation of mass requires the two rates to be equal:

$$\rho_0 u_s = \rho(u_s - u).$$

Consider now the rate of change of momentum per unit area in the shock. The rate at which momentum is generated in the shock is equal to the rate of flow of mass through the shock, namely $\rho_0 u_s$, multiplied by the velocity acquired, u, and it is also equal to the difference of pressure across the front. Thus

$$p - p_0 = \rho_0 u_s u.$$

Finally, the rate at which work is done on the material passing through the shock is equal to the rate of flow of mass through the shock ($\rho_0 u_s$)

Figure 4.5. Velocity, u, pressure, p, and density, ρ, in a shock front.

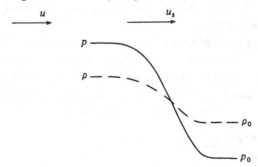

multiplied by the change of kinetic energy per unit mass ($\frac{1}{2}u^2$) plus the change of internal energy per unit mass ($E - E_0$). The rate is also equal to the rate at which the pressure does work on the material. Thus

$$pu = \rho_0 u_s(\tfrac{1}{2}u^2) + \rho_0 u_s(E - E_0).$$

The three equations give

$$u_s = V_0 \frac{(p - p_0)^{1/2}}{(V_0 - V)}$$

$$u = \tfrac{1}{2}\{(p - p_0)(V_0 - V)\}^{1/2}$$

$$E - E_0 = \tfrac{1}{2}(p + p_0)(V_0 - V),$$

where V_0, V are the specific volumes equal to ρ_0^{-1} and ρ^{-1} respectively. Since p_0, ρ_0, u_s and u are known, these equations may be solved for p, ρ and ($E - E_0$). In many cases, the Hugoniot equation takes a simple form, for it is often found that there is a linear relation between u_s and u, namely

$$u_s = c_0 + \lambda u$$

where c_0 is the velocity of sound, to which u_s reduces when the shock is weak and u very small.

When that relation is satisfied, the Hugoniot equation is found to be

$$p = p_0 + \frac{c_0^2(V_0 - V)}{V_0 - \lambda(V_0 - V)^2}.$$

The reduction of the Hugoniot equation to an isotherm depends on the use of the Mie–Grüneisen equation:

$$p - p_0 = \frac{\gamma}{V}(E - E_0)$$

for two states (p, E) and (p_0, E_0) having the same volume.

Various schemes of calculation have been used to effect these reductions (Takeuchi and Kanamori, 1966; Knopoff and MacDonald, 1960; Shapiro and Knopoff, 1969; Ahrens, Anderson and Ringwood, 1969; Davies and Anderson, 1971).

The main difficulty is to obtain a value of γ at high pressures (Knopoff and Shapiro, 1969). As will be realized from the earlier theoretical discussion, γ is a somewhat empirical quantity and there is no sound theoretical basis on which to predict its variation. Empirical relations for the dependence of γ on volume have been proposed:

$$\gamma = -\tfrac{1}{2}V \frac{\partial^2 p/\partial V^2}{\partial p/\partial V} - \frac{2}{3} \quad \text{(Slater, 1940)}$$

or

$$\gamma = -\tfrac{1}{2}V\frac{\partial^2(pV^{2/3})/\partial V^2}{\partial(pV^{2/3})/\partial V} - \tfrac{1}{3} \qquad \text{(Dugdale and MacDonald, 1953)}$$

but the most reliable calculations are probably those which employ an experimentally determined value of γ.

Now γ is equal to $K\alpha V/C_p$, and it should be possible to determine it from shock-wave studies on material of the same K, α and C_p, but of different specific volume or density. In fact, if the Mie–Grüneisen equation of state is substituted into the Hugoniot equation, it is found that

$$\gamma = \frac{V(p_H - p_T)}{\tfrac{1}{2}(p_H + p_{H0})(V_0 - V) + (E_{H0} - E_T)}.$$

In this expression p_H is the pressure on the Hugoniot curve, p_T the isothermal compression giving the same specific volume V, p_{H0} is the zero pressure on the Hugoniot curve and E_{H0} and E_T have corresponding meanings.

It is possible to reduce the specific volume of a material while keeping the other parameters constant by preparing it in a porous sintered form. Thus Altshuler, Krupnikov and Brazhnik (1958) and Altshuler *et al.* (1958*a*) carried out experiments on sintered iron and found a value for γ of about 1.6 at 10^{11} Pa and, more recently, experiments have been done on sintered magnesium oxide, a material of considerable importance in studies of the mantle of the Earth (Carter, Marsh, Fritz and McQueen, 1971). The latter experiments show that $\rho\gamma$ is nearly constant.

Comprehensive reviews of the methods of shock-wave experiments have been given by Rice, McQueen and Walsh (1958) and by Duvall and Fowles (1967). Shock waves are generated by driving a metal plate at high velocity against specimens of the material, for in that way it is possible to generate stronger shocks than by detonating an explosion against the specimen. The plate is driven by an explosive charge (Figure 4.6), the composition of which determines the speed of the plate and hence the strength of the shock.

In order that the plate may move with a uniform velocity the explosive charge must burn uniformly across its area and this means that it must be ignited at the same time all over. That condition is answered by igniting the main charge from a lens-like charge itself ignited by a detonator. The lens charge is made up in such a way that the time taken for the ignition wave to travel from the detonator to the main charge is everywhere the same.

In the earliest experiments the time of arrival of the shock was shown by the specimen being forced into contact with a pin, so making an electrical circuit. By drilling holes of different depths in the specimen and inserting pins in them, the arrival of the shock at different distances through the specimen could be detected and thus the speed calculated. The method is applicable only to metals and a more versatile method uses the flash of light generated by the shock heating of gas (usually argon) trapped between the surface of the specimen and the surface of a block of leucite. A number of specimens with a range of thicknesses is carried on a support against which the driver plate is forced (Figure 4.6) and, by placing leucite blocks close to the support and the rear surface of the specimens, flashes of light are generated at times corresponding to the times taken for the shock to pass through the different specimens. Measurements of the times of the successive flashes taken with different thicknesses of the blocks yield the shock velocity. The mass velocity behind the shock is found from the time taken by a specimen to travel across a known gap to a leucite slab. The flashes are recorded photographically on a streak camera placed either behind the specimens or to one side.

Most experiments have been done on minerals and metals, but some results are available for liquids, in particular water and mercury. Shock pressures of up to 5×10^{11} Pa have been achieved, beyond the pressures attained at the centres of the terrestrial planets, but less than those attained in the major planets. One difficulty in using the results has

Figure 4.6. Generation of shock waves in solid specimens.

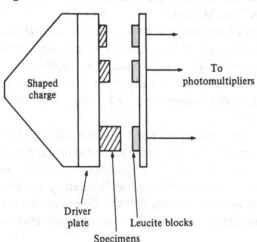

already been discussed, namely the uncertainty in the reduction of the Hugoniot curve to an isotherm or adiabat. There are two other objections which have been advanced against the uncritical use of shock-wave data in planetary studies. The first is that the shock wave does not set up a hydrostatic system of stresses and it is therefore a question of how far the Hugoniot equations of state may depart from a hydrostatic equation: what influence in fact does the strength of the material have on the behaviour of a shocked solid? Carter, Marsh, Fritz and McQueen (1971) have discussed the point and conclude that the Hugoniot curve is a hydrostat, at least if a small constant correction is made for the effects of strength.

The second possible difficulty lies in the time it takes for a material to take up its new density as the shock wave passes. Is there an effective bulk viscosity which prevents the density attaining its ideal value? The question is most serious, probably, when shock waves are used to study transformations of minerals to forms stable at high pressure, for, if the transition is too slow, its occurrence would be missed. The times in question are a few microseconds. It does indeed seem to be the case that some transitions observed in static experiments are not found in shock studies (Ahrens, Anderson and Ringwood, 1969). Fortunately studies of phase transformations can be made statically. Shock-wave studies do then give the properties of those high pressure phases which result from the shock. These questions will be considered further when high pressure transformations are discussed.

4.4   Pressure–density relations

The chemical and mineralogical compositions of the terrestrial planets are not known. Arguments based on the composition of the Earth's mantle and of meteorites lead to plausible suggestions but not to certainty, so that the composition of the zones of the Earth must be inferred from comparisons of the pressure–density relation established from seismic data with the results of experiment, while, for the other planets, for which no seismic data exist, a whole range of models may be constructed that fit the values of size, mean density and moment of inertia (when known). The composition of the Earth was discussed briefly in Chapter 2 and models of the Moon are considered in Chapter 5 and of the terrestrial planets in Chapter 6. A planet might be composed of a great variety of metal silicate minerals, and the range of possibilities is such that it would be pointless to construct all possible models, for many would no doubt be very similar. Even for the Earth the available data would not allow of other than broad distinctions to be drawn between possible models, and far less for the other planets. It turns out that the density of a silicate mineral, at zero pressure, is determined almost entirely by its mean atomic weight. Thus, pure silicon itself is the least dense whilst the densest silicates are those with high proportions of iron. Anderson (1967) found that the density at zero pressure and mean atomic weight were related to the seismic parameter Φ, equal to $(\alpha^2 - \frac{4}{3}\beta^2)$ or K_0/ρ_0, by the empirical formula

$$\rho_0/\bar{M} = a\Phi^n,$$

where ρ_0 is the density at zero pressure in kg/m^3, $\bar{M}$ is the mean atomic weight, and Φ is in $(km/s)^2$.

Anderson gave two sets of values for a and n. Either a is 48 and n is 0.323 or, for more compact minerals, a is 49.2 and n is $\frac{1}{3}$ (Anderson, 1969).

Anderson's relation, it will be noted, is somewhat different from the indication of the Thomas–Fermi–Dirac equation, which would suggest a relation between mean atomic *number* and density and elastic moduli. However, on the range of atomic weight and atomic number encountered in silicates, a relation in terms of atomic number would probably be as satisfactory as one in terms of atomic weight.

Anderson's relation may be used in one of two ways. If ρ_0 and Φ are known, as they are for the Earth, it is possible to infer their values at zero pressure and, hence, with the help of Anderson's relation, to establish the mean atomic weight. On the other hand, if values of ρ_0 and $\bar{M}$ are postulated, a value of K_0 is entailed by Anderson's formal relation, so that the change of density with pressure for low pressures may be calculated. Formally

$$K_0^n = \frac{\rho_0^{n+1}}{a\bar{M}}.$$

Thus

$$\frac{d\rho}{\rho_0} = \frac{dp}{K_0} = \frac{dp}{\rho_0^{1+1/n}}(a\bar{M})^{1/n}$$

or

$$\rho_0^{1/n}\,d\rho = (a\bar{M})^{1/n}\,dp.$$

The hydrostatic equation gives

$$dp = -g\rho\,dr,$$

with

$$g = \frac{G}{r^2} \int 4\pi \rho r^2 \, dr$$

to complete a set of equations that may be integrated to give ρ and p as functions of radius.

If a value of ρ_0 is postulated, integration of the equations yields the mass of a planet of specified radius.

Here it is assumed that K_0 is the value at zero pressure. However, K does not remain constant as the pressure increases, and the bulk modulus at a pressure of 10^{11} Pa may be much greater than at zero density. Table 4.1 contains values for K_0 and $\partial K_0/\partial p$ taken from Ahrens, Anderson and Ringwood (1969). The values given are the isothermal ones; the adiabatic bulk modulus is somewhat greater, but the pressure gradients of the isothermal and adiabatic bulk moduli are hardly distinguishable.

It was suggested in section 4.2 that the value of $\partial K/\partial p$ would tend at very high pressures towards the value of $\frac{5}{3}$, or 1.67, the value for the Thomas–Fermi–Dirac equation of state; the values of $\partial K/\partial p$ for the three silicates given in Table 4.1 are all less than 1.67, and that for fayalite is negative. Values for oxides are, however, all greater than 1.67.

$\partial K/\partial p$ itself is not constant but falls with increase of pressure, at least for MgO, Al_2O and stishovite (SiO_2) as shown in Figure 4.7.

Experiments at high pressure give equations of state directly, but information about the derivatives at low pressures comes also from the

Table 4.1. *Values of K_0 and $\partial K_0/\partial p$ for some silicates and oxides* (from Ahrens, Anderson and Ringwood, 1969)

Mineral	$K_0(10^{11}$ Pa$)$	$\partial K_0/\partial p$
Oxides		
Al_2O_3 corundum	2.901	3.24
Al_2O_3 ceramic	2.532	3.94
MgO periclase	1.648	4.06
SiO_2 α-quartz (stishovite)	3.627	3.04
MnO_2 pyrolusite	3.390	1.67
Fe_3O_4 magnetite	4.483	1.79
Fe_2O_3 haematite	3.814	1.88
Silicates		
Mg_2SiO_4 forsterite	4.307	1.07
Fe_2SiO_4 fayalite	3.953	−1.66
$MgSiO_3$ enstatite	3.277	1.50

elastic constants of solids. Many observations have been made by acoustic methods, the elastic constants being obtained from the times of travel of sound waves through crystals. Data relating to minerals of geophysical interest have been reviewed by Anderson, Schreicher, Liebermann and Soga (1968).

Some of the first shock-wave experiments were done on metals (Rice, McQueen and Walsh, 1957; Altshuler *et al.*, 1958*a*; Altshuler, Krupnikov and Brazhnik, 1958). Takeuchi and Kanamori (1966) have reduced Hugoniot equations of state to isotherms and adiabats and, according to their results, the bulk modulus of iron is well represented by

$$K = 2 + 3.6p$$

below a pressure of 4×10^{11} Pa, but, at pressures of 10^{12} Pa or more, the slope $\partial K / \partial p$ falls to about 2.8.

The high pressure equations of state for oxides can well be fitted by the lattice dynamic type of equation dismissed in section 4.2 with values of n and m about 6 and 1, respectively, but it is difficult to see how the behaviour of olivines and enstatite can be represented in a similar way.

Quantum mechanical calculations have been carried out for aluminium and iron by Berggren and Fröman (1969). The calculated values of the density of aluminium are quite close to shock-wave values, and the

Figure 4.7. Behaviour of $\partial K / \partial p$ at high pressures.

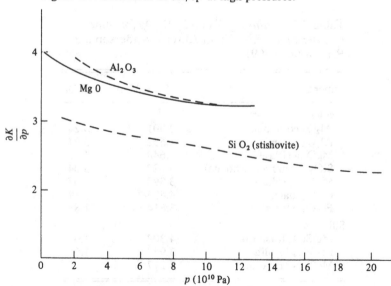

calculated values of bulk modulus are closely represented by

$$K = 1.2 + 2.9p \quad (\text{in } 10^{11} \text{ Pa}),$$

which, over the range from $1\text{–}3 \times 10^{11}$ Pa is not very different from the shock-wave result,

$$K = 2.4 + 2.4p \quad (\text{in } 10^{11} \text{ Pa}).$$

The calculations for iron are in less satisfactory agreement with the shock-wave data. New calculations for iron have recently been carried out by Bukowinski and Knopoff (1976).

According to Bullen's extended incompressibility–pressure hypothesis, materials of terrestrial planets all show, at sufficiently high pressures, a compressibility that varies as

$$K = K_0 + bp,$$

where b is a constant close to 3. The value for the lower mantle and core of the Earth is very close to 3.2 (Chapter 2). Well though such a rule represents the compressibility within the Earth, it is clearly not of universal application. The slopes, b, equal to $\partial K / \partial p$, of a number of possible planetary materials are plotted in Figure 4.8 and it will be seen that they vary widely with composition and substantially with pressure. Clearly the Bullen rule as derived from the Earth may not necessarily apply to the other planets. So far as the Earth is concerned the major question raised by Bullen's rule is why the bulk moduli of such very different materials as those comprising the core and lower mantle of the Earth should agree as closely as they do, and it has been argued (Cook, 1972) that the coincidence is related to the particular size of the core and the pressure that happens to be attained at the boundary between core and mantle. According to a plausible model, the lower mantle consists of the oxides of iron, magnesium, aluminium and silicon, and the bulk moduli of all these oxides happen to be rather close to each other at the pressure at the boundary of the core (1.5×10^{11} Pa); they are also close to the bulk moduli of iron and of the lower mantle and core at the same pressure. Were the pressure at the boundary of the core and mantle to differ by 0.5×10^{11} Pa from its actual value, the coincidence between all these values of the bulk modulus would be appreciably less close.

4.5   Changes of crystal structure

Polymorphic forms of ionic crystals have been known for some time, and Bernal (1936) suggested that common olivine might transform under the pressures and temperatures established in the mantle of the

Earth to a spinel form with a density 9 per cent greater. His suggestion, which was based on the observation (Goldschmidt, 1931) that the analogous compound Mg_2GeO_4 could exist, even at atmospheric pressures, in olivine and spinel forms, was employed by Jeffreys (1937) as the basis for an explanation of the rapid increase in seismic velocities between the upper and lower mantles. Olivine–spinel transitions of a number of analogous minerals are now known, the increase of density in most of them being about 10 per cent and the transition pressures and temperatures ranging from zero to 1.2×10^{10} Pa and 820 to 1000 °C (Ringwood, 1975, p. 839).

Figure 4.8. Values of $\partial K/\partial p$ for planetary materials. (From Cook, 1972.)

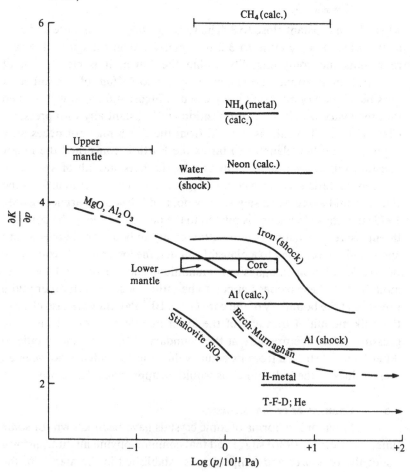

At first it was not possible to study the (Fe, Mg) olivine system directly under static conditions, but that has now been made possible by improvements in experimental apparatus, as indicated in section 4.3, and extensive studies of the olivine and other systems have been carried out. At the same time not all possible planetary materials are susceptible of direct study, and so the investigation of analogous materials remains important, while shock-wave experiments enable the properties of high pressure forms of minerals to be investigated over a wider range of pressure and temperature than do static experiments.

Along the temperature–pressure boundary between two forms of a single substance, such as the boundary between ice and water or olivine and spinel, the Gibbs free energy at constant pressure must be equal for the two forms. The Gibbs free energy G is defined by

$$G = E - TS + pV.$$

Now the differences in the internal energies and entropies of two forms do not depend much on pressure, whereas the product $p \, \Delta V$, where ΔV is the difference of specific volumes, is proportional to pressure. Thus, for a sufficiently high pressure, the Gibbs free energy of the denser form must be less than that of the less dense form of the substance, even though the internal energy is greater, and so a transition to the denser form will take place.

The pressure at which the transition occurs will depend on temperature according to Clapeyron's equation

$$\frac{dp}{dT} = \frac{1}{T} \frac{\Delta H}{\Delta V},$$

where ΔH is the difference of the heat functions $(U + pV)$ for the two forms and ΔV is the difference of specific volumes; ΔH is the latent heat of the transition.

Changes of crystal form are not the only possible transitions, and Birch (1952) suggested that the lower mantle of the Earth might be a mixture of oxides having much the same overall composition as olivine, basing his argument on the agreement of values of the ratio K/ρ for such oxides with the value for the lower mantle.

Within recent years, static experiments have yielded a great deal of information about transitions in the olivine, $(Mg, Fe)_2SiO_4$, pyroxene, $(Mg, Fe)SiO_3$, and other systems. The transition from the olivine form which is hexagonal, to the spinel form, which is cubic, is well established, and a diagram showing the dependence of the transition pressure on composition is shown in Figure 4.9. However, when the proportion of iron is less than 20 per cent, the high pressure crystal structure is not the true spinel form but a birefringent form of lower symmetry known as the β-phase.

Ringwood (1975, Chapter 11) has discussed at length the possible transformations of olivines, initially to the spinel and β-forms, then to a denser form having the structure of strontium plumbate; at yet higher pressures the mineral of olivine composition splits up ('disproportionates') into simple compounds, possibly into (Mg, Fe)SiO (aluminite structure) and (Mg, Fe)O (rock salt structure) and perhaps at still higher pressures into the isochemical mixture of oxides, MgO, FeO and SiO_2.

Lastly, at the highest pressures attained in the mantle of the Earth, it has been suggested that a post-oxide phase may be formed, consisting of silicates denser than the isochemical oxides. One indication of this possibility is found in shock-wave studies, where the densities at pressures greater than 7×10^{10} Pa appear to be some 5 per cent greater than those of the isochemical oxides (McQueen and Marsh, 1966).

Evidence for the transformations at very high pressures has become available quite recently from static experiments. Thus Bassett and Ming (1972) (see also, Ming and Bassett, 1975) studied the disproportionation

Figure 4.9. Dependence of the olivine–spinel–oxide transition on composition at $T = 1000$ K.

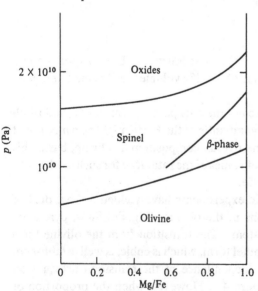

of fayalite up to 2.5×10^{10} Pa and 3300 K, but Liu (1976) found that within that range forsterite (Mg_2SiO_4) did not decompose but adopted a different structure. Liu (1975a) has also investigated the post-oxide phases of forsterite and enstatite and has argued (Liu, 1975b) that the lower mantle is composed of a mixture of perovskite $(Mg, Fe)SiO_3$ and the oxides FeO and MgO.

The possible transformations of olivines are summarized in Table 4.2. Pyroxenes are found to undergo somewhat similar transformations to those of olivines (Ringwood, 1975, Chapter 12), initially to spinel and stishovite according to an equation such as

$$2ABO_3 \rightarrow A_2BO_4 + BO_2$$
(pyroxene)   (spinel) (oxide)

(see Figure 4.10 and Table 4.2).

In the presence of aluminium oxide, garnets are formed:

$$3MgSiO_3.x\,Al_2O_3 \rightarrow x\,Mg_3Al_2Si_3O_{12} + 3(1-x)MgSiO_3$$
(aluminium enstatite)   (pyrope garnet)   (enstatite)

(Ringwood, 1975, Chapter 12; Paprika and Cameron, 1976).

It seems from the experiments of Liu (1974a, b) that, at pressures of about 3×10^{11} Pa, pyroxenes transform to crystals of ilmenite structure (perovskite) which are denser than the isochemical mixed oxides, a conclusion supported by shock-wave results (McQueen and Marsh, 1966).

Table 4.2. *Transformations of forsterite and enstatite*
(from Liu, 1975a)

Formula	Phase	Density (kg/m^3)
Forsterite		
Mg_2SiO_4	olivine	3214
	β-phase	3480
	spinel	3549
$2MgO + SiO_2$	periclase + stishovite	3852
$MgO + MgSiO_3$	periclase + perovskite	3927
Enstatite		
$MgSiO_3$	enstatite	3190
$\frac{1}{2}SiO_2 + \frac{1}{2}Mg_2SiO_4$	spinel + stishovite	3741
$MgSiO_3$	ilmenite	3813
$MgO + SiO_2$	periclase + shishovite	3972
$MgSiO_3$	perovskite	4083

The changes of density in a number of transitions are summarized in Table 4.3.

The transformations which have just been outlined have been established mainly by static experiments at high temperatures, but shock-wave experiments supplement the static ones, for they can attain higher pressures and also some phases produced at high pressures in static experiments may not be observed because they revert to low pressure forms when the pressure is released. On the other hand, some phases may not have time to form in a shocked specimen. The information about equations of state of high pressure forms comes almost entirely from shock-wave studies.

Ringwood (1975, Chapter 3) has argued that the mantle of the Earth is predominantly a mixture of pyroxene and olivine with perhaps some garnet, to which (Ringwood, 1962) he gave the name *pyrolite*, and he has constructed, on the basis of experimental evidence summarized above, the transformations of crystal structure and composition that pyrolite may be expected to undergo. At a depth of about 400 km in the Earth, pyroxene would transform to garnet and olivine to the spinel or β-phase, followed, at about 700 km depth, by transformations to a mixture of oxides and perovskite, with possibly a recombination of oxides to perovskite at yet higher pressures. The changes at 400 and 700 km correspond rather well with major changes in density and elastic properties in the upper mantle, but the evidence for further changes at great depths is less convincing: the expected changes of density are themselves only a few per cent and the seismic evidence, especially the models inferred from free oscillations of the Earth, suggest very smooth changes of density and elastic moduli below 1000 km.

Table 4.3. *Some changes of density in phase transformations*

Low pressure form	High pressure form	$\Delta\rho$ (kg/m^3)	V_l/V_h
Mg_2SiO_4 forsterite	β-phase	+264	1.078
Fe_2SiO_4 fayalite	spinel	+440	1.100
Mg_2SiO_4 β-phase	oxides	+372	1.112
Fe_2SiO_4 spinel	oxides	+460	1.096
$MgSiO_3$ enstatite	oxides	+750	1.245
$MgO + SiO_2$ oxides	perovskite	+110	1.028

V_l: specific volume of low density form.
V_h: specific volume of high density form.
$\Delta\rho$: change of density.

While rather convincing models of the Earth's mantle can be constructed on the basis of known transformations of minerals likely to constitute the mantle, the models cannot be applied in a straightforward manner to the other planets. In the first place, the composition may not be the same as in that of the Earth's mantle, and, as has been seen, the olivine–spinel transformation, as an example, occurs at a pressure which varies with composition (Figure 4.9). Secondly, the transition pressure depends on temperature, according to the Clapeyron equation (p. 123).

The specific volumes of some of the different structures of silicate minerals are known (Table 4.4), but the latent heats are in general poorly known. Some examples are given in Table 4.4, calculated from the slope of the phase boundary. It will be noticed that no direct information is available for forsterite nor for solid solutions of fayalite and forsterite. Only one estimate of dp/dT is available for the spinel disproportionation; it was obtained from calorimetric data for the enthalpy of formation of the tin analogues, Mg_2SnO_4 and Co_2SnO_4, from component oxides, and by argument based on analogy the value of dp/dT is thought to be less than 10^5 Pa/deg for silicates (Navrotsky and Kasper, 1976).

Figure 4.10. Dependence of the pyroxene–spinel–oxide transition on composition at $T = 1000$ K.

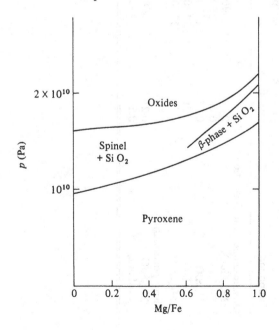

Table 4.4. *Thermodynamic properties of olivine–spinel transformation (after Ringwood, 1975) and disproportionation (after Navrotsky and Kasper, 1976)*

Compound	Transition pressure (10^{10} Pa)	Transition temperature (K)	ΔV (m^3/kg mole)	L (J/kg mole)	dp/dT (10^6 Pa/deg)
Olivine–spinel transformation					
Mg_2GeO_4	0	1090	3.5×10^{-3}	1.3×10^6	3.3
Fe_2SiO_4	0.49	1270	4.4×10^{-3}	1.6×10^6	2.8
Ni_2SiO_4	0.31	1270	3.3×10^{-3}	0.7×10^6	1.6
Co_2SiO_4	0.70	1170	4.0×10^{-3}	1.5×10^6	3.2
Mg_2SiO_4	1.25	1270	4.2×10^{-3}	$1.6\text{–}2.7 \times 10^6$	3–5
Disproportionation					
Mg_2SnO_4	0.26	400	3.85×10^{-3}	4.7×10^{-6}	3.04
Co_2SnO_4	0.12	400	3.80×10^{-3}	-1.6×10^{-6}	-1.03

Knowledge of transformations of minerals to polymorphic forms or constituent oxides may be summarized as follows: the transitions likely to occur at pressures encountered within the Earth and planets have been identified, changes of density at the transition are known and high pressure equations of state are available. Thermodynamic data are, however, sparse, and the values of dp/dT along the phase line are not well known.

4.6   Melting

By Clapeyron's equation, the variation of melting temperature with pressure is given by

$$\frac{\mathrm{d}T}{\mathrm{d}p} = \frac{T\Delta V}{L}$$

where ΔV is the difference of specific volume and L the latent heat.

Values of dT/dp are known for planetary materials (in particular iron) only at low pressures, if at all, and a problem which has been extensively discussed is how to estimate melting temperatures at high pressures in the absence of the necessary thermodynamic data measured at high pressures in excess of 10^{11} Pa. Even the difference of specific volumes is not known, for although shock-wave experiments have been done on iron (Section 4.4) they do not distinguish between solid and liquid. The change of latent heat along the phase boundary is given by Clausius's equation:

$$\frac{\mathrm{D}L}{\mathrm{D}T} - \frac{L}{T} = \Delta C_p - \frac{L}{\Delta V}\Delta\left(\frac{\partial V}{\partial T}\right)_p ;$$

the symbol Δ, as before, denotes the difference between the values of a quantity for the solid and the liquid, while $\mathrm{D}/\mathrm{D}T$ indicates differentiation along the phase boundary. It will be seen that, in addition to the difference of specific volumes, it is necessary to know the coefficients of expansion $(\partial V/\partial T)$ and the specific heat at constant pressure. Now

$$C_p - C_V = \frac{T}{VK_T}\left(\frac{\partial V}{\partial T}\right)_p^2 \quad \text{(Wilson, 1966, p. 39)}$$

which shows that ΔC_p also involves ΔV and $\Delta(\partial V/\partial T)$ as well as the difference ΔC_V.

In the absence of thermodynamic data at high pressure, recourse must be had to empirical or quasi-theoretical approaches to estimate the melting temperature at high pressure. A well-known empirical rule is that of Simon (1937) which reads

$$p = A\left(\frac{T_m}{T_0}\right)^c - 1,$$

where T_m is the melting temperature at pressure p, T_0 that at atmospheric pressure and A and c are numerical constants.

c is related to Grüneisen's parameter:

$$c = (6\gamma + 1)/(6\gamma + 2).$$

Kraut and Kennedy (1966) found that Simon's rule gave melting temperatures generally higher than those measured at high pressures. They observed that for most metals there is a linear relation between melting temperature and the volume compression of the solid:

$$T_m = T_0\left[1 + a\left(\frac{\Delta V}{V_0}\right)\right].$$

Most metals, as distinct from non-metals, do appear to follow such a linear law, but some, for example mercury and bismuth, do not (see Chan, Spetzler and Meyer, 1976). Furthermore, as Birch (1972) has pointed out, phase transformations in the solid, known to occur in iron, will affect the melting temperature.

If, then, empirical relations do not seem to be generally applicable, and in any case cannot be checked at 10^{11} Pa, quasi-theoretical estimates are scarcely more satisfactory. The prime difficulty is that we do not know how to describe what happens when a solid, specifically a metal, melts. The solid of course has a structure of ions in a lattice with well-defined mean parameters about which it executes thermal vibrations, but the order in a liquid is statistical; the average distance of nearest neighbours is known, but not the separation of a particular pair at a particular instance. It is not so far possible to describe how the one structure changes to the other. Lindemann (1910) early suggested that melting occurred when the amplitude of lattice vibrations was comparable with the distance between nearest neighbours and, more recently, melting has been related to dislocations in the solid, melting being defined as the state when the density of dislocations becomes infinite. Other contributions have been made by Leppaluota (1972), using the idea (John and Eyring, 1971) that liquids have an essentially solid structure through which ions move. Another approach (Boschi, 1974a) is to use the results of computer calculations on models of solids and liquids consisting of hard spheres (see Ree, 1971); the results appear to be inconsistent with the Kraut–Kennedy relation. Boschi (1974b) has also pointed out that the Kraut–

Kennedy relation may be regarded as the linear approximation to a more general relation of the form

$$T_m = T_0\left(1 - \frac{\Delta V}{V_0}\right)^{-n/3},$$

where n is between 8 and 9.

Since $\Delta V/V_0$ exceeds 10 per cent within the core of the Earth, terms of order $(\Delta V/V_0)^2$ may be quite significant. Boschi finds much higher melting temperatures for iron than those given by the Kraut–Kennedy rule.

5

The Moon

5.1 Introduction

After the Earth, the Moon is much the best known body of the solar system. Almost all physical measurements that have been made on the Earth have also to some extent been made on the Moon. Artificial satellites have been placed in orbit about the Moon and have enabled the components of the gravitational potential to be estimated. The physical librations, the equivalent of the luni-solar precession of the Earth, have been observed, especially by laser ranging to the retroreflectors left on the Moon by Apollo astronauts. The Apollo astronauts took with them seismometers that have recorded impacts of meteorites and rockets on the surface and moonquakes within the Moon. The flow of heat through the surface of the Moon was measured.

The magnetic field of the Moon has been studied intensively, globally by satellites at a distance from the surface and in detail by others close to it, while the magnetization of rock samples brought back by the Apollo astronauts has been studied in the laboratory. In addition, electro-magnetic induction in the Moon has been studied. Thus, there is some prospect of being able to construct models of the interior of the Moon using much the same methods as are followed for the Earth, whereas there is at present no such prospect for any of the planets. However, there are major gaps in our knowledge of the Moon as compared with the Earth, and the principal one is that seismic data are comparatively very sparse because there are only four seismic stations on the Moon and all of them are on the same hemisphere and, furthermore, because free oscillations of the Moon have never been observed.

We shall see that the Moon is in some respects very different from the Earth. Thus it provides an alternative to the Earth as a basis for models of the terrestrial planets: we shall see that Mars may be more like the Moon in constitution than the Earth. A close study of the Moon is, therefore, a

132

necessary prelude to the study of the terrestrial planets as well as being of great interest in its own right.

5.2   The mass and radii of the Moon

The mass of the Moon is obtained in two ways with comparably high precision. It may, in the first place, be derived from the period and semi-major axis (or orbital velocity) of a lunar satellite by using Kepler's third law. In this way, Michael and Blackshear (1972) using Explorers 35 and 49, obtained the result

$$GM = 4.902\,84 \times 10^{12}\,\mathrm{m}^3/\mathrm{s}^2.$$

The other way involves finding the ratio of the mass of the Moon to that of the Earth. Just as the Moon describes an orbit about the centre of mass of the system of the Earth and the Moon, so does the Earth, but in an orbit smaller in the ratio of the mass of the Moon to that of the Earth. In consequence, all stars appear to move in circular orbits with a monthly period and an amplitude of some $6''$. It was from this phenomenon that the mass of the Moon was determined until quite recently (Chapter 3). Another consequence of the same motion of the Earth is that a body, such as a space probe moving with a steady velocity in space, will appear to have a variation in velocity with a monthly period and an amplitude of about 12.4 m/s. That variation can be found with high accuracy from the Doppler shifts of radio transmissions from such vehicles, and the observations lead to the result (Anderson, Efrom and Wong, 1970) that the ratio of the mass of the Earth to the mass of the Moon is 81.3008 ± 0.0008, and consequently that GM for the Moon is $4.902\,82 \times 10^{12}\,\mathrm{m}^3/\mathrm{s}^2$.

The radii of the Moon have been found traditionally from telescopic observations from the Earth, but the current best values are derived from radio echo sounding from satellites of known orbits. The results for the principal axes of the Moon's figure are (Sjogren and Wollenhaupt, 1976)

	Maria (km)	Highlands (km)
a	1736.6	1738.1
b	1735.0	1738.2
c	1733.0	1738.0

5.3   The gravitational field of the Moon from orbiting satellites

As a preliminary to the landing of the Apollo astronauts, the gravitational potential of the Moon was investigated through the changes in the orbits of artificial satellites of the Moon. Five such satellites were established (Michael and Blackshear, 1972; Table 5.1) and were tracked by the Doppler shift of radio signals.

The motions of artificial satellites of the Moon are more complex than those of the Earth. Just as the main deviation of the Moon's orbit about the Earth from a Keplerian ellipse results from the attraction of the Sun, so in a similar way the main deviation of the orbit of a satellite about the Moon is the consequence of the attraction of the Earth. When the second zonal harmonic of the Moon is also taken into account, orbits of lunar satellites fall into two groups according to whether the perilune circulates (that is, has a steady secular motion) or librates about a longitude which is an odd multiple of $\frac{1}{2}\pi$. Into which group an orbit falls depends on the initial conditions (Felsentreger, 1968).

Another important difference from terrestrial satellites is that the second degree zonal harmonic does not dominate the potential as it does for the Earth, because the tesseral and sectorial harmonics are comparable in magnitude with the principal zonal harmonics. The motions of node and perigee are therefore not controlled by the second degree zonal harmonic as for terrestrial satellites. The orbits of the five satellites of the Moon have been interpreted by numerical integration of the equations of motion, adjusting the values of the coefficients of a number of harmonics until agreement is obtained with the observed orbits.

The estimates of the coefficients cannot be independent, derived as they are from the orbits of only five satellites. It was seen in Chapter 3 that the estimates of the coefficients of the Earth's gravity field depend on fewer independent observations than there are significant coefficients. The lower harmonics of the Earth are, however, reasonably well determined because the coefficients in general die away fairly rapidly as the

Table 5.1. *Artificial satellites of the Moon*

Satellite		Semi-major axis (km)	Inclination to lunar equator (degree)
Lunar Orbiter	1	2670	12
	2	2702	18
	3a	2688	21
	3b	1968	27
	4	3751	84
	5	2832	85
Explorer	35	5980	170
	49	2803	62

degree increases, so that errors arising from correlation between estimates, arising as they do from neglect of coefficients beyond a certain degree, are relatively more serious for the high than for the low degree harmonics. Things are not so clear cut for the Moon. The coefficients do not die away so rapidly as the degree increases as do those for the Earth (see Figure 9.1) so that errors due to neglect of higher harmonics are probably relatively more important in coefficients of second and third degree than they are for the Earth.

A list of harmonic coefficients is not very instructive: square roots of the sum of squares of the coefficients of a given order provide a useful summary of the behaviour; maps are still more convenient. Two ways of representing the lunar field have been used; one a map of an equipotential surface (see Chapter 2 for the procedure for the Earth) and the other a map of the acceleration due to gravity at the surfaces. Examples of such maps of equipotentials for the near and far side of the Moon, taken from Michael and Blackshear (1972) are shown in Figures 5.1(a) and (b).

An alternative description of parts at least of the lunar gravity field is available. Doppler tracking provides directly the line-of-sight velocities of a satellite and so it is straightforward to derive the accelerations of the satellite normal to the surface of the Moon over much of the visible hemisphere, accelerations which can be interpreted in terms of mass excess or defect near the surface below the satellite. In this way concentrations of mass (mascons) related to certain large craters were identified and the gravity field has been represented by a set of point masses near the surface. While instructive for the near surface structure of the Moon, this representation does not help much in obtaining a value of J_2 or C_{22} and thus the moments of inertia of the Moon. The representation is also ambiguous because the depths at which the masses lie cannot be specified (the same is true of the spherical harmonic representation which describes the external field but does not specify the location of the sources).

The best values of J_2 and C_{22} have been obtained from the Explorer 35 and 49 satellites which have greater semi-major axes than the other lunar orbiters, so that the effects of harmonics of low degree are more significant relative to those of high degree than for the orbiters which have orbits close to the surface of the Moon. The effect of a harmonic term of degree n is proportional to $(a/a')^n$, where a is the mean equatorial radius of the Moon and a' is the semi-major axis of the satellite. Thus, if a and a' are nearly equal, harmonics of many degrees will have effects of similar relative magnitude, whereas, if a' is much

Figure 5.1. Maps of gravity at the surface of the Moon: (*a*) near side
(—— positive deviation, – – – negative deviation); (*b*) far side. (From
Michael and Blackshear, 1972.)

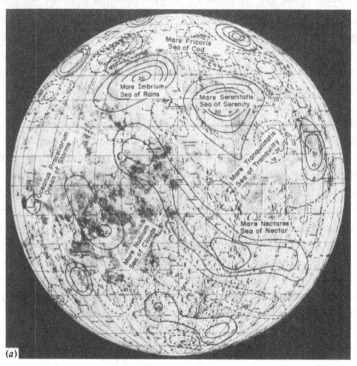

(*a*)

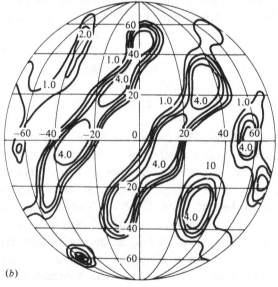

(*b*)

greater than a, only the lowest harmonics will have much effect on the orbit. Using then satellites for which a' was respectively $1\frac{1}{2}$ and more than 3 times a, Gapcynski, Blackshear, Tolson and Compton (1975) obtained values for J_2 and C_{22} that are probably less dependent on other harmonics than are the estimates derived from all lunar orbiters. Bryant and Williamson (1974) have also used Explorer 49 while Sjogren (1971) has used the data for Orbiter 4 which has the greatest semi-major axis of the Orbiter series. The most recent analysis is that of Ferrari (1977) whose results are given in Table 5.2.

Other evidence about the harmonics of third order is now available from observations of the physical librations, to which, therefore, we now turn.

Table 5.2. *Coefficients of spherical harmonics in the lunar gravitational potential*

A	$J_2 = 2.0272 \pm 0.0148 \times 10^{-4}$
	$J_4 = 6.276 \pm 5 \times 10^{-6}$
	(Gapcynski, Blackshear, Tolson and Compton, 1975)

B   Some of the lower degree normalized coefficients (Ferrari, 1977)

n	m	$C_{nm} \times 10^4$	$S_{nm} \times 10^4$
2	0	−0.91	—
2	1	0.02	−0.002
2	2	0.34	−0.004
3	0	−0.05	—
3	1	0.27	0.06
3	2	0.12	0.01
3	3	0.25	−0.09
4	0	0.02	—
4	1	−0.05	0.03
4	2	0.14	−0.03
4	3	−0.01	−0.21
4	4	−0.13	−0.01
5	0	−0.04	—
5	1	−0.04	−0.04
5	2	0.04	0.07
5	4	0.06	0.18
5	5	−0.05	0.16

Material also from Blackshear and Gapcynski (1977), Anauda (1977), Ferarri (1977).

5.4 The gravitational potential and moments of inertia from the physical librations

The Moon has three unequal moments of inertia, A, B and C. The axis of least inertia (A) points nearly in the direction of the Earth, the axis of greatest inertia (C) is nearly perpendicular to the ecliptic and to the plane of the Moon's orbit about the Earth, while the axis of intermediate inertia (B) is nearly tangential to the orbit. The Moon's mean angular velocity in its orbit is equal to its spin angular velocity, so that, on the average, the Earth lies always in the same direction as seen from the Moon. But the Moon's orbit is not circular, nor does its plane coincide with the equator of the Moon as defined by the A and B axes of inertia, and, therefore, as seen from the Moon, the position of the Earth oscillates about its mean direction, the axis of least inertia (A). The Earth accordingly exerts periodic torques upon the Moon and, in consequence, the Moon is driven into forced oscillations about the mean directions of its axes of inertia. These oscillations are the physical librations of the Moon (for an outline of the theory see Appendix 3). The principal manifestation of the physical librations is the inclination of the axis of greatest inertia (C) to the normal to the ecliptic at an angle of about 1 ° 30 '. The C axis of inertia rotates about the normal to the ecliptic at the angular velocity of the Moon in its orbit about the Earth so that the axis of inertia, the normal to the ecliptic and the normal to the Moon's orbit, remain nearly coplanar, a phenomenon first described by Cassini.

The angle between the C axis and the normal to the ecliptic is relatively large because of a near resonance in the motion; it is proportional to

$$\frac{6\beta \sin \frac{1}{2}i}{2(g'-n_0/n)-3\beta}$$

where β is the ratio of moments of inertia, $(C-A)/B$, i is the inclination of the Moon's orbit to the ecliptic (about 7 °), n_0 is the mean angular velocity of the Earth about the Sun, n is the mean angular velocity of the Moon about the Earth, and g' is the rate at which the node of the Moon's orbit rotates about the Earth.

β is about 6×10^{-4} and $\sin \frac{1}{2}i$ is about 0.045 but, because $(g'-n_0/n)$ is about 5×10^{-3}, the amplitude is about 1 ° 30 '.

The inclination term in the libration corresponds to synchronous rotations about the A and B axes of intertia. Another important term is the libration in longitude, corresponding to a rotation about the C axis; it is proportional to γe, where γ is the ratio $(B-A)/C$ (about 2×10^{-4}) and e

is the eccentricity of the Moon's orbit, and it is small because there is no small divisor to amplify it.

Most terms in the physical libration are indeed of order 10^{-6} or less, and only very few of them were detected by telescopic observation of the Moon. The ability to measure distances from the Earth to the retroreflectors placed on the Moon by the Apollo astronauts (Figure 5.2) has now made it possible to observe the physical librations in very much greater detail. This has led to estimates of β and γ in which considerable confidence can be placed; when combined with values of J_2, C_{22} and S_{22} from lunar satellites, they lead to estimates of the ratios C/Ma^2, B/Ma^2 and A/Ma^2 reliable enough to be used in the study of the interior of the Moon.

The unequal moments of inertia of the Moon correspond to the gravitational coefficients J_2, C_{22} and S_{22}. The harmonics of third and fourth order in the gravitational field of the Moon also give rise to torques exerted by the Earth upon the Moon and thus to components of the physical librations (Appendix 3). Indeed the best estimates of certain harmonic coefficients are at present obtained from librations rather than from orbits of lunar satellites, and so to some extent provide a check on results from satellites. An account of laser ranging to the Moon has been given by Bender *et al.* (1973).

The principle of the determination of the physical librations by laser distance measurements is shown in Figure 5.3. C_1 and C_2 are respectively the centres of mass of the Earth and the Moon, O is the position of an

Figure 5.2. Positions of cube-corner reflectors on the Moon. A11, A14 and A15 are Apollo sites and LK2 is Lunakhod 2.

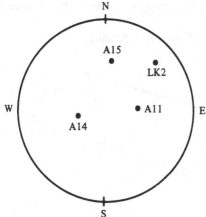

observatory on the Earth and L that of a retroreflector on the Moon. Denote the vector C_1C_2 by R, C_1O by r_1 and C_2L by r_2. The vector OL, denoted by D, is then given by

$$D = R - r_1 - r_2.$$

The quantity that is measured is D equal to the absolute value of D. Now

$$D^2 = D \cdot D = (R - r_1 - r_2) \cdot (R - r_1 - r_2).$$

Thus

$$D^2 = R^2 - 2R \cdot (r_1 + r_2) + r_1^2 + r_2^2 + 2r_1 \cdot r_2.$$

or approximately

$$D = R[1 - R \cdot (r_1 + r_2)/R].$$

Now $D \cdot r_1/D$ and $D \cdot r_2/D$ are the respective projections of r_1 and r_2 on R, p_1 and p_2 say, so that the approximate result is

$$D = R - p_1 - p_2.$$

Of course, this is not a sufficiently accurate formula for measurements with a precision of much better than 1 m, but it does show the factors on which the measured distance, D, depends.

In the first place, there is the distance, R, between the centres of mass of the Earth and the Moon which is determined by the orbital motion of the Moon about the Earth. Secondly, p_1, which depends on the angle between R and r_1, as well as on the geocentric position of O, varies with a period of one day as the Earth spins about its axis. Thirdly, p_2 depends on the selenocentric position of L and also on the rotation of the Moon relative to the Earth. Now the Moon rotates relative to the Earth in part through the physical librations, but also, in greater part, because the orbit of the Moon about the Earth is not circular, giving rise, as has already been said, to oscillations of the position of the Earth as seen from the

Figure 5.3. Principle of lunar laser ranging.

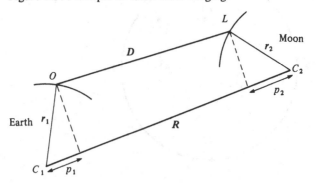

Moon about its mean position along the A axis of the Moon. These relative rotations of the Moon are the geometrical librations; with amplitudes of up to 7° they are much larger than the physical librations, but may be calculated from the Moon's orbit. Like the principal term in the physical libration, they have a period of 1 month, so that, if there were only one reflector on the Moon, they could not be separated from variations in R on account of the eccentricity of the lunar orbit. However, with three or more reflectors fairly symmetrically disposed around the intersection of the line of centres $C_1 C_2$ with the lunar surface, it is easy to make the separation, for, when p_2 increases for one reflector, it will decrease for a reflector on the other side of the intersection. Thus it is that it is possible to make a clear separation of the librations from other motions.

The measurements of distances to the lunar retroreflectors from the MacDonald Observatory (Bender *et al.* 1973) have been interpreted by numerical solutions for the lunar orbit and for the physical librations. The librations depend, as seen above, on the ratios β and γ, and also on the ratios of the harmonic coefficients of third order in the lunar potential to a coefficient of second order, J_2 say. Thus β, γ and the third order harmonic coefficients were adjusted until the librations determined from the numerical integration agreed with the measurements of distance. The values of β and γ so found are (Williams, 1976) $\beta = 6.3126 \times 10^{-4}$ and $\gamma = 2.277 \times 10^{-4}$.

The retroreflectors were not the only ranging equipment left on the Moon. Radio transmitters formed part of the so-called ALSEPs (Apollo Lunar Surface Experimental Packages) with power provided from nuclear sources. The positions of the transmitters are determined by very long baseline radio interferometry (VLBI); the measurements are most sensitive to relative rotations of the transmitters and not very sensitive to variations in the Earth–Moon centre-to-centre distance nor to the co-ordinates of the observing sites (King, Counselman and Shapiro, 1976). Again the observations have been compared with a numerical integration of the equations of libration.

King, Counselman and Shapiro (1976) combined lunar laser ranging to the reflectors with VLBI observations to the radio transmitters and obtained the following results:

$$10^6 \beta = 631.27 \pm 0.03 \qquad 10^6 C_{31} = 26 \pm 4 \qquad 10^6 S_{31} = -1 \pm 30$$

$$10^6 \gamma = 227.7 \pm 0.07 \qquad 10^6 C_{32} = 4.7 \pm 0.2 \qquad 10^6 S_{32} = 1.8 \pm 0.3$$

$$10^6 J_3 = 3 \pm 20 \qquad 10^6 C_{33} = 2 \pm 2 \qquad 10^6 S_{33} = -0.3 \pm 1.$$

The harmonic coefficients significantly determined from the librations are C_{31}, C_{32} and A_{32}.

Now

$$J_2 = \frac{C - \frac{1}{2}(A+B)}{Ma^2},$$

while

$$\frac{A}{C} = \frac{1 - \beta\gamma}{1 + \beta};$$

and

$$\frac{B}{C} = \frac{1 + \gamma}{1 + \beta};$$

thus

$$\frac{C}{Ma^2} = \frac{2J_2(1 + \beta)}{2\beta - \gamma + \beta\gamma}.$$

With the above values of β and γ and the value of J_2 derived by Gapcynski *et al.* (1975), namely $(202.72 \pm 1.48) \times 10^{-6}$, it follows that $C/Ma^2 = 0.392 \pm 0.003$.

5.5 Lunar seismology

The Apollo astronauts left on the Moon at their landing sites groups of seismometers with which natural and artificial seismic events have been observed. The first, at the Apollo 11 site, was powered by solar cells and did not operate for long, but the others, at the sites of Apollos 12, 14, 15 and 16, had nuclear power supplies and are still operating. The 12 and 14 sites are close together (181 km), the other two are at about 1100 km from 12 and 14 and with them form a roughly equilateral triangle on the near side of the Moon facing the Earth. The need to have clear line of sight for the radio transmission of data to the Earth means of course that seismometers cannot be placed on the far side of the Moon.

The seismometers at each site comprised a three-component set of long period seismometers and a single vertical short period seismometer, the response curves of which are shown in Figure 5.4 (Lammlein *et al.*, 1974; Toksöz, Dainty, Solomon and Anderson, 1974). The long period seismometers can operate in a peaked mode or, with feedback, in a flat response mode. Both the long period and short period instruments can detect ground motion of the order of 0.3 nm. The sensitivity is much greater than for terrestrial seismometers because the natural noise on the Moon is far less than on the Earth. Of the sixteen instruments, only two failed, the short period vertical at Site 12 which was damaged on installation and the long period vertical at Site 14 which has operated only intermittently since January 1972.

Signals are received at the seismometers from artificial sources on the surface (impacts of Saturn 4B boosters and lunar modules), from meteorite impacts on the surface and from internal moonquakes. The characters of lunar signals are very different from those of terrestrial signals. The latter consist of a number of distinct pulses, corresponding to waves travelling by different paths through the body of the Earth as longitudinal or shear waves, followed by long trains of surface waves showing dispersion. The pulses to appear first on lunar seismograms are small, those from surface sources are followed by a slow increase of amplitude followed by an even slower decay over a very long time. The long trains of waves show neither dispersion nor coherence between the components in different directions. Some examples are shown in Figure 5.5.

It is now generally accepted that elastic waves are trapped in a relatively thin layer close to the surface of the Moon, within which they

Figure 5.4. Apollo seismometer response curves. (After Lammlein *et al.*, 1974.)

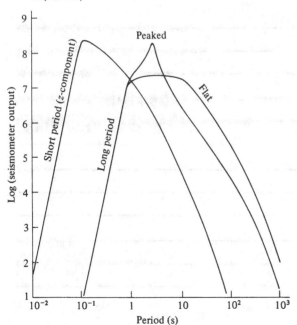

Figure 5.5. Lunar seismograms: (a) from impact of lunar module; (b) from internal moonquake. (From Toksöz et al., 1974.)

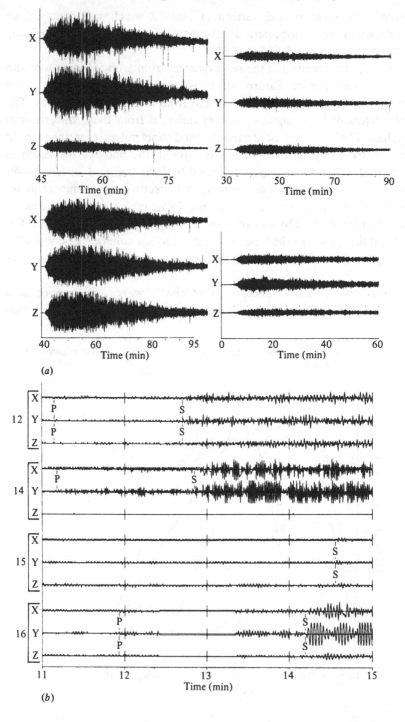

suffer random scattering. Theory and experience (Toksöz *et al.*, 1974) show that, in such circumstances, the amplitude of a signal received at a detector from a distant surface source will build up slowly and then decay even more slowly. For such phenomena to be observed, as they are on the Moon, two factors are necessary: the existence of a wave trap and very low attenuation, so that waves may be scattered many times within the trap. The trap is formed by the lunar surface material which is a loose powder at the surface and becomes rapidly more compact with depth with a corresponding increase of seismic velocity (Cooper, Kovack and Watkins, 1974). Ray paths are, in consequence, concave towards the lunar surface so that rays originating in the surface are refracted strongly back towards it, whilst rays coming up from below are of course reflected at the free surface. Thus a wave trap is formed, which is of the order of 1 km thick. The material at the lunar surface, the regolith, formed by the smashing of lunar material by meteorite impacts, is very dry, indeed it seems that the Moon as a whole is very dry compared with the Earth. It is on that account that it is believed that the attenuation of seismic waves within the Moon is much less than within the Earth. The combination of the scattering in the wave trap and the low attenuation means that only the first one or two signals arriving by the most direct path can be seen on seismograms, for pulses arriving later by longer paths, if they exist, are swamped by the reverberation in the wave trap.

The points of impact of spent boosters and lunar modules are all quite close to the seismometer sites, and so signals from them provide information about seismic velocities in the upper parts of the Moon. Table 5.3 gives a list of the impacts that were used. No unique relation between seismic velocity and depth can be obtained from the travel-time data, but bounds can be set to the range within which the velocity must lie at any depth. Toksöz *et al.* (1974) have used methods developed by McMechan and Wiggins (1972) and Bessanova *et al.* (1974) which make use of the ray parameter $dT/d\Delta$, denoted by p (T is the time taken along a ray joining two points on the surface that subtend an angle Δ at the centre of the Moon). The first step is to derive from the travel-time curves limits on p, considered as a function of Δ and considered as a function of a residual time τ equal to $T - \Delta \, dT/d\Delta$. From those bounds and the travel-time curve, limits can be set to the velocity.

The information in the travel-time curves does not exhaust that in the signals recorded at the seismometer, and the velocity–depth relation can be refined through comparisons of the observed seismograms with those calculated theoretically for postulated velocity–depth relations.

The lunar module impact data determine the velocity to a depth of about 80 km. Within the uppermost 1 km, the velocity increases from about 200 m/s to 4 km/s, presumably corresponding to compaction of the lunar regolith. There is then a smooth increase to about 6.7 km/s at a depth of about 21 km, a value which is maintained to some 54.5 km. The velocity then increases to about 8.9 km/s at 60 km depth and remains roughly constant below that, so far as can be determined from the impact data. There is, however, a discordance between data from close events, the lunar modules and boosters, and more distant events such as meteorite impacts, for, as will be seen below, the latter indicate that the P-wave velocity below the crust is about 8.1 km/s. The velocity of 6.7 km/s is similar to velocities encountered in the crust of the Earth and that of 8.9 km/s to velocities encountered in the upper mantle and so the outermost 60 km of the Moon is called the crust and the region below it the upper mantle; the correspondence is not of course exact and, in particular, crustal material on the Moon is much thicker than on the Earth: approximately twice as thick as the continental crust on the Earth. The results refer to the region occupied by the seismometer sites, the nearside region most directly opposite the Earth, and nothing is known about possible differences in other parts of the Moon.

The records of impact signals contain S-wave arrivals which can sometimes be identified with the help of diagrams of particle motion. It is found that the travel-time data for S-waves can be reproduced by taking the P-wave velocities and assuming that Poisson's ratio is 0.25, i.e. the

Table 5.3. *Particulars of lunar impact seismic sources*
(Toksöz *et al.*, 1974)

Impact	Kinetic energy (10^3 MJ)	Distance from seismometer at site (km)			
		12	14	15	16
LM-12	3.4	73	—	—	—
S4B-13	46.3	135	—	—	—
S4B-14	45.2	172	—	—	—
LM-14	3.4	114	67	—	—
S4B-15	46.1	355	184	—	—
LM-15	3.1	1130	1048	93	—
S4B-16	not known	132	243	1099	—
S4B-17	47.1	338	157	1032	850
LM-17	3.2	1750	1598	770	995

LM: lunar module.
S4B: Saturn 4 booster.

S-wave velocities are taken to be the P-wave velocities divided by $3^{1/2}$. The S-wave data are of poor quality and cannot be used to make an independent determination of velocity with depth.

Knowledge of the seismic velocities at depths greater than about 100 km comes from meteorite impacts, from moonquakes and from high frequency teleseismic events (HFT events, Nakamura *et al.*, 1974), which may lie at any distance from the seismometers. Meteorite impacts are of course surface sources, and the high frequency teleseismic events originate close to the surface, whereas the moonquakes lie deep within the Moon. The problem with these signals is that, in contrast to the artificial impacts, the positions and times of the sources are unknown. Suppose the source lies in the surface and its time is known; then times of arrival at two seismometers will locate it if the wave velocities are known (with an ambiguity as to the hemisphere in which it lies). With three seismometers, the time of origin can also be determined, and with four seismometers, as on the Moon, something may be said about velocities. Because moonquakes do not lie in the surface, five seismometers are needed if an average velocity to the surface is to be determined, as well as position and time, from waves of one type. The Apollo series provides only four seismometers. Hence, a sequence of successive approximations is used in which a variation of velocity with depth is assumed, locations and times are calculated, and then changes (it is hoped, improvements) are made to the velocities. Toksöz *et al.* (1974) point out that there is no guarantee that the process will converge to the correct structure; indeed it is not clear that it will converge at all. When both S and P arrivals can be read, something more definite can be derived, but it is rare for both to be clearly seen on the same record.

It is a striking feature of the strong signals from internal sources that they fall into groups such that, at any one seismometer, the signals in one group from different events are all very similar, but distinct from members of other groups. Now the details of a seismogram after the first arrival depend both on the source function and on the scattering, diffraction, reflexions and so on, suffered by a wave in passing from source to detector, and, if a number of seismograms at one station have similar form, the inference to be drawn is that in each case the waves have followed the same path and have therefore come from the same source. Thus, all events of one class can be ascribed to sources within a relatively small volume. Further, because the signals are the same, the signal-to-noise ratio can be improved by adding all the seismograms of a class together to produce a composite seismogram for sources in that particular small volume. The data are then sufficient to enable the sources to be located and a mean velocity to the seismometer to be determined. A map of the principal moonquake epicentres on the near side of the Moon is shown in Figure 5.6 from which it will be seen that the epicentres form two quite well-defined belts.

The occurrence of moonquakes appears to be influenced by the tidal stresses in the Moon set up by the variable torques exerted by the Earth as it librates about its mean position relative to the principal axes of inertia of the Moon. Figure 5.7 shows how the frequency of events at a particular location depends on the position of the Earth relative to the Moon.

Moonquakes differ from earthquakes in two clear ways: many of them recur within a relatively small volume and the frequency of occurrence is controlled by tidal stresses in the Moon. There are two other very clear differences. Moonquakes occur deep within the Moon, between depths of 300 and 800 km, and far below the lunar crust, whereas the great majority of earthquakes occur in or close to the crust of the Earth and relatively few at great depth. The energy released in moonquakes is also far less than in earthquakes, the total annual rate being 10^4 J for moonquakes and 5×10^{17} J for earthquakes. It is only because the Moon is seismically very quiet that any of these events is detectable.

Figure 5.6. Location of moonquakes. (From Lammlein *et al.*, 1974.)

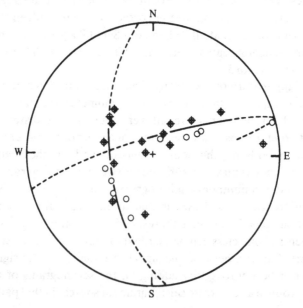

Travel-time curves for various sources are shown in Figure 5.8. The location of moonquakes at depths of 300 to 800 km depends on the behaviour of the differences between arrival times of P-waves and S-waves. When signals from a particular group of events can be identified at all four seismometers, it is found that the S–P intervals are large (not less than 90 s) and comparable on all four seismometers, whereas the differences of S arrivals between separate seismometers are less. It follows that the sources must be roughly equidistant from all four seismometers and, given that the seismometers are about 1100 km apart, the sources cannot lie on the surface but must lie deep below the near side of the Moon. It is possible to reproduce the observed times if the P velocity of 8.1 km/s is supposed to extend to 800 km. The interval between S and P arrivals is related to the P arrival time by the differential equation

$$\partial(T_S - T_P)/\partial T_P \sim 1,$$

a result clearly distinct from that for surface sources, namely

$$\partial(T_S - T_P)/\partial T_P \sim 0.73.$$

The relation for the deep sources entails a value of the S-wave velocity of 3.8 ± 0.2 km/s over the range of 300–800 km.

It may be noted that the values of depths and S-wave velocities given here follow from the most recent analyses (Toksöz et al., 1974; Nakamura et al., 1974) whereas earlier analyses (Lammlein et al., 1974) gave greater velocities and greater depths.

In the analyses just described, the P-wave and S-wave velocities are supposed not to vary with depth down to 880 km, but Nakamura et al. (1974) find that, when they determine velocities from distant meteorite

Figure 5.7. History of moonquake activity showing dependence upon distance of the Moon from the Earth. (From Lammlein et al., 1974.)

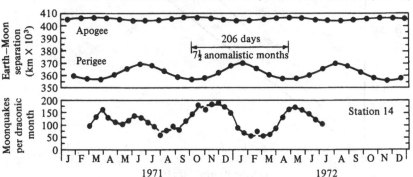

Figure 5.8. (*a*) Travel-time curves for surface sources on the Moon. A is the single point which suggests the presence of a core. (*b*) S–P intervals for different types of source. (After Nakamura *et al.*, 1974.)

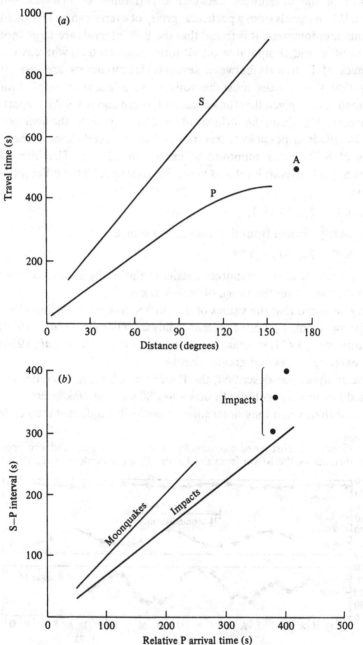

impacts and high frequency teleseismic events, making the assumption that the velocity does not vary with depth, then the velocities so found vary in a systematic way with range; a constant P-wave velocity is not, they claim, an adequate model, and they suggest that the velocity decreases with depth, either with a discontinuity at 300 km or continuously to a minimum of 7.8 km/s to around 500 km, or perhaps both types of behaviour may occur.

There is some other evidence for a discontinuity at 300 km. S-waves from surface and HFT events extend, albeit weakly, to distances of 150 °, although high frequency shear waves from internal sources are not received in circumstances which suggest they do not propagate in the Moon below 800–1000 km. The S signals from the surface sources are therefore thought to be refracted at or diffracted round the discontinuity at 300 km, although there is the third possibility that they are refracted in a low velocity zone below 300 km. So far, then, the seismic evidence reveals to us a thin outer layer of smashed up rock in which waves are trapped and scattered, a crust about 60 km thick, the properties of which are determined from signals from impacts of lunar modules, and a region (the upper mantle) from 60 to 300 km depth in which the P-wave and S-wave velocities are about 8.9 and 5.1 km/s, respectively, the properties of which are determined from signals received from distant meteorite impacts and near surface high frequency teleseismic events; and finally a region (the middle mantle) from 300 to 800 km depth where the P and S velocities, of about 8.1 km/s and 3.8 km/s, respectively, are derived from signals from deep moonquakes that occur within or below the middle mantle. The seismic velocities within the different zones are shown in Figure 5.9. A somewhat different interpretation is, however, possible, involving a smaller difference between the upper and middle mantle. Toksöz *et al.* (1974) suggest that all observations might be consistent with constant velocities (about 8.1 km/s for P and 4.0 km/s for S) throughout the upper and middle mantle, although recent interpretations do indicate a decrease in the S-wave velocity with depth from about 4.7 km/s just below the crust to 4.0 km/s at 700 km depth (Dainty, Goins and Toksöz, 1975; Toksöz, 1975).

Little is known of the zone below 800 km (the lower mantle) save for evidence from certain moonquakes that S-waves in it are strongly attenuated.

There is some evidence for a small inner core (Nakamura *et al.*, 1974). A signal has been received from a source at 168 °, nearly at the antipodes of the seismometers, with a large delay on the P-wave arrival compared with signals from near sources (Figure 5.8). Although only one such signal has been identified, Nakamura *et al.* (1974) interpret it as coming from a wave which has passed through a central zone or core in which the P-wave velocity is much less than in the mantle outside it. They set the following limits on the size of and velocity in a possible core:

radius: 170–360 km

P-wave velocity: 3.7 – 5.1 km/s.

The interpretations placed upon the seismic signals so far analysed may be summarized in the following list of zones of the Moon:

(*a*) *The crust*, 50–60 km thick, at least in the area of the seismometers. It is probably composed of material rich in plagioclase and the top few hundred metres are pulverized, forming the regolith.

(*b*) *The upper mantle*, 250 km thick, extending to a depth of 300 km. The P-wave velocity is 8.1 km/s, Poisson's ratio is 0.25, and comparisons with laboratory determinations of seismic wave velocities suggest that the material is a mixture of olivine and pyroxene with the ratio $Mg/(Mg + Fe)$ about 4 to 5.

Figure 5.9. Seismic velocities in the Moon. (After Nakamura *et al.*, 1974.)

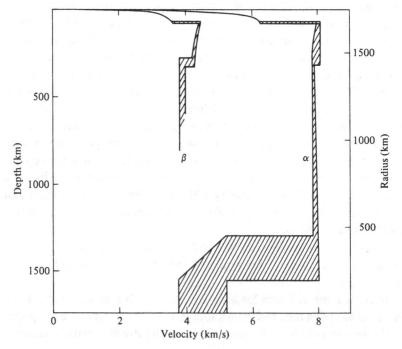

(c) *The middle mantle*, extending from 300 to 800 km depth. The P-wave velocity is about 8.1 km/s or possibly slightly less, but the S-wave velocity (3.8 km/s) is much less than in the upper mantle and Poisson's ratio is between 0.33 and 0.36, an unusually large value. The deep moonquakes occur mainly towards the base of this zone.

(d) *The lower mantle*, below 800 km. Velocities in this zone are not known; all that is known about it is that S-waves do not penetrate it.

(e) A possible *core* of radius between 170 and 360 km in which the P-wave velocity may lie between 3.7 and 5.1 km/s.

It must be emphasized that this division of the Moon is by no means so clearly based as that of the Earth into crust, upper and lower mantle and inner and outer core. The number of seismometers is almost never sufficient to enable velocities as well as locations to be determined, and the only possible assumption to make for a given source is that mantle velocities are constant. There is no possibility, as there is on the Earth, of deriving a travel-time curve for P or S, or any other phase, from signals from a single source observed at a wide spread of seismometer stations. Furthermore, because of scattering in the outer wave trap, it is only very rarely that signals can be seen that might be interpreted as having arrived by paths other than the direct P or S rays. There are of course no observations of free oscillations, but then the division of the Earth into zones was established before any observations of free oscillations on the basis of signals arriving at well-spaced seismometers by a multiplicity of paths. There are no doubt deficiencies in the distributions of earthquakes and seismometers on the Earth, which prevent a close analysis of regional variations, but the travel times and the velocity distributions derived from them are for the most part unambiguous, whereas unique solutions are not possible for the Moon.

5.6 Lunar magnetization, heat flow and electrical conductivity

The magnetization of the Moon, the heat flow through its surface and the electrical conductivity of its deep interior all provide information about the internal structure of the Moon, although the interpretation to be placed on any of them is far from clear.

The magnetic field of the Earth has provided three lines of evidence about internal structure. In the first place, the properties of the main dipole field, coupled with those of the secular variation, have, through the concepts of the dynamo theory of the main field, clarified some of the questions we should ask about the core, about its electrical conductivity and viscosity, the motions in it and the sources of energy it contains (see

Gubbins, 1974). Unfortunately it is perhaps true to say that at present the theory of the geomagnetic dynamo (see Moffat, 1978) is not sufficiently developed to enable us to derive, in more than the sketchiest of ways, answers to those questions from the properties of the main field. There are also major problems, as yet unapproached in dynamo theory, such as why the main dipole is inclined to the spin axis of the Earth, just as it is in Jupiter although not in Saturn. Secondly, the field generated around and within the Earth by the magnetosphere and ring currents induces currents in the outer parts of the Earth which themselves generate time-varying magnetic fields that can be detected at the surface. By comparing the field of the induced currents with the fields that induce the currents, it is possible to determine how the electrical conductivity of the Earth varies with depth. Then, using the measured behaviour of conductivity with temperature in semiconductors such as olivine, some idea of the temperature within the outer parts of the Earth may be obtained. Lastly, the remanent magnetization of rocks at the surface has been used to show how parts of the surface of the Earth have moved relative to each other, so that it has contributed to the development of the concepts of plate tectonics, the movements corresponding to which are seen as surface manifestations of the outward flow of energy from the interior of the Earth. Does the magnetization of the Moon seem likely to give us any comparable information about its interior?

The general magnetization of the Moon is extremely feeble, as has been shown by a succession of observations with magnetometers carried in satellites in orbit about the Moon. Those that went closest to the Moon were the sub-satellites of the Apollo 15 and 16 missions at altitudes of some 100 km above the surface. They showed that the Moon's permanent dipole moment is not more than 4.4×10^8 T m^3, whereas that of the Earth is 1.8×10^{16} T m^3; the surface field corresponding to the Moon's permanent moment would not exceed 10^{-10} T compared with the Earth's field of the order of 5×10^{-5} T. However, closer inspection shows that the Moon has local magnetic fields. The astronauts of Apollos 12, 14, 15 and 16 made measurements with magnetometers that showed fields of up to 3×10^{-7} T, and similar results were obtained in traverses by Lunakhod 2. The magnetometers in the Apollo 15 and 16 sub-satellites were able to map fields which are less than 10^{-9} T, but extend in a coherent way over appreciable areas of the surface of the Moon. Then again, studies of the flow of the solar wind past the Moon by the Explorer 35 satellite have indicated magnetized regions at the surface which compress the stream of protons and electrons, while most recently electron probes in orbiting satellites have provided further evidence of magnetized regions. Thus, the picture of the magnetization of the Moon is quite different from that of the Earth where the main field dominates and most of the magnetization of rocks is induced by the main dipole field and is not, as it appears to be on the Moon, mainly permanent magnetization. The question to answer for the Moon is, how did the permanent magnetization, varying comparatively rapidly from place to place as it does, come into existence?

Further evidence of the permanent magnetization of lunar material has been provided by extensive studies of rocks brought back in the Apollo sample programme. Natural remanent magnetization is widespread but, whereas on the Earth such magnetization is carried by ferrimagnetic oxides of iron and titanium, that of the lunar rocks resides in very fine particles of iron. Attempts have been made to use the demagnetization curves of samples under various conditions to estimate the field in which the magnetization was established. For that to be possible it must be established that the remanent magnetization is thermal or chemical remanent magnetization and that is something difficult to do for lunar samples. Thus, very few estimates of the intensity of the inducing field have been made and they range from 2×10^{-6} T to 10^{-4} T (Fuller, 1974).

Local studies of samples and wider surveys of the surface of the Moon show that the rocks at the surface are permanently magnetized, and that the magnetization maintains its sign over areas of some $10°$ (300 km) in extent. Some of the areas of magnetization appear to be associated with large craters though that is not always so; the strengths of the fields appear to be greater on the far hemisphere than on the near hemisphere. Yet the Moon itself has no main dipole field, so how have the surface materials become magnetized? It may be pointed out that, because the surface fields change sign every 300 to 500 km, it is not sufficient simply to have a dipole field. Some way is needed of producing in addition remanent magnetization of variable direction relative to the main field. Another difficulty is that it cannot be said that any of the lunar rocks are in the positions they were in when magnetized. On the Earth various tests can be applied to check that rocks have indeed not been disturbed since they were magnetized, but that has not been possible for lunar samples, and so the directions of the fields in which they were magnetized cannot be reconstructed.

A wide range of explanations has been advanced to account for the magnetizing fields; they fall into three categories: fields generated in the Moon and since decayed; fields of the Sun or Earth that were once much stronger; and fields generated by solar or other plasmas, again once stronger than now, streaming past the Moon. It has also been suggested that the shock of meteorite impacts might magnetize the surface material. One of the more complete attempts at an explanation is that of Runcorn who has postulated that the crust of the Moon (outside the Curie point of iron) became magnetized in the dipole field of a lunar dynamo maintained in a core which was liquid early in the history of the Moon but which has now solidified. The field of a shell magnetized by a dipole field vanishes outside the shell after the dipole field is removed, but the crust is still magnetized and, where it is damaged, by craters for example, field lines are discontinuous and external fields appear.

Unfortunately it has to be admitted that we do not know how the lunar surface was magnetized. We therefore do not know if the Moon did once have a dipole field and we do not know how strong it was. At present then, intriguing as are the issues it raises, we cannot draw on the remanent magnetization of the Moon to tell us about the structure of the interior. One thing does seem clear: whatever the origin and history of the Moon's field, it is not simply a small scale model of the Earth's field (see also Runcorn, 1977).

The Apollo 15 and 17 astronauts carried with them instruments for measuring the temperature gradient in the lunar soil and its thermal conductivity and hence they were able to estimate the flow of heat out through the surface of the Moon (Langseth *et al.*, 1972; Langseth, Keihm and Chute, 1973).

The measured value of the flow at the site of the Apollo 15 landing was $2.2 \times 10^{-2} \, \text{W/m}^2$, while at the Apollo 17 site it was $1.6 \times 10^{-2} \, \text{W/m}^2$ (Schubert, Young and Cassen, 1977); the difference between the two values is within the uncertainty of both. It is found from a study of the surface brightness temperature, made on the Apollo 15 flight, that the thermal conductivity increases rapidly with depth down from the surface of the Moon (Keihm *et al.*, 1973). The implications of these values will be discussed in Chapter 9.

As with the Earth, it is not possible to estimate the temperature within the Moon solely from the rate of flow of heat out through the surface. To do so, we would need to know the original temperature, the rate of generation of heat within the Earth or Moon and the processes of heat transfer. None of them do we know. We can, however, gain some idea of the variation of temperature within both the Earth and the Moon from the electrical conductivity of the material, and that in turn can be estimated from the magnetic fields generated by currents induced by external fields. The materials of the outer parts of the Earth and the Moon are generally supposed to be semi-conductors, such as olivine, and it is well known that the electrical conductivity of a semi-conductor depends on temperature, T, according to the exponential formula

$$\sigma = \sum \sigma_i \exp\left(-E_i/kT\right)$$

where σ_i is a constant independent of temperature and E_i is an activation energy. There are distinct values of σ_i and E_i for impurity, intrinsic and ionic conductivity and those modes of conduction may dominate at different temperatures according to the values of σ and E_i.

Suppose now that σ_i is known as a function of radius in the Moon and suppose also that the constants σ_i and E_i are known for the lunar interior; then it would be possible to calculate temperature as a function of radius. In fact we do not know what we need to know to make the calculation; as will be seen, σ_i is not known as a function of radius, rather some general indication of its behaviour can be given, and also we do not know for sure of what the lunar interior is composed. It will be argued below that it is probably not closely comparable to any material in the Earth. Despite the lack of precise information, it does seem nonetheless possible to set useful constraints to the possible behaviour of temperature as a function of radius in the Moon.

The estimates of electrical conductivity depend on the way in which the presence of the Moon affects magnetic fields convected past it in the solar wind. Those fields induce currents in the Moon which, in turn, generate magnetic fields that change those moving with the solar wind, so that the response of a magnetometer carried in a satellite on the side of the Moon remote from the Sun differs from that of an instrument on the sunward side of the Moon. A number of such studies have been made, and the variations of conductivity with depth as proposed by different authors are shown in Figure 5.10. Two different types of analysis have been used: one, applicable to variations on the night side of the Moon, deals with the transient response of the Moon (Dyal and Parkin, 1973; Dyal, Parkin and Daily, 1974*a*, *b*), while the other, more applicable to the day side, employs a Fourier decomposition of the response (Sonnett *et al.*, 1972; Kuckes, 1971). Hobbs (1973) has pointed out that unique solutions cannot be obtained and has applied inversion theory to indicate the range of conductivity entailed by the data, although since then the scope of the available data has widened and correspondingly tighter limits might now be set to the conductivity. Notice that the conductivity cannot be estimated at a radius less than 0.4 of the surface radius.

The electrical conductivity of olivine at high pressure has been determined by Duba *et al.* (1974) and, by using it, temperatures can be estimated from the conductivity. It is a general feature of all estimates that the temperature rises rapidly to about 1200 K at a radius of about 0.8 of the surface radius and thereafter more slowly to about 2000 K at 0.4 of the surface radius. The particular values depend on the material of the Moon being olivine and too much credence should not be placed on particular numerical values of the temperature, but it may well be that the qualitative feature of a rapid rise with depth followed by a much slower one to just below half the radius (where estimates fail) does represent the behaviour in the Moon.

Figure 5.10. Electrical conductivity in the Moon.

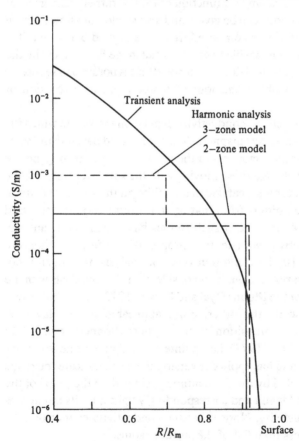

5.7   **The crust and mantle of the Moon**

Until recently (see, for example, Lyttleton, 1963; Cole, 1971; Cook, 1972) models of the density and elastic properties within the Moon were constructed on the basis that the equation of state of lunar material was similar to that of the upper mantle of the Earth. There seemed good justification for so doing, for the mean density of the Moon (3340 kg/m^3) is very close to that of the upper mantle at zero pressure, the moment of inertia appeared to be very nearly $0.4Ma^2$, indicating an almost uniform density, and the pressure was so low relative to the bulk modulus of mantle material that very little change of density with radius could occur. These considerations are still valid in a first approximation, but need to be refined. Thus the moment of inertia is distinctly less than $0.4Ma^2$, so that some central condensation or some lighter surface layer, or both, are required. Then, as will shortly be seen, the seismic data, not available to earlier workers, clearly show that the density and bulk modulus do change with temperature as well as with pressure within the lunar mantle, and in fact have different properties from the Earth's mantle. Finally, and it is to this point we now turn, there are clear indications from studies of the chemistry of lunar samples that the material of the Moon is not identical with that of any zone in the Earth.

The surface of the Moon has been shown to consist of two distinct types of rock (see Burnett, 1975), the dark material of the *maria* and the light material of the highlands. The former has the higher content of iron and titanium (higher in fact than terrestrial basalts) and has a low content of alumina, and it is composed of clinopyroxene, ilmenite and various iron oxides. The highlands on the other hand are primarily composed of plagioclase with a high content of calcium, they have a high content of alumina but are low in iron oxides. It is supposed that both are derived from the original mantle material of the Moon, whether by differentiation in a molten magma, or by melting and recrystallization under the impact of meteorites, although it is possible that the crust accreted separately onto the Moon, from meteorites, after the formation of the mantle.

It seems that, if the *mare* and highland materials were derived from the mantle, the mantle is unlikely now to have the same constitution as the upper mantle of the Earth.

Further evidence for differences between the Moon and the Earth comes from a study of trace elements. Thus, uranium and thorium appear to be concentrated in the outer parts of the Moon, for, if the Apollo sample concentrations were maintained throughout the Moon, the generation of heat would be too great for the observed outward flow of heat. Studies of lead/uranium and $^{87}Sr/^{86}Sr$ ratios show that the chemistry of lunar material is not the same as of terrestrial material. Generally, the volatile trace elements (Na, Pb for example) are depleted relative to refractory ones (U, Th) in the Moon, and the siderophile elements (those with metallic association, for instance Ni, Ir, Au, Ge) have abundances in lunar material that are low compared with the average solar composition. These observations appear to indicate that the chemistry of the Moon differs from that of the Earth and from that of chondritic meteorites (often supposed to represent the composition of the solar system). In particular, the low abundances of siderophile elements may indicate that metallic components were separated from the material that eventually formed the Moon before that formation took place, and thus one might speculate that the lunar mantle is also deficient in metals relative to the terrestrial mantle.

It should be noted that the density of the *mare* basalts is between 3300 and 3400 kg/m^3 whereas that of the highland material is 2800 to 3100 kg/m^3. It has been suggested on the basis of laboratory experiments that the density of the basalts would increase by phase transformation to between 3500 and 3700 kg/m^3 at the pressures encountered in the mantle of the Moon; if so, *mare* basalt cannot compose the mantle.

Let us now see if such seismological evidence as there is about the Moon's interior is consistent or not with the inferences drawn from the chemical and isotopic constitution of surface samples.

It was seen that the P-wave velocity is almost constant over depths from about 60 km (the base of the crust) to 700 or 800 km, while over the same range the S-wave velocity decreases somewhat. The implications of this behaviour are now developed. It will be shown first that density and elastic moduli must depend upon temperature as well as upon pressure and cannot be a function of pressure alone.

Let α and β stand as usual for the P-wave and S-wave velocities and let K be the bulk modulus, μ the shear modulus and ρ the density. Then

$$\alpha^2 = (K + \tfrac{4}{3}\mu)/\rho, \qquad \beta^2 = \mu/\rho,$$

and, as in Chapter 2,

$$K/\rho = \alpha^2 - \tfrac{4}{3}\beta^2.$$

The adopted values of the P-wave and S-wave velocities at 60 and 800 km are given in Table 5.4. From them there follow the values of K/ρ and μ/ρ which are also given in that table: K/ρ increases from 36 km^2/s^2 to 43.2 km^2/s^2 and μ/K, which is equal to $\beta^2/(\alpha^2 + \tfrac{4}{3}\beta^2)$, decreases from 0.61 to 0.36.

Let us compare these values for the lunar mantle with the corresponding values for the Earth at the same pressure. To calculate the pressures at 60 and 800 km, it is sufficient to treat the Moon as a sphere of uniform density; if the surface radius is a and the density ρ, the pressure at radius r is

$$\frac{2\pi}{3} G\rho(a^2 - r^2),$$

where G is the constant of gravitation.

Taking a to be 1738 km and ρ to be 3340 kg/m^3, the pressure at a depth of 60 km will be found to be 0.4×10^9 Pa, while at 800 km it will be 3.45×10^9 Pa.

The Adams–Williamson argument may now be used to set limits to the change of density from 60 to 800 km, assuming that it comes solely from the increase of pressure. We have

$$\frac{d\rho}{\rho} = \frac{dp}{K}$$

so that

$$d\rho = \Psi \, dp,$$

where Ψ stands for ρ/K.

An integration by parts gives

$$\Delta\rho = [\Psi p]_{p_1}^{p_2} - \int_{p_1}^{p_2} p \, d\Psi.$$

Here p_1 is the pressure at 60 km (0.4×10^9 Pa) and p_2 that at 800 km (3.45×10^9 Pa). The integral depends on the behaviour of Ψ, that is, on

Table 5.4. *Properties of the lunar mantle*

Depth (km)	Pressure (10^9 Pa)	α (km/s)	β (km/s)	K/ρ (km/s)^2	μ/ρ (km/s)^2	K_{min} (10^{11} Pa)	μ_{min} (10^{11} Pa)
60	0.4	8.1	4.7	36.16	22.09	1.15	0.76
800	3.45	8.1	4.1	43.20	16.81	1.37	0.53

K_{min}, μ_{min}: values calculated with the minimum density of 3180 kg/m^3.
If α is 8.0 km/s, the corresponding values of K_{min} are:

Depth below the surface (km)	K_{min} (10^{11} Pa)
60	1.10
800	1.37

whether the seismic wave velocities change steadily or in jumps. Limits may, however, be placed upon it if we assume that Ψ changes monotonically. Then

$$\left|\int p\,\mathrm{d}\Psi\right| \leqslant |(p_2 - p_1)(\Psi_2 - \Psi_1)|,$$

where Ψ_1 and Ψ_2 are the values of Ψ at p_2 and p_1.

In fact $\int p\,\mathrm{d}\Psi$ is less than zero, for $p_2 > p_1$, but $\Psi_2 < \Psi_1$. The numerical values are

	p (10^9 Pa)	K/p (km^2/s^2)	Ψ (10^{-8} s^2/m^2)
1	0.4	36.16	2.76
2	3.45	43.20	2.31

Hence

$$[\Psi p]_{p_1}^{p_2} = 68.6 \text{ kg/m}^3,$$

while

$$|(p_2 - p_1)(\Psi_2 - \Psi_1)| \leqslant 14.3 \text{ kg/m}^3.$$

Thus, the increase of density from 60 to 800 km should lie between 69 and 83 kg/m^3, provided that pressure is the only factor causing the density to change.

The seismic data show that K/ρ increases by about 20 per cent over the range 1–3.5×10^9 Pa in the Moon, while the argument from the Adams–Williamson relation suggests that the increase of density is not more than about 2.5 per cent. The value of the density itself may be estimated from the dimensionless angular momentum of the Moon, the current estimate of which is 0.392 (see above). The density between 60 and 800 km will take its least value if the density below 800 km is constant from there to the centre, and the least value between 60 and 800 km in those circumstances is then 3180 kg/m^3. Thus we can calculate the minimum values of K by multiplying the observed values of K/ρ by 3180 kg/m^3. The values so found are given in Table 5.4, together with corresponding values of μ.

The terrestrial values of K, μ and ρ to compare with the foregoing estimates of lunar properties are taken from the models of Gilbert and Dziewonski (1975). Gilbert and Dziewonski constructed two models, 1066A and 1066B, the differences between which are for the present purpose insignificant, and the values given in Table 5.5 are the means of K, μ and ρ from the two models. The comparison between terrestrial and lunar properties is contained in Table 5.6.

The density within the Earth over the relevant range of pressure (that is, to a depth of about 100 km, or within the lithosphere) is very close to the mean density of the Moon (3340 kg/m^3) and increases by less than the amount estimated above for the Moon. Not only are the values of K/ρ at the same pressure different for the Earth and the Moon, but K is greater for the Moon than for the Earth. The values of K/ρ are independent of estimates of density, whereas the lunar values of K are proportional to the assumed density; nonetheless, the minimum values exceed the terrestrial values by substantial amounts. The increases of K, in the Moon as well as the Earth, are of the same order as suggested by laboratory measurements of the variation of K with pressure for olivine. Kumagawa and Anderson (1969) found $\partial K/\partial p$ to be 5.2 for polycrystalline olivine, so that the change of K from 1×10^9 to 3.5×10^9 Pa would be 13×10^9 Pa as compared with the estimated increases of 20×10^9 Pa for the Earth and 22×10^9 Pa for the Moon. At the same time, the temperature will increase with depth, both in the Earth and in the Moon, leading to smaller changes of K than estimated from the effect of pressure alone. It follows that the

Table 5.5. *Values of density ρ, bulk modulus K, shear modulus μ and the ratios K/ρ, μ/K for the outermost parts of the Earth*

Depth (km)	Pressure (10^9 Pa)	Density (kg/m^3)	K (10^{11} Pa)	μ (10^{11} Pa)	K/ρ (km/s)^2	μ/K	σ
42	1.0	3330	1.04	0.72	31.2	0.70	0.20
73	2.0	3342	1.12	0.70	33.4	0.62	0.24
100	3.0	3346	1.20	0.68	35.4	0.57	0.26
123	3.7	3357	1.26	0.66	37.6	0.52	0.28
168	5.2	3375	1.36	0.64	40.5	0.46	0.30

The values are approximately means of the values given by the two models 1066A and 1066B constructed by Gilbert and Dziewonski (1975).

Table 5.6. *Comparison between properties of the Earth and the Moon*

Pressure (10^9 Pa)	K (10^{11} Pa)		μ (10^{11} Pa)		μ/K		σ	
	Earth	Moon	Earth	Moon	Earth	Moon	Earth	Moon
1	1.04	1.15	0.72	0.70	0.69	0.61	0.20	0.25
3.5	1.24	1.37	0.67	0.53	0.54	0.39	0.28	0.32

bulk modulus of the lunar mantle, supposing it to have a uniform composition, increases more rapidly with pressure than does that of olivine. The bulk modulus of olivine decreases by 1.5×10^7 Pa/deg (Kumagawa and Anderson, 1969). Thus the difference between the increase of K in the Earth and in the Moon could be explained if the temperature between 40 and 120 km in the Earth increased by 200 K more than that in the Moon between 60 and 800 km. However, if a difference in temperature is to be invoked to explain the differences in K itself, the temperature in the Earth at 40 km would need to exceed that in the Moon at 60 km by 700 K, and that seems unlikely.

It is possible that the changes of bulk modulus with depth in both the Earth and the Moon may be accounted for by the actual increase in pressure and plausible differences of temperature, but the differences in the values of K themselves indicate a difference in composition of lunar and terrestrial material.

There is an evident difference between the ratio μ/K for the Earth and the Moon and, as may be seen from Table 5.6, not only is K greater in the Moon than the Earth, but μ is lower, and, whereas K *increases* with pressure more rapidly in the Moon than the Earth, μ *decreases* more rapidly. The two effects together lead to great differences between the ratios μ/K.

μ, like K, increases with pressure and decreases with temperature. If a temperature effect *could* account for the different values of K in the Earth and the Moon, could it account for the different values of μ as well? The signs of the differences show that it cannot. K increases more rapidly in the Moon than in the Earth, implying a lower temperature at a pressure of 3.5×10^9 Pa in the Moon than in the Earth, but μ decreases more rapidly in the Moon, implying a higher temperature in the Moon. The validity of this argument depends only upon the assumption that the changes of K and μ are due solely to pressure and temperature and that the coefficients of temperature have the same signs for K and μ, as is the case.

We must conclude that the changes of K and μ with pressure and temperature cannot account for the differences of K and μ and μ/K between Earth and Moon. It seems clear that there are real differences of composition, for, not only is the bulk modulus greater in the Moon, but also, it seems likely, the density is less. The minimum density, as set by the dimensionless moment of inertia, is considerably less than the terrestrial density over the same range of pressure, and the maximum possible density must also be less, for the terrestrial density is very close to the *mean* density of the Moon, while the density of the outer part of the Moon must be less than the mean density for the dimensionless moment of inertia to be less than 0.4.

Consideration of the shear modulus introduces other factors. Kumagawa and Anderson (1969) give the following values for the pressure and temperature coefficients of the moduli of olivine:

$\partial K/\partial p$: 5.2 $\partial \mu/\partial p$: 1.8

$\partial K/\partial T$: -1.5×10^7 Pa/deg $\partial \mu/\partial T$: -1.3×10^7 Pa/deg.

If one were to suppose that these values apply to the upper mantle of the Earth, then the change in temperature between 40 and 110 km would be estimated to be

from the change of K: -470 K

from the change of μ: $+730$ K.

The corresponding estimates for the lunar temperature at 800 km depth are

from the change of K: -600 K

from the change of μ: $+1730$ K.

It is clear from these estimates that, in the Earth and Moon alike, μ decreases more rapidly with depth than can be accounted for by a simple decrease proportional to the increase of temperature. Furthermore the lunar and terrestrial values of $\partial K/\partial p$ appear to be significantly greater than found in laboratory experiments, namely at least 8 to 9 instead of about 5 for olivine. The greatest values found experimentally are about 6 (Cook, 1972) and this suggests that the change of K is not solely due to pressure, but to a change of composition as well.

Two conclusions seem to follow from the foregoing analysis. The first is that the values and changes of K and μ in the Earth and Moon cannot be reconciled if it is supposed that the material is the same for each. The material of the lunar mantle appears to be less compressible than the terrestrial mantle. The other conclusion is that in the Earth as in the Moon the decrease of shear modulus shows that the behaviour of the shear modulus with temperature is more complex than laboratory work indicates.

5.8   **Lunar models**

Any model of the Moon or a planet must satisfy the constraints of the known mass and moment of inertia, but in themselves those constraints are far from being sufficient to determine the internal structure and in general some assumptions must be made about the present state of the body or of the conditions that gave rise to it. As has been shown in Chapter 4, a great deal is known in a general way about the behaviour of planetary materials, but the complexity of the chemical constitution is such that it is not possible to predict the state of a particular body in the absence of a detailed chemical specification. There is little direct evidence on which such a specification can be based and so we are faced with two possibilities; either to draw such general conclusions as are possible in the absence of detailed chemical knowledge, or to call upon theories of the origin of the planets for a chemical recipe. In this book I emphasize the first approach; I am concerned to see how far it is possible to go without making use of theories of origin. That is not to say that such theories are to be ignored: they have much of value to tell us about the planets, but one aim of this book is to see how far it is possible to use our knowledge of the present state of the planets to check the validity of such theories, and in consequence we must proceed as far as possible independently of them.

Our knowledge of the Moon consists then of the mean density, 3340 kg/m^3, the moment of inertia factor, 0.392, and the seismic data as analysed in the previous section. Because it seems that the material of the outer parts of the Moon is not identical with that of the upper mantle of the Earth, we are not allowed to construct models of the interior of the Moon using terrestrial equations of state.

The moment of inertia factor is somewhat less than $\frac{2}{5}$ so that some increase of density towards the centre occurs. Our knowledge of the Earth indicates that there are three possible causes: compression of material of uniform composition under self-gravitation; the olivine–spinel or similar change of phase; and the presence of a heavy core. Of these possibilities, the second, a phase transition, may be eliminated because the pressure is never great enough for it to occur. The moment of inertia factor shows that the density is nearly constant throughout the Moon, so that the central pressure may to a first approximation be taken to be that in a sphere of constant density, namely

$$\frac{2\pi G\rho^2}{3} a^2,$$

where G is the constant of gravitation, ρ the density and a the radius. With ρ equal to 3340 kg/m^3 and a equal to 1738 km, the central pressure so estimated is 4.7×10^9 Pa, which is much less than the pressure ($\sim 1.5 \times 10^{10}$ Pa) at which the olivine–spinel transition occurs. It may therefore be taken that the outer parts of the Moon have the low pressure crystal structure, which is consistent with the mean density of the Moon being close to that of olivine and to that of the upper mantle of the Earth.

We therefore consider two possibilities: either the Moon is of more or less uniform constitution throughout (apart from the crust of 60 km) or it has a small central core of heavy material of different chemical constitution, and we ask first if the limits given in Appendix 1 provide any guide to the choice. Parker's limit is that, if there is a core, its density must exceed 1.03 times the mean density and its radius must be less than 0.99 times the Moon's radius, hardly stringent conditions. If the variance of density is to be minimized, the radius of the core must be 0.75 times the lunar radius and the outer density must be 0.95 and the core density 1.06 times the mean density. These values do not suggest the presence of a small dense core. Similarly Rietsch's (1978) bound on the maximum density, namely 3472 kg/m^3, within a radius of 0.9 times the lunar radius, does not indicate the existence of a dense core. These limits tell us that, unless there are any strong reasons for supposing a small heavy core to exist, we should first consider the effect of self-compression on material of uniform composition.

In considering the Moon we can take advantage of the fact that the maximum pressure is relatively low and so adopt a simplified treatment. The relevant parameter here is the ratio of the central pressure to the bulk modulus of the material; the former has just been estimated to be about 4.7×10^9 Pa, whilst the latter was seen in the previous section to be of the order of 1.5×10^{11} Pa, so that the ratio is about 3×10^{-2}.

This means that the maximum change of density due to self-compression is of the order 3×10^{-2} and accordingly an approximate algebraic treatment of the problem is possible in which the pressure is calculated as if the density were constant. The pressure at any radius, r, is accordingly taken to be

$$\frac{2\pi}{3} G\rho^2(a^2 - r^2),$$

where ρ is understood to be the mean density of the Moon.

Further, K is assumed to be given by the linear expression

$$K = K_0 + bp,$$

where, for the Moon, K_0 is 1.5×10^{11} Pa and b is 8.

The change of density with pressure is given by

$$\frac{d\rho}{\rho} = \frac{dp}{K};$$

with the foregoing expressions,

$$dp = -\frac{2\pi}{3} G\rho^2 \, d(r^2)$$

and

$$K = K_0 + \frac{2\pi b}{3} G\rho^2 (a^2 - r^2).$$

Thus

$$d\rho = -\frac{2\pi}{3} \frac{G\rho^3}{K_0} \left[1 - \frac{2\pi b G\rho^2 (a^2 - r^2)}{3K_0} \right] d(r^2),$$

to the approximation adopted here, leading to

$$\rho = \rho_0 + \frac{2\pi}{3} \frac{G\rho_0^3}{K_0} (a^2 - r^2) - \frac{2\pi^2}{9K_0^2} G^2 b \rho_0^5 (a^2 - r^2)^2.$$

Thus

$$\rho = \rho_0 \left[1 + \frac{\xi}{a^2} (a^2 - r^2) - \frac{\eta}{a^4} (a^2 - r^2)^2 \right],$$

where

$$\xi = \frac{2\pi}{3} \frac{G\rho_0^2 a^2}{K_0}$$

and

$$\eta = \frac{2\pi^2}{9} \frac{b G^2 \rho_0^4 a^4}{K_0^2} = \tfrac{1}{2}\xi^2 b.$$

ξ is p_c/K_0, where p_c is the central pressure and η is $\tfrac{1}{2}b(p_c/K_0)^2$. The mass is equal to

$$\int_0^a 4\pi\rho r^2 \, dr$$

and the moment of inertia to

$$\int_0^a \frac{8\pi}{3} \rho r^4 \, dr.$$

Thus

$$M = \frac{4\pi}{3} a^3 \rho_0 (1 + \tfrac{2}{5}\xi - \tfrac{8}{35}\eta)$$

and

$$I = \frac{8\pi}{15} a^5 \rho_0 (1 + \tfrac{2}{7}\xi - \tfrac{8}{63}\eta)$$

whence

$$\frac{I}{Ma^2} = \tfrac{2}{5}[1 - \tfrac{4}{35}\xi + \tfrac{8}{1575}(9 + 10b)\xi^2].$$

Since ξ is 3×10^{-2}, the third term is negligible and we find

$$\frac{I}{Ma^2} = 0.3986.$$

The result is significantly greater than the observed value of 0.392 and we conclude that self-compression is inadequate to account for the observed moment of inertia.

So far the effect of the 60 km crust has not been considered. We do not know what the density is, but suppose it to be 3000 kg/m^3. Then the moment of inertia factor of a uniform interior covered by such a crust, the whole having a density of 3340 kg/m^3, would be 0.397. The effects of the crust and self-compression may be combined according to the formula given in Appendix 2. Let γ be $5I/2Ma^2$. If γ_1 is the value corresponding to one factor and γ_2 that to another and, if both are close to 1, the two factors together give a value of γ equal to $\gamma_1\gamma_2$. If then we combine a crust as just specified with an interior in which the density increases by self-compression, the moment of inertia ratio is found to be 0.396.

This value is still in excess of the observed value and it therefore seems likely that the Moon possesses a small heavy core. Let the core in an otherwise uniform Moon have a value of γ equal to γ_3. We must then have

$$\gamma_{\text{obs}} = \gamma_1\gamma_2\gamma_3$$

or

$$\gamma_3 = \gamma_{\text{obs}}/\gamma_1\gamma_2,$$

where γ_{obs} is the observed value for the Moon and γ_1 and γ_2 are respectively the values for the crust and self-compression.

We have just seen that $\gamma_1\gamma_2 = 0.99$ ($I/Ma^2 = 0.396$). Since γ_{obs} is 0.98, we require that γ_3 should be 0.99. Suppose, for example, that the core is not allowed to have a density in excess of 8000 kg/m^3 (an iron-like material). The required radius of the core would then be 340 km and the mean density of the mantle material between the core and the crust would be 3270 kg/m^3.

To summarize, a model with the following characteristics has a moment of inertia ratio very close to the observed value:

(*a*) a crust of 60 km thickness and a density of 3000 kg/m^3, (*b*) a mantle of mean density 3270 kg/m^3 and bulk modulus of 1.5 × 10^{11} Pa, under self-compression,

(*c*) a core of density 8000 kg/m^3 and radius 340 km.

All these are consistent with our knowledge of the Earth and the Moon, but other possibilities are by no means excluded. The nature of the crust is in fact rather uncertain. It is not known if it extends over the whole Moon (Lammlein, 1977) and there is no independent evidence for the density. A value of 3000 kg/m^3 seems reasonable in the light of lunar sample properties. If the crust were less dense, the core would need to be smaller, and vice versa. The fact that the very sparse seismic data seem to set a limit of about 300 km on the radius of the core (Nakamura *et al.*, 1974) may be seen as support for a crustal density not less than 3000 kg/m^3. As to the material of the mantle, it does seem to differ significantly from that of the upper mantle of the Earth, for the density appears to be less than that of the upper mantle of the Earth over a comparable range of pressure, while, as seen above, the bulk modulus is rather greater.

More elaborate models than the foregoing have been constructed, necessarily based on more assumptions (see also Bills and Ferrari, 1977). No allowance has been made in the present model for any increase of temperature towards the centre and that, by reducing the increase of density somewhat, would increase the moment of inertia ratio.

6

Mars, Venus and Mercury

6.1 Introduction

Mars, Venus and Mercury form with the Earth and the Moon a group of rather similar bodies. By comparison with the giant planets on the one hand and the small satellites on the other, the sizes lie in a relatively restricted range, while the mean densities are higher than those of most other bodies in the solar system. It is natural to think that their compositions are similar and that the structures of Mars, Venus and Mercury might be inferred from what is known of the Earth and the Moon.

Seismological data are, of course, not available for any of the planets other than the Earth, so that the structures of the terrestrial planets must be derived from the dynamical data, together with such inferences as may be drawn from the magnetic and electrical properties, together with analogies with the Earth and the Moon.

Unfortunately, the dynamical data themselves are less informative for Mars, Venus and Mercury than they are for the Earth and the Moon or for the major planets. The solar precession of Mars has not so far been observed and, in consequence, the moment of inertia cannot be derived from the value of J_2 without making the assumption of hydrostatic equilibrium. Yet it is clear that Mars is not in hydrostatic equilibrium. The theory of the errors likely to be committed by making the assumption of hydrostatic equilibrium was given in Chapter 3 and subsequently in this chapter (section 6.6) it will be applied to Mars. Our knowledge of Venus and Mercury is even less than that of Mars. It was seen in Chapter 3 that the value of J_2 of a planet in hydrostatic equilibrium is proportional to m, and so to the spin of the planet. Now Venus and Mercury rotate very slowly, so that m and therefore J_2 are small, so small in fact that for neither planet is J_2 greater than the current limit of detection set by the observations of the paths of space probes that have passed close to the

171

planets. Thus, with m very small, and J_2 not detectable, Darwin's formula (Chapter 3) cannot be used to estimate C/Ma^2, even supposing the planets to be in hydrostatic equilibrium, a supposition that may not be too erroneous for Venus, but may well be unsatisfactory for Mercury.

A seismometer was placed on Mars by the Viking mission of 1976 (Anderson *et al.*, 1977). The disturbances caused to the instrument by winds prevent seismic signals being detected in the daytime, and only a few events have been detected at night. The few records obtained at a single site provide no information about internal structure.

The magnetic field of the Earth is associated with the metallic core, while the absence of an external dipole field of the Moon is consistent with the absence of a core of significant size. Magnetic fields of the other terrestrial planets might therefore provide clues to the presence or absence of cores, but the fields of internal origin are at the most very small, so that no clear conclusions can be drawn about the existence of cores.

The variation of density in the mantle of the Earth is controlled both by composition and pressure. There is a steady increase of density owing to pressure, and there are additional increases of density within the upper mantle, both because of an increase of the ratio of iron to magnesium and also because of polymorphic changes of silicates under pressure. It may be expected that the same three factors will determine the densities of silicates in the mantles of the Moon, Mars, Venus and Mercury and, although it was seen in Chapter 5 that the pressure in the Moon never attains that required for the olivine–spinel transformation, the pressure in Mars and Venus does exceed that transition pressure. Thus, it may be supposed that, as in the mantle of the Earth, so in the mantles of Mars and Venus, the increase of density with depth will be determined by hydrostatic compression, changes of composition and polymorphic transitions. There is a further difficulty: the pressure at which a polymorphic transition occurs depends both on composition (that is, the iron–magnesium ratio in olivine and pyroxene) and on temperature and so, since both composition and temperature will probably vary with radius and pressure otherwise than they do in the Earth, the radii or pressures at which the polymorphic transitions occur in Mars and Venus are unlikely to be the same as in the Earth.

Given such complexities, inherent in the variability of iron–magnesium silicates, it is hopeless to attempt to derive detailed models from such dynamical data as there are, together with analogies with the Earth or Moon to resolve the very wide range of possibilities consistent with the

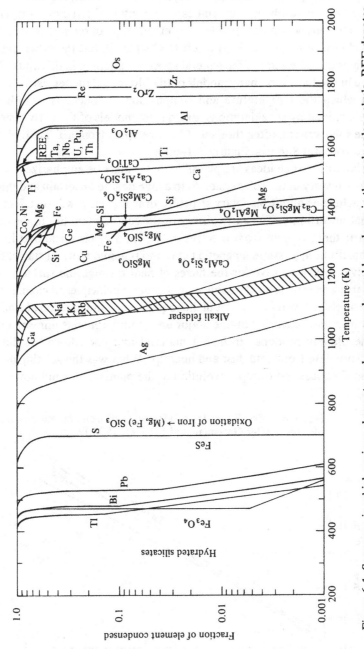

Figure 6.1. Sequence in which various elements and compounds would condense in the early solar nebula. REE denotes rare earth elements. (From Grossman and Larimer, 1974.)

mechanical properties. Are the magnetic fields, which are very small, of any help in resolving ambiguities in the internal structure, or can any conclusions be drawn from ideas about the origin of the solar system? The evidence of magnetic fields is left to Chapter 9, but the other question brings up the matter of the general approach adopted in this book. There are in the literature many models of the Moon and the terrestrial planets in which the temperature and composition as functions of radius are determined by considerations drawn from models of the early history of the solar system, often thought of as developing from a T-Tauri phase of the youthful Sun (see Figures 6.1–6.3).

No doubt such ideas are plausible, but it must be emphasized that the early history of the solar system is to a high degree uncertain and that any conclusions about planetary structure drawn from it will be affected by that uncertainty. The early history of the solar system lies in the distant past; there remain from it traces in the form of chemical and isotopic constitution, possible magnetic fields and possible thermal states, themselves unknown save for the fluxes of heat through the surfaces of the Earth and the Moon. To some extent general principles of chemistry can be applied to work out in detail the sequence of condensation of a mixture of iron and silicates, but one major uncertainty remains unresolved: did the planets condense from hot material and cool down, or did they accumulate from cold dust and heat up? What was the starting point of the chemical and isotopic evolution of the planets? It is unknown.

Figure 6.2. Conditions for condensation of iron and magnesium silicates. (After Grossman and Larimer, 1974.)

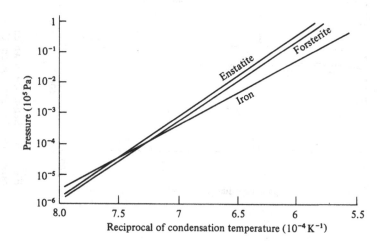

A different attitude is adopted in this book, the aim of which is to find out as much as possible about the planets from their present state, so far as it is observable, and to use only established physical principles. When, together, these fail, the failure is explicitly recognized and no attempt will be made to fill the gap by inference from an assumed past state. From such a self-denying approach it will yet be possible to draw some conclusions (Chapter 10) about the origin of the solar system.

6.2   Investigation of Mars, Venus and Mercury

Of these three planets, it had been possible to learn much of Mars by telescopic observations from the Earth. The surface can be seen through the tenuous atmosphere, so that the period of rotation and the size and polar flattening of the geometrical surface can all be estimated with adequate precision. Furthermore, Mars has two small natural satellites, Phobos and Deimos, the orbits of which can be determined from the Earth, so enabling the mass of Mars and the value of J_2 to be found (Kovalevsky, 1970; Sinclair, 1972). Venus and Mercury, on the other hand, are less accessible to telescopic observation from the Earth, Venus

Figure 6.3. Mean atomic weights of the planets predicted from the fractionation of metals and silicates in the early solar nebula. (From Lewis, 1972.)

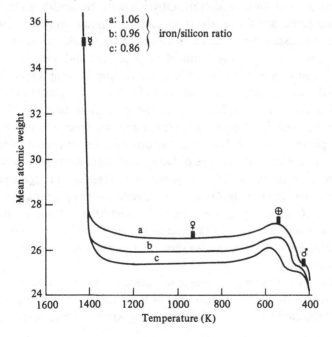

because the solid surface is concealed by the thick atmosphere and Mercury because it is always close to the Sun. Thus, whereas the older estimates of the mechanical properties of Mars have been refined only in detail by the new means of radar and space craft, quite radical changes have taken place in the knowledge of Venus and Mercury. A convenient summary of the results of telescopic observations (together with some of the newer methods) will be found in the compilation edited by Dollfus (1970).

Telescopic observations from the Earth have not been without their development in recent years. It has been possible to improve the resolution of photographs very considerably by using telescopes at high altitude, such as that at the Pic-du-Midi, and the measurements of diameters was much improved by the split-image micrometer devised by Dollfus (see, for example, Dollfus 1972*a*, *b*). Nonetheless, it has been the application of radar and various observations from space craft that have so greatly improved our knowledge of Venus and Mercury in the last two decades or so.

Radar studies of the planets began with the detection of echoes from Venus, followed by echoes from Mercury in 1962 and Mars in 1963. (Radar returns from the Moon had been obtained as early as 1946.) A radar system to observe echoes from the planets differs in important ways from one that detects ships or aircraft. Obviously the transmitters must generate more power and the receivers operate at optimum sensitivity for echoes from such distant bodies to be received. The aerial also must be as large as is reasonable from the point of view of cost and mechanical practicability, but it does not have to follow a rapidly moving target. It should indeed operate under computer control to follow the planet being observed. Perhaps more fundamental is the difference between the character of the echo received from a planet and one from an aircraft. Because the planet is large and rotating, the observed returns come from a range of distances and with a range of Doppler shifts of frequency, and, in fact, from the distribution of echo power with respect to range and frequency, it is possible to determine both the radius of the planet and its rate of rotation (Green, 1968). Analyses of the echo pattern or scattering function of a planet are greatly facilitated by the fact that successive echoes are effectively identical, unlike those of a ship or aircraft which are continually changing.

The obvious use of radar is of course to determine the distance of the planet. The ratios of the semi-major axes of the orbits of the planets are very well established by Kepler's laws from the known orbital periods, and so the scale of the solar system, namely the astronomical unit or the semi-major axis of the Earth's orbit, follows, in principle, from a single determination of the distance between two planets. A number of careful observations have led to international agreement on a value to be adopted for the astronomical unit (Ash, Shapiro and Smith, 1967).

There are two ways of estimating the radius of a planet from radar observations. Besides that just mentioned, involving the analysis of the echo pattern, the radius may be found by comparing the least distance of the planet from the observatory with the distance calculated dynamically from the orbital period. The former gives the distance to the closest part of the surface of the planet, the latter the distance to the centre of mass, and the difference between the two is the radius along the line of sight to the observatory. As explained by Shapiro (1968) a process of successive approximations has to be followed to estimate all the parameters that determine the echo time from the planet.

A second method is also available to establish the period of rotation of a planet, complementary to that just mentioned, which depends on the Doppler shift of the frequency of an echo from a moving surface. That method works well for a planet rotating relatively fast, but is less satisfactory for the slow rotators, Venus and Mercury. However, for those it is possible to detect specific features of the echo which come from some particular part of the surface and move in time through the echo pattern as the planet rotates; by following such factors, the period of the planet's rotation may be established. It was in this way that the current values of the rotation periods of Venus and Mercury were first obtained.

Most radar observations of the planets have been made in the United States where there are some very powerful observatories, particularly the steerable systems at Millstone Hill operated by the Lincoln Laboratory of the Massachusetts Institute of Technology and at Goldstone, operated by the Jet Propulsion Laboratory of the California Institute of Technology. The Lincoln Laboratory also operates the Haystack system. Cornell University operates a system with a very large antenna at Arecibo (Puerto Rico), where the parabolic reflector is formed in the ground and has a fixed vertical axis, although by moving the aerial feed it can operate within 20° of the zenith. Outside the United States the Mark I radio telescope at Jodrell Bank (England) has been used for radar work, as has the Crimean Deep Space Tracking Station in the U.S.S.R.

Accounts of the theory of radar astronomy, of installations and of methods and results of observations of the planets will be found in the co-operative work edited by Evans and Hagfors (1968).

Powerful as radar methods are, they have their limits. The radii which are determined are essentially equatorial radii; the radar echo is insensitive to polar flattening, but might reveal, or set limits to, any ellipticity of the equator. More important, radar gives no information about the mass or gravity field of the planets, and it is there that space craft have made fundamental contributions. The principal results have come from the Mariner series of space craft (see, for example, Rea, 1970) which are stabilized with respect to three axes and carry a variety of instruments according to the special emphasis of a particular flight, but which have always carried equipment for accurate tracking, whether by radio Doppler shift measurements or by radio time delay measurements or both, and have often carried television equipment. The first determination of the mass of a planet from the perturbations it produced on the trajectory of a space craft was that of Venus by means of Mariner 2 (Kovalevsky, 1970). Much more recently the flight of Mariner 10 past Venus and Mercury in 1974 and 1975 has led to accurate estimates of the masses of those planets (Howard *et al.*, 1974*a, b*; Deane, 1976).

The first space craft to fly by Mars was Mariner 4, followed by Mariners 6 and 7, which carried television equipment (Smith, 1971; Kliore, Fjeldbo and Seidel, 1971) and subsequently by Mariner 9, which was successfully placed in orbit round Mars in 1971. The detailed study of the orbit of Mariner 9 has enabled very detailed description of the gravity field of Mars to be given (Jordan and Lorell, 1975).

The major aims of the Mariner space craft, and the corresponding craft launched from the U.S.S.R., especially Mars 1 to 6 (Marov and Petrov, 1973) and the Venera space craft, have been the study of the surface features, the geology and morphology, and the ionospheres and magnetospheres of the planets, but the instruments carried by the Mariner space craft in particular have allowed the determination of the sizes and rotation rates in certain cases. The size can be found from occultations by the planet of radio signals received on Earth from the space craft. Thus the polar as well as the equatorial radii of Mars have been determined and radii of Mercury at orientations of 2 ° and 68 ° to the equator.

Important observations have been made with the television cameras carried on Mariners 9 and 10. The former photographed the moons, Phobos and Deimos, enabling their orbits to be established. The data so obtained could be combined with tracking of the space craft itself, leading to a very detailed description of the gravity field (Born, 1974). Photographs from Mariner 10 of the surface of Mercury led to an accurate determination of the period of rotation, concordant with other methods (Table 6.4).

Space craft of the U.S.S.R. Mars series have landed on Mars and were followed by the soft landings of the Viking Landers 1 and 2 (Soffen and Young, 1972). From the point of view of dynamics, the important point about the Viking landings is that they took place from space craft in orbit about Mars, the tracking of which led to further improvements in the knowledge of the Martian gravity field (section 6.3).

6.3 Dynamical properties

The principal dynamical properties of the terrestrial planets have been reasonably well known for quite a long time. All three planets are close enough for their dimensions to be established from telescopic observations from the ground. The polar flattening and other properties of Mars are also well determined from telescopic observations; estimation of the period of rotation is straightforward, whilst the mass and the gravitational coefficient, J_2, are derived from the orbital motions of the natural satellites, Phobos and Deimos. Mercury and Venus present some difficulties for ground based observations. They have no natural satellites, so that it was necessary to estimate their masses from the perturbations that they produced of the orbits of other planets, an inherently less accurate procedure than the more direct one utilizing the size and period of a satellite orbit. Again, in the absence of natural satellites, it was not possible to estimate J_2, while telescopic observations failed to detect any polar flattening of the geometrical figures. Furthermore, estimates of the periods of rotation were very uncertain. Valuable summaries of the knowledge of the terrestrial planets up to 1970 will be found in the book edited by Dollfus (1970). Radar measurements, begun in the 1950s, contributed to improved knowledge of some properties of the terrestrial planets, in particular of the periods of rotation of Venus and Mercury.

Knowledge of the dynamics of the terrestrial planets has been very greatly advanced through observations made with the Mariner series of space craft and the Viking Landers on Mars. The periods of rotation of Venus and Mercury have been firmly established and their masses carefully determined from the behaviour of space craft in their neighbourhood. The sizes have been found from occultations of very high frequency radio transmissions from space craft and limits placed upon the geometrical flattening. A detailed knowledge of the gravity field of Mars has been obtained from the orbits of Mariner 9 and Vikings 1 and 2 as well as from the orbits of Phobos and Deimos derived from photographs of them taken from Mariner 9. Finally, limits have been placed upon values of J_2 for Venus and Mercury.

The rotational period of Mars is 1.025 96 d, corresponding to an angular velocity of $7.088\,22 \times 10^{-5}$ rad/s (Dollfus, 1970; Allen, 1963).

Recently, de Vaucouleur, Davies and Sturms (1973) have obtained a somewhat improved value, from telescopic observations of surface markings. They find that the period of rotation is 24 h 37 min 22.655 s or 88 642.655 s, corresponding to an angular velocity of $7.088\,21 \times 10^{-5}$ rad/s.

The most recent observations of the geometrical figure of Mars are given in Table 6.1(*a*).

Three means have been used to determine the geometrical figure of Mars; visual observations from telescopes on the ground (see Dollfus, 1972*b*); radar observations; and occultations of radio transmitters carried in space craft, especially in Mariner 9. Dollfus devised a double-image micrometer in which two images of the planet are arranged so that their opposite limbs just touch, and has used it at telescopes at Pic-du-Midi and Meudon and in Greece, working with a number of collaborators over some years. The values he obtained for the equatorial and polar radii, as given in Table 6.1(*a*), are the results of combining the optical observations with material from the Mariner 4, 6 and 7 flights. Dollfus, like Born (1974) who used radar data and observations of the natural satellites made from Mariner 9, did not distinguish between equatorial radii: he took the equator to be circular. Cain *et al.* (1972) and Christensen (1975), on the other hand, found the equator to be distinctly elliptical. Cain *et al.* analysed radio occultations from Mariner 9, while Christensen combined the occultations with radar and optical observations of the surface.

The agreement between the various estimates of the Martian radii is satisfactory, although the estimates are not independent, for some of them use common data. Nonetheless, the polar radius is probably known to within ± 2 km, while it seems likely that the equator is definitely elliptical.

The polar flattening of the geometrical surface is comparatively large and greater than the dynamical flattening (see below), indicating that Mars is not in hydrostatic equilibrium. The indication is strengthened by the implication of the large geometrical flattening. Were Mars in hydrostatic equilibrium, Darwin's formula (Chapter 3) could be applied in the form

$$\frac{C}{Ma^2} = \frac{2}{3}\left[1 - \frac{2}{5}\left(\frac{5}{2}\frac{m}{f} - 1\right)^{1/2}\right]$$

to give C/Ma^2. If the value of m and the geometrical flattening are inserted, it will be found that $C/Ma^2 = 0.45$. Since the largest possible value of C/Ma^2 is that for a uniform sphere, namely 0.4, it is evident that the assumption of hydrostatic equilibrium is incorrect (see Dollfus, 1972b).

The mass of Mars is well determined from the tracking of a number of space craft, namely Mariners 4, 6 and 9, by means of the Doppler shifts of frequencies of radio transmissions from them. The result from the tracking of Mariner 4 (Null, 1969) was especially precise. Mariner 9 was

Table 6.1. *Dynamical properties of Mars*

(a) Geometrical figure

	Radii (km)			Polar flattening $\times 10^{-3}$
	a	b	c	
Born (1974)	3397.2 ± 1		3375.5 ± 1	6.38 ± 0.03
Dollfus (1972b)	3398 ± 3		3371 ± 4	7.94
Christensen (1975)	3399.1	3394.1	3376.7	5.9
Cain et al. (1972)	3400.8	3394.7	3372.5	7.42

(b) Mass

	GM (10^{13} m^3/s^2)	M (10^{23} kg)
Mariner 4 (Null, 1969)	$4.282\,83 \pm 0.000\,01$	6.4191
Mariner 6 (Anderson, 1970)	$4.282\,80 \pm 0.000\,20$	6.419 06
Mariner 9, approach		
(Lorell et al., 1973)	$4.282\,82 \pm 0.000\,10$	6.419 09
S. K. Wong (quoted in Born, 1974)	$4.282\,85 \pm 0.000\,04$	6.419 13
Mariner 9, Phobos and Deimos (Born, 1974)	$4.282\,81 \pm 0.000\,5$	6.419 07

(c) J_2

	$10^3 J_2$
Natural satellites	
(Sinclair, 1972)	1.9655 ± 0.0014
Mariner 9	
(Jordan and Lorell, 1975)	1.9606
(Sjogren, Lorell, Wong and Downs, 1975)	1.9577
Mariner 9 and Vikings 1 and 2	
(Gapcynski, Tolson and Michael, 1977)	1.9557 ± 0.0004

more versatile, for it carried cameras which took photographs of Phobos and Deimos, so that better orbits of those satellites were determined than could be done from the ground, and Born (1974) combined the data from the orbits of the satellites with Doppler tracking to improve the estimate of the mass of Mars as derived from the tracking alone (Sjogren, Lorell, Wong and Downs, 1975). It seems likely that the best result for the mass of Mars is to be obtained by taking the average of the Mariner 4 result and that of Born (1974) and it is that value, namely $6.419\,09 \times 10^{23}$ kg that is given in the summary in Table 6.5. In fact, none of the values in Table 6.1(b) differs from this mean value by more than its own quoted standard deviation.

Estimates of the value of J_2 had long been obtained from the motions of Phobos and Deimos as observed from the Earth. Both satellites are close enough to Mars for the perturbations due to J_2 to be appreciable, and, since Mars is relatively close to the Earth, quite accurate observations of the perturbations are available (Sinclair, 1972). However, because there are only two satellites, neither of which can be observed in detail from the ground, it is in effect only J_2 that can be determined. Much greater detail of the field is derived from the very detailed observations of the motions of Mariner 9. Two approaches have been adopted. In one (Sjogren *et al.*, 1975) the accelerations of the space craft were determined over short segments of the orbit and were represented by the attraction of 92 point masses distributed over the planet, a method that has also been adopted to represent the gravity field of the Moon (Chapter 5). Then the spherical harmonic representation of the potential of the point masses was calculated. The coefficients of the spherical harmonics so determined are listed in Table 6.2. The Viking Landers have provided new data, which have been analysed by Gapcynski, Tolson and Michael (1977). Like Mariner 9, Vikings 1 and 2 were followed by observations of the Doppler shifts of radio signals from transmitters that they carried, so that it was possible to study three orbits. The parameters of the sections of the orbits that were analysed are as follows:

	Mariner 9	Viking 1	Viking 2
a (km)	1600–9000	1500–11 700	
e	0.6	0.75	0.75
i	64°	38°	75°
T (h)	12	22–27	

a is the semi-major axis, e the eccentricity, i the inclination and T the period.

It will be noted that in particular a good range of inclination is attained.

The accelerations were found over short sections of the orbits and the potential was represented by an expansion in spherical harmonics up to degree 6 and order 6. The coefficients are listed in Table 6.2 for

Table 6.2. *Gravity field of Mars*

		Normalized coefficients $\times 10^5$					
		C^a			S^a		
n	m	1	2	3	1	2	3
2	0	87.68	87.55	87.46	—	—	—
	2	−8.46	−8.27	−8.50	4.93	5.06	4.97
3	0	1.12	1.30	−1.26	—	—	—
	1	0.47	0.39	0.34	2.51	2.46	2.43
	2	−1.60	−1.68	−1.57	0.81	1.01	0.80
	3	3.53	3.60	3.52	2.52	2.69	2.50
4	0	−0.97	−0.42	0.61			
	1	0.50	0.48	0.44	0.33	0.32	0.33
	2	−0.20	−0.17	−0.09	−0.98	−0.91	−0.88
	3	0.69	0.66	0.68	0.00	0.12	−0.01
	4	0.00	−0.11	−0.02	1.30	−1.12	−1.21
5	0	—	—	−0.23	—	—	—
	1	—	—	−0.02	—	—	0.25
	2	—	—	−0.44	—	—	−0.07
	3	—	—	0.30	—	—	0.00
	4	—	—	−0.48	—	—	−0.36
	5	—	—	−0.50	—	—	0.31
6	0	—	—	0.20	—	—	—
	1	—	—	0.24	—	—	−0.01
	2	—	—	0.14	—	—	0.20
	3	—	—	0.09	—	—	−0.07
	4	—	—	0.23	—	—	0.31
	5	—	—	0.21	—	—	0.02
	6	—	—	0.28	—	—	0.02

[a] C denotes the coefficient of the term proportional to $\cos \phi$, S that of the term proportional to $\sin \phi$.

1 Jordan and Lorell (1975)
2 Sjogren, Lorell, Wong and Downs (1975)
3 Gapcynski, Tolson and Michael (1977).

The normalization is such that the integral of the square of a harmonic term on the unit sphere is 4π. Unnormalized coefficients were obtained by multiplying by

$$\left[\frac{(2n+1)(n-m)!(2-\delta_{nm})}{(n+m)!} \right]^{1/2},$$

this factor is $5^{1/2}$ for J_2.

comparison with those of Sjogren, Lorell, Wong and Downs (1975). However reliable or unreliable estimates of coefficients of individual harmonics may be, it is evident that Mars is far from being in hydrostatic equilibrium. The significance of the coefficients, and in particular of their dependence on degree, will be considered later in this chapter.

The gravity field of Mars thus now appears to be known in almost as much detail as that of the Moon; the fields of Venus and Mercury, however, remain almost completely unknown. The dynamical properties of Venus are listed in Table 6.3. Because Venus is covered by a thick cloudy atmosphere, optical observations of the solid surface cannot be made and, consequently, it has not been possible until recently to obtain reliable values of the rate of rotation and the surface radius. Radar observations of the surface have changed that situation. A reliable value for the (retrograde) spin rate was derived from radar reflexions from the surface by Shapiro (1967). The period is very close to the value of 243.16 d required if Venus is to present the same aspect to the Earth at inferior conjunction; and the implication is that Venus is constrained to move commensurably with the Earth by a couple generated through departures of Venus from perfect sphericity. It has been estimated that for the commensurable condition to be maintained

$$\frac{B-A}{C} \nless 10^{-4},$$

where A, B and C are the moments of inertia, with A the least and C the greatest.

Table 6.3. *Dynamical properties of Venus*

Spin period (d)[a]	243.09 ± 0.18
Radius (km)[b]	6052.1 ± 2
Mass (kg)[c]	$4.868\,96 \times 10^{24} \pm 3 \times 10^{18}$
J_2[d]	$< 2 \times 10^{-5}$

[a] Shapiro (1967).
[b] There are two values from radar observation: 6050.5 ± 5 km (Ash *et al.*, 1968), 6053.7 ± 2.2 km (Melbourne, Muhlmann and O'Handley, 1968) (see also Dollfus, 1972a).
[c] Howard *et al.* (1974a); the mass is given as the reciprocal of the solar mass, namely $M_\odot/408\,523.9 \pm 1.2$.
[d] Howard, *et al.* (1974a).

The radius of the solid surface of Venus is also obtained from radar observations. The visual observations made by Dollfus (1972*a*) refer to a layer of cloud at a radius of 6115 ± 13 km. Other estimates of the radius (6085 ± 10 km) were derived from space craft (Venera 4 and Mariner 5) approaching the planet, but it seems (Ash *et al.*, 1968) that the motions of the space craft have been misinterpreted, for the inferred radius is inconsistent with a radius obtained from radar observations extending over some years made from a number of observatories. The radar radii estimated by Ash *et al.* (1968) are effectively confirmed by observations made from the Goldstone observatory (Melbourne *et al.*, 1968). The radar values are those given in Table 6.3.

The mass of Venus was originally found from the perturbations it produced in the orbits of other planets, but the current values derive from perturbations of tracks of space craft near Venus. The value given in Table 6.3 (Howard *et al.*, 1974*a*) comes from the observation of radio transmissions from the Mariner 10 space craft. No departure of the gravity field from spherical symmetry was detectable, and the authors conclude that J_2 lies between 10^{-6} and 2×10^{-5}. They also estimate that the relative differences of any pairs of moments of inertia must be less than 1 in 10^4, which on the face of it appears to conflict with the argument based on the commensurable spin of Venus (see above), according to which one of those differences should exceed 1 in 10^4.

Mercury, being free of an atmosphere, is in some ways less difficult to study than Venus. The spin period is well established by three different methods: Goldstein (1971) used radar reflexions from the surface; Murray, Dollfus and Smith (1972) mapped the surface of Mercury by observing it through a telescope on the surface of the Earth; and finally Klaasen (1975) found the spin period from photographs of the surface taken from Mariner 10. The second and third results agree within their joint uncertainty.

The radius of Mercury was found from observations with Mariner 10 by Howard *et al.* (1974*b*). The value of the mass given in Table 6.4 was derived by the same authors, who also studied the departures of the gravity field from sphericity. As with Venus, it has not so far been possible to detect any departures, and the authors state that all harmonic coefficients of the gravity field must, in non-dimensional form, be less than the corresponding coefficients of the lunar field, that is to say, less than about 2×10^{-4}.

The foregoing dynamical data are all the mechanical information we have from which to infer internal structures for Mars, Venus and

Mercury. A seismometer was placed on Mars by the Viking Lander (Anderson *et al.*, 1977), but the results have so far been somewhat disappointing. During the Martian day, the strong winds that blow generate strong noise and it is only at night that true seismic events might be detected. At the time of the publication just referred to, only one event that might represent a Martian earthquake had been observed. Evidently Mars is seismically very quiet compared to the Earth. Anderson *et al* (1977) suggest that the characteristics of that one possible event may indicate that there is a crustal layer on Mars some 30 km thick.

6.4 Models of Mars

It is now time to attempt to assemble such data as we have about the internal density of Mars and to see what may be inferred about the structure of the planet. We begin by noting that the mean density of Mars, 3935 kg/m^3, lies between the densities of the lower and upper mantles of the Earth when reduced to zero pressure (3300 and 4000 kg/m^3 respectively) and is greater than the mean density of the Moon. Recalling, also, that the difference in density between the upper and lower mantles of the Earth is in part the consequence of polymorphic transitions in olivine and pyroxene, it is likely that Mars is composed in larger part of material similar to that of the mantle of the Earth, with polymorphic transitions somewhere within the planet. The argument for a structure of this type was first set out by Jeffreys (1937) and the considerations leading to it are now discussed in more detail.

The central pressure in a sphere of radius a and uniform density ρ is $\frac{2}{3}\pi G\rho^2 a^2$. Using the values of ρ and a for Mars in Table 6.5, the central

Table 6.4. *Dynamical properties of Mercury*

Spin period (d)	
Klaasen (1975)	58.661 ± 0.018
Goldstein (1971)	58.65 ± 0.25
Murray *et al.* (1972)	58.644 ± 0.009
Radius (km)	
Howard *et al.* (1974*b*)	2439.0
Mass (kg)	
Howard *et al.* (1974*b*)	$M = M_\odot/(6\,023\,600 \pm 600)$
	$= 3.302\,16 \times 10^{23} \pm 3 \times 10^{19}$
J_2	
Howard *et al.* (1974*b*)	Not so far detected ($<10^{-4}$)

pressure in Mars will be found to be of the order of 2.5×10^{10} Pa (compare Lyttleton, 1965).

The polymorphic and compositional changes which occur in the Earth's mantle lie in the zone which Bullen (in, for example, Bullen, 1975) calls Zone C, a zone which lies between the depths of 410 and 980 km; at greater depths the density and elastic moduli increase steadily with depth in the *lower mantle*. The pressure within the Earth at 410 km is 1.4×10^{10} Pa and that at 980 km is 3.8×10^{10} Pa. We would therefore expect that the initial polymorphic changes in the olivine and pyroxene series would occur in Mars (where the maximum pressure is 2.5×10^{10} Pa), but that those which take place at the higher pressures would not be found in Mars. Experimental evidence on the transitions of olivine and pyroxene is in general agreement (Chapter 4). The central pressure in Mars would allow the spinel transition and that to the β-phase to occur, but not disproportionation (if it does indeed take place) nor the final perovskite stage.

It has to be remembered that the pressures at which the phase transitions occur depend on temperature and composition. It is very likely that the temperature in Mars at a pressure of say 1.5×10^{10} Pa is not the same as that in the Earth at the same pressure, whilst the example of the Moon warns us that, although the composition of Mars is probably generally similar to that of the Earth, it may well not be identical. Accordingly, it would probably be incorrect to construct a model of Mars on the basis that the density was the same function of pressure as it is in the Earth.

Table 6.5. *Summary of dynamical properties of Mars, Venus and Mercury*

	Mars	Venus	Mercury
Spin period (d)	1.025 96	243.09	58.65
Equatorial radius (km)	3397.4	6052.1	2439.0
Polar radius (km)	3373.9		
Mass (kg)	$6.419\,09 \times 10^{23}$	4.869×10^{24}	3.3022×10^{23}
Mean density (kg/m³)	3935.2	5243.7	5433.5
$m\,(a^3\omega^2/GM)$	4.600×10^{-3}	6.107×10^{-8}	1.012×10^{-6}
J_2	1.9606×10^{-3}	$<2 \times 10^{-5}$	$<10^{-4}$
Polar flattening			
geometrical	6.92×10^{-3}	—	—
dynamical	5.24×10^{-3}	—	—
C/Ma^2	0.3752	—	—

The simplest model to take is one in which there is just a single polymorphic transition in Mars, a plausible model in view of the rather moderate central pressure. It will also be supposed to begin with that there is no heavy metallic core. With a mean density less than 4000 kg/m^3 it is not obvious that a heavy core is needed. It may be that it is needed to reproduce the value of C/Ma^2, and one aim of the study of models without a heavy core must be to see if they can yield the observed value of C/Ma^2 or whether a core must be added to do so. The first models to discuss, therefore, are those with two zones, the material in the inner zone being chemically the same as that in the outer zone but with a polymorphic transition.

The moment of inertia factor of Mars as calculated from the actual value of J_2 and the rotational angular acceleration is 0.3752, but it was shown in Chapter 3 that C/Ma^2 should be calculated from the hydrostatic part of J_2. Let us therefore now estimate the effect of non-hydrostatic distributions of density upon J_2. There is of course no direct determination of the non-hydrostatic part of J_2, and it must be inferred from the behaviour of other non-hydrostatic terms in the gravitational potential.

The mean square coefficients for degrees 3 to 6 are listed in Table 9.1 and plotted against degree in Figure 9.1. The harmonics of degree 3 and greater are produced by non-hydrostatic distributions of density and can be seen to vary in a regular way with degree. Accordingly we suppose that the non-hydrostatic terms of degree 2 follow the same empirical rule and, if so, we find that the expected mean square non-hydrostatic coefficient of degree 2 is 1.259×10^{-9} whilst the contribution of the observed harmonics of degree 2, other than J_2, is, from Table 6.2, 1.9×10^{-9}. This result suggests that J_2 contains no significant non-hydrostatic part and we therefore use the observed value of J_2 in order to calculate C/Ma^2, recognizing that that may be a somewhat uncertain thing to do (compare Cook, 1977). Different arguments led Binder and Davis (1973) and Reasenberg (1977) to estimate larger corrections to J_2 and C/Ma^2.

The value of γ, equal to $5C/2Ma^2$, corresponding to the value of 0.3752 for C/Ma^2, is 0.938. Since γ is not very far from 1, the effects of different types of departure from a constant density may be combined according to the prescription of Appendix 2, just as was done for the Moon. Three effects will now be considered: a polymorphic transition; the increase of density by self-compression; and a possible crust. So far as the crust is concerned, let us suppose that its effect is comparable with that of the lunar crust, entailing a value of γ of 0.993.

Consider next the effect of self-compression and use the formula that was derived for the Moon in Chapter 5, namely

$$\gamma = 1 - \tfrac{4}{35}\xi + \tfrac{8}{1575}(9 + 10b)\xi^2,$$

where ξ is the ratio of central pressure to K_0 and K_0 and b are the constants in the linear formula for the bulk modulus

$$K = K_0 + bp.$$

If K_0 is taken to be 1.5×10^{11} Pa, the same value as for the Moon, then ξ is 0.17.

The third term in the expression for γ is negligible, and the value of γ for self-compression turns out to be 0.98.

On the hypothesis of a planet with a polymorphic transition, the observed value, γ_{obs}, will be the product of three factors, that corresponding to the crust (γ_1 say), that corresponding to self-compression (γ_2 say), and that corresponding to the two zones into which the polymorphic transition divides the planet. Calling this γ_3, we should have

$$\gamma_3 = \gamma_{obs}/\gamma_1\gamma_2.$$

Now γ_{obs} is 0.938, γ_1 is 0.993 and γ_2 is 0.98. Thus γ_3 is 0.964, and the question to be addressed is whether any two-zone model with a specified difference of density can give such a value of γ_3.

Because the pressure at which a polymorphic transition takes place depends both on the temperature and the chemical composition (in olivine and pyroxene, the iron–magnesium ratio), we have to contemplate, as Lyttleton (1965) did, that the transition may occur anywhere within Mars. Of course, if the transition occurs near the surface or almost at the centre, the density will be uniform and the value of γ_3 will be 1. For transitions at intermediate radii, the value of γ_3 will be less, and we seek the minimum value to compare with our estimate of what γ_3 should be. If the least value for the two-zone model exceeds γ_3, we shall conclude that there is in addition a heavy core as we found to be required in the Moon.

The formal expressions for a two-zone model of constant density in each zone are given in Appendix 2. The densities are taken to be constant in each zone because the effect of self-compression has already been allowed in the factor γ_2.

If σ_1 is the relative density in the outer zone, σ_2 that in the inner zone and if α is the ratio of the radius of the inner zone to that of the planet, then

$$\alpha^3\sigma_2 + (1 - \alpha^3)\sigma_1 = 1,$$
$$\alpha^5\sigma_2 + (1 - \alpha^5)\sigma_1 = \gamma.$$

We suppose that σ_2 exceeds σ_1 by the factor $1+\varepsilon$, where ε is specified, and seek the least value of γ. (The greatest value of γ will be 1 when α is 0 or 1.) Now

$$\gamma = \frac{1+\alpha^5\varepsilon}{1+\alpha^3\varepsilon},$$

and the minimum value occurs when

$$\alpha = 0.703 + 0.033\varepsilon.$$

If we take ε to be 0.1, α is 0.706 and the least value of γ is 0.983.

This is a value so much greater than the required value of 0.964 that our preliminary conclusion is that a core is indeed required in Mars. The minimum size of a core may be estimated by using the same device of multiplying values of γ. Suppose γ_4 represents the effect of inserting a core into an otherwise uniform planet. We must have

$$\gamma_{obs} = \gamma_1\gamma_2\gamma_3\gamma_4.$$

We have just estimated the least value of γ_3 for a 10 per cent polymorphic change and, if that is inserted in the above expression, the value of γ_4 that results is the greatest value consistent with the assumptions, and so leads to the minimum size of core. With γ_3 as just estimated and the values of γ_1, γ_2 and γ_{obs} as before, the greatest value of γ_4 is found to be 0.980. As for the Moon, let the density of the supposed core be 8000 kg/m^3. Then, using the same formulae as for the lunar core, the minimum radius of a Martian core is found to be 925 km.

The methods so far followed are rather crude and, as for the Moon, no account has been taken of the effects of an inward increase of temperature. One refinement is to determine the self-compression exactly for a two-zone model by integrating the equations of hydrostatic equilibrium throughout the model instead of assuming, as was done above, that the effect in a two-zone model would be taken into account by using the analytical expression for a model with uniform composition.

Within each zone an equation of state must be adopted and as before it will be taken to be Bullen's relation

$$K = K_0 + bp,$$

which experimental evidence shows to be generally obeyed over moderate ranges of pressure.

Let the outer zone be labelled 1 and the inner zone 2, so that in either

$$K_i = K_{0i} + b_ip$$

where i stands for 1 or for 2. If the material in each zone is supposed to be chemically homogeneous, and if the effect of temperature may be ignored, density depends only on pressure within each zone and

$$\frac{\mathrm{d}\rho}{\mathrm{d}p} = \frac{\rho}{K}$$

or

$$\frac{\mathrm{d}\rho}{\rho} = \frac{\mathrm{d}p}{K_{0i} + b_i p}.$$

The differential equation for ρ integrates to give

$$\ln \frac{\rho}{\rho_0} = \frac{1}{b_i} \ln \left(\frac{K_{0i} + b_i p}{K_{0i}} \right),$$

where ρ_0 is the value of ρ at zero pressure. Thus

$$p = \frac{K_{0i}}{b_i} \left\{ \left(\frac{\rho}{\rho_0} \right)^{b_i} - 1 \right\}.$$

Now suppose that Mars is in hydrostatic equilibrium. The non-hydrostatic terms in the gravity field show that that is clearly not true, but it is likely (Chapter 9) that the density anomalies corresponding to the non-hydrostatic terms lie close to the surface. Thus, bearing in mind other simplifications that are being made, the assumption of hydrostatic equilibrium will be sufficient. The relation between pressure and radius is then

$$\frac{\mathrm{d}p}{\mathrm{d}r} = -g\rho = -\frac{GM_r \rho}{r^2},$$

where M_r is the mass within the radius r. In addition

$$\mathrm{d}M_r = 4\pi\rho r^2 \, \mathrm{d}r.$$

These two equations, together with the $p(\rho)$ relation and appropriate boundary conditions will determine the distribution of density with radius.

Because

$$\frac{\mathrm{d}\rho}{\mathrm{d}p} = \frac{\rho}{K}$$

and

$$\frac{\mathrm{d}K}{\mathrm{d}p} = b_i,$$

we get

$$K = C\rho^{b_i},$$

where

$$C = K_{0i}\rho_0^{-b_i}.$$

Further, the equation of hydrostatic equilibrium together with the definition of bulk modulus, K, give

$$\frac{d\rho}{dr} = \frac{d\rho}{dp}\frac{dp}{dr} = -\frac{\rho}{K}\frac{GM_r\rho}{r^2}$$

or

$$M_r = -\frac{Kr^2}{G\rho^2}\frac{d\rho}{dr}.$$

But

$$dM_r = 4\pi\rho r^2\, dr$$

and so

$$\frac{d}{dr}\left(\frac{Kr^2}{\rho^2}\frac{d\rho}{dr}\right) = -4\pi G\rho r^2.$$

On putting $K = C\rho^{b_i}$ it follows that

$$\frac{d}{dr}\left(r^2\rho^{b_i-2}\frac{d\rho}{dr}\right) = -A^2 r^2\rho,$$

where

$$A^2 = 4\pi G/C.$$

Now let

$$\rho^{b_i-2}\, d\rho = d\eta$$

or

$$\eta = \rho^{b_i-1}/(b_i - 1).$$

Then

$$\frac{d}{dr}\left(r^2\frac{d\eta}{dr}\right) = -A^2 r^2\{(b_i-1)\eta\}^{1/(b_i-1)},$$

or, if $\nu = 1/(b_i - 1)$,

$$\frac{d}{dr}\left(r^2\frac{d\eta}{dr}\right) = -A^2 r^2 (\eta/\nu)^\nu.$$

With the further substitution of

$$\xi = Ar\nu^{-\nu/2}$$

it will be found that

$$\frac{d}{d\xi}\left(\xi^2\frac{d\eta}{d\xi}\right) + \xi^2\eta^\nu = 0$$

or

$$\frac{d^2\eta}{d\xi^2}+\frac{2}{\xi}\frac{d\eta}{d\xi}+\eta^\nu=0.$$

This is Emden's equation (Lyttleton, 1965; Bullen, 1975) and admits of analytical solution only for $\nu = 1$ and 5. In the terrestrial planets b_i is about 3.2 so that ν is about 0.45 and Emden's equation must be solved numerically.

Because Emden's equation is of second order, two boundary conditions are required for each zone. If a planet were supposed to consist of a single zone, the boundary conditions could be adjusted to give the correst mass and C/Ma^2 for a specified radius. If a two-zone model is considered, pressure and gravity must be continuous at the boundary.

An extensive numerical study of the solutions of Emden's equation for a two-zone model of Mars was made by Lyttleton (1965). He took equations of state that were the same as those for the upper and lower mantles of the Earth, as known at that time, and constructed a sequence of models in which the polymorphic transition was allowed to occur at any radius from the centre to the surface of Mars. The sequence has a minimum value of C/Ma^2, as shown in Figure 6.4; the minimum occurs when the transition occurs at 0.84 times the surface radius and is 0.381. It

Figure 6.4. Values of C/Ma^2 and density for Lyttleton's (1965) models of Mars.

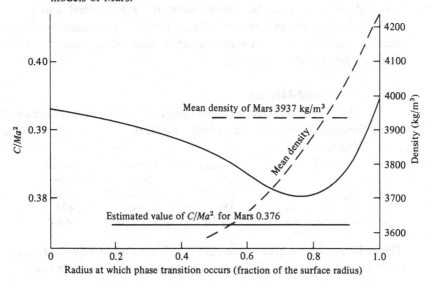

Mean density of Mars 3937 kg/m³

Mean density

Estimated value of C/Ma^2 for Mars 0.376

Radius at which phase transition occurs (fraction of the surface radius)

follows (Cook, 1972, 1977) that, because the minimum value of C/Ma^2 exceeds the observed value of 0.375, the presence of a heavy core must be postulated.

As for the Moon, sets of more detailed models have been constructed based on detailed models of the chemical composition and of the thermal state of the interior. One group of such models is based on Lewis's (1972) study of the fractionation of metals and silicates in the early history of the solar system. Figures 6.1 and 6.2 show the calculated variation of temperature in the early solar nebula, together with the temperatures at which various components of the planets would condense. Figure 6.3 shows the resulting mean densities of bodies formed in the positions of the terrestrial planets, and it will be seen that this model does indeed reproduce well the mean densities of the planets.

Johnston and Toksöz (1977) (see also earlier work in Johnston, McGetchin and Toksöz, 1974) have adopted Lewis's results and have followed Solomon and Toksöz (1973) in their procedures and assumptions in calculating the temperature in Mars (see further Chapter 9). They assume that Mars has a crust 50 km thick with a density of 3000 km (as in Chapter 5 and above) and that an olivine–spinel transition occurs within a mantle. The core is supposed to be a mixture of iron and iron sulphide at a temperature in excess of 2000 K so that it would be molten. Some examples of their models are shown in Figure 6.5. Johnston and Toksöz (1977) take a lower density for the core than that adopted above and in consequence obtain larger radii for the core.

A different approach has been followed by Ringwood and Clark (1971) who suppose that the metal abundances are the same in all the planets, but that their different densities may be accounted for by different degrees of oxidation.

6.5 Venus and Mercury

All we know about Venus and Mercury are the radii, masses and mean densities. They compare as follows with the corresponding values for the Earth:

	Earth	Venus	Mercury
Radius (km)	6378	6052	2439
Mass (kg)	6×10^{24}	4.87×10^{24}	3.30×10^{23}
Mean density (kg/m³)	5517	5244	5434
Central pressure (Pa)	3×10^{11}	2.5×10^{11}	3×10^{10}

The estimates of the central pressure which are also listed require some explanation. All exceed the values calculated from the formula $\frac{2}{3}\pi G\rho^2 a^2$

using the mean density. The value for the Earth is the actual pressure obtained by integrating throughout the Earth using the known values of g, a and ρ (Chapter 2). The values of Venus and Mercury have been increased *pro rata* to allow for the fact that the formula for a body of uniform density underestimates the pressure when the density increases

Figure 6.5. Examples of models of Mars with different thermal regimes: (*a*) conduction model; (*b*) convection model. The four different models correspond to four different models of the core. (From Johnston and Toksöz, 1977.)

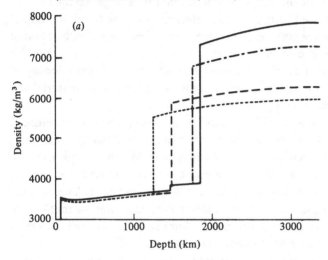

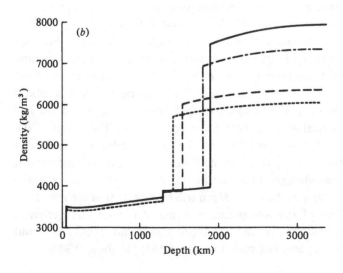

towards the centre. The actual values are not too significant: what matters is that the pressures at the centres of Venus and the Earth are much the same and that at the centre of Mercury it is an order of magnitude less.

If the bulk modulus throughout much of the planet is taken to be $(3 \times 10^{11} + 3.2p)$ Pa, as it is within the core and much of the mantle of the Earth, it can be seen that the increase of density due to self-compression will be of the order of 25 per cent or so throughout a substantial part of the Earth and Venus, but nowhere more than 10 per cent in Mercury. The conclusion to be drawn is that the mean density of Venus, like that of the Earth, is determined to a large extent by self-compression. In the Earth, it is largely determined by self-compression of the mantle since the volume of the core is only about one-eighth of that of the Earth as a whole so that the density of the core does not have a major effect on the mean density. We may see that the lower mean density of Venus is consistent with self-compression of similar mantle material at a slightly lower maximum pressure.

We know also that part of the increase of density towards the centre of the Earth is produced by the sequence of polymorphic changes of silicates (Chapter 4). The maximum pressure in Venus, like that in the Earth, is about ten times greater than the pressures at which those changes occur and so we infer that they do occur in Venus. So similar is Venus to the Earth in position, size and density, that it is natural to suppose that Venus, like the Earth, has a core composed predominantly of iron, with a density ranging from about 8000 to about 11 000 kg/m^3.

The high density of Mercury cannot be accounted for by self-compression of mantle material, for the greatest pressure is far too low to account for the difference between 5434 kg/m^3 and 3300 kg/m^3: the greatest density that mantle material could attain in Mercury, with allowance made for a polymorphic transition is about 4000 kg/m^3. Mercury must therefore have a substantial heavy core. If it be supposed to be of the same composition as the Earth's core, its mean density will be about 8300 kg/m^3. With some allowance for self-compression, the density of the mantle material may be taken to be 3500 kg/m^3. The radius of the core will then be found to be 1800 km. The core boundary thus may lie at a depth of 600 km; at that depth in the Earth the pressure is great enough for polymorphic changes to occur in olivine and pyroxene, but the pressure in Mercury at the same depth will be less, so that the mantle of Mercury will be of the low pressure forms of olivine and pyroxene throughout. Consequently, the assumption that the density is about 3500 kg/m^3 is justified. For such a model, C/Ma^2 is about 0.335.

6.6 Conclusion

The models of Mars, Venus and Mercury considered in this chapter have been, like those of the Moon discussed in Chapter 5, cold models, in that the effects of temperature on the density, and hence on the ratio C/Ma^2, have been neglected. The general justification for the neglect was given in Chapter 4; namely that the coefficient of expansion of any material decreases with pressure. Of course, the pressures are less in the Moon and Mars, so that the decrease would be expected to be less than in the Earth or Venus. On the whole, allowance for temperature would not be expected to modify the models adopted here in any serious way. If it is desired to obtain an indication of the effect of a specified distribution of temperature, T, it may be done by calculating the corresponding factor γ for a sphere of uniform composition. Let

$$\rho = \rho_0(1 + \alpha T),$$

where α and T are functions of r, known or postulated. Then the mass is

$$M = M_0 + 4\pi\rho_0 \int_0^a r^2 \alpha T \, dr,$$

and the moment of inertia

$$I = I_0 + \frac{8\pi}{3}\rho_0 \int_0^a r^4 \alpha T \, dr.$$

Here $M_0 = \frac{4}{3}\pi\rho_0 a^3$ and $I_0 = \frac{8}{15}\pi\rho_0 a^5$. Thus

$$\gamma = 1 + \frac{8\pi\rho_0}{3I_0} \int_0^a r^4 \alpha T \, dr - \frac{4\pi\rho_0}{M_0} \int_0^a r^2 \alpha T \, dr.$$

The maximum value of αT is likely to be less than 1 per cent and consequently γ is unlikely to be less than about 0.99.

All the models we have considered have heavy cores, supposed to consist of iron-like material. Table 6.6 summarizes the sizes and relative volumes of the cores.

Table 6.6. *The cores of the Moon and the terrestrial planets*

	Radius	Volume of core (10^{11}m^3)	Volume of core / Volume of planet
Moon	340	4.8	0.0075
Mercury	1800	135.7	0.40
Mars	925	35.8	0.02
Venus	—	—	(supposed same as the Earth)
Earth	3400	484.2	0.15

There is a wide range in both the absolute volumes of the cores (1 to 100) and in the relative volumes (1 to 53). The implications of these ranges will be taken up again in Chapter 10.

7

High pressure metals

7.1   Introduction

The possibility that, at the pressures encountered in the planets, materials ordinarily non-metallic at low pressures might transform into metals has been discussed for more than forty years. Two main ideas have been considered: one, that metal silicates, such as olivine, might become metallic at pressures developed in the core of the Earth, and the other, that hydrogen, helium and other light elements might transform to metals at pressures encountered in the major planets. Sufficient is now known about changes of density in metallic transformations under high pressure to be sure that the jump of density between the mantle and the core of the Earth is too great to be explained by such a transformation and in the preceding chapters on the terrestrial planets it has been assumed that the difference between the core and mantle of the Earth is one of composition (see also Anderson, 1977). It is otherwise with Jupiter and Saturn. The mean densities of those planets are too low for them to be composed of anything but hydrogen, helium and other materials of low atomic number, and the likelihood of a metallic transformation of hydrogen in particular is crucial to a discussion of their internal structures. One of the first studies of the metallic transformation in hydrogen (Kronig, de Boer and Korringa, 1946) was prompted by the idea of Kuhn and Rittman (1941) that the inability of the core of the Earth to support shear waves might be because it was of solar composition, that is, mainly of hydrogen, and by the subsequent suggestion of van der Waals that at core pressures the hydrogen might be metallic. It is now clear that the density of the core of the Earth is about ten times that of metallic hydrogen at the same pressure; on the other hand, it is almost certain that hydrogen in metallic form makes up large proportions of Jupiter and Saturn.

The pressure at which molecular hydrogen may transform to a metal is probably greater than 2×10^{11} Pa and is therefore beyond the reach of

199

present experimental methods using steady pressures. Until recently, the only experimental studies with which theoretical calculations of the equation of state of molecular hydrogen could be compared were those of Stewart (1956). Another set of data is now available (Anderson and Swenson, 1974) which is, in fact, in close agreement with that of Stewart and which extends to about the same pressure as Stewart's, namely 2.5×10^9 Pa. Further, because the compressibility of hydrogen is great, it is not possible to develop high pressures in it by shock waves. Thus, no one has yet certainly demonstrated the metallic transformation in hydrogen, although there have been claims that it has been observed (Grigorev *et al.*, 1972; Mlynek, 1974; Kamarad, 1975; see also Hawke, 1974). The properties of metallic hydrogen can therefore at present be found only by calculation, but, as will appear in this chapter, the structure of metallic hydrogen seems now to be well understood, and different methods of calculation appear to give concordant equations of state. It is in fact more difficult to calculate the properties of molecular hydrogen at high pressure and, in consequence, the pressure at which the transformation to the metallic form occurs cannot be estimated to better than an order of magnitude. In so far as this is because the equations of state of the molecular and metallic forms are thought to be very similar, it does not affect the study of the internal constitution of Jupiter and Saturn too seriously.

The primary aim of theoretical studies of metallic (and molecular) hydrogen is to calculate an equation of state, that is to say, to calculate the density as a function of pressure. It is also desirable to calculate transport properties, heat capacity, latent heat of transformation, the melting curve of metallic hydrogen and to predict whether metallic hydrogen may be superconducting or superfluid. The first step is to determine the density of the metallic form at zero temperature. The internal energy is calculated as a function of the separation of the protons in the metal, and the minimum is found as the separation is varied. That minimum is the internal energy, and the corresponding separation gives the atomic volume (or density).†
To find the relation between pressure and density, the internal energy, E, is calculated as a function of the proton separation, and hence of the specific volume, and the pressure is calculated as the differential

$$p = -\frac{\partial E}{\partial V}.\ddagger$$

† It is uncertain whether the metallic form of hydrogen is stable at zero pressure.

‡ At zero temperature, the free energy, F, is equal to the internal energy, E.

Most calculations are carried out for a lattice of fixed protons, that is, for a solid at zero temperature, but some work has been done on the solid at high temperature and on the liquid. The heart of the theory is, however, the calculation of the internal energy as a function of proton separation.

It is well known that many of the properties of simple metals like sodium are very close to those of a gas of free electrons, and accordingly the first step in the calculation of the properties of metallic hydrogen is to regard it as a Fermi gas with purely kinetic energy. Allowance has to be made for the correlation of the electrons or, put in another way, for the energy of collective plasma oscillations. The gas is not, however, free for it lies in the periodic potential of the lattice of protons, and so it is necessary to add to the kinetic energy of the free gas the potential energy of the electrons in the proton lattice and also that of the repulsion of the protons. The calculations of the various components of the energy will be discussed in the next sections of this chapter, and will be summarized in comparisons of the results of different methods of calculation. The means of estimating the properties of molecular hydrogen will also be described, and the transition pressure will be discussed.

The first calculations on metallic hydrogen were those of Wigner and Huntington (1935) who used the methods developed by Wigner and Seitz (1933, 1934) (see also Wigner, 1934) for the study of sodium. Further calculations were undertaken by Critchfield (1942) and, as already mentioned, by Kronig, de Boer and Korringa (1946), who incorporated work by Bardeen (1938). Subsequently, de Marcus (1958) re-examined the theory with particular reference to the constitution of Jupiter and Saturn, but following the same lines as Wigner and Huntington (1935) and Kronig, de Boer and Korringa (1946). In recent years, interest in metallic hydrogen has increased considerably (see Caron, 1975) and, in particular, new methods for the calculation of the electron energy have been employed that derive from developments in the theory of metals generally, while, in addition, the effects of different possible lattice structures for the protons have been investigated.

In theoretical calculations of equations of state it is common to use atomic units.

The unit of length is the Bohr radius, a_0, namely $\hbar^2/me^2$, where m is the mass and e the charge of the electron ($a_0 = 5.2918 \times 10^{-11}$ m).

The atomic unit of energy is the Rydberg, Ry, namely $me^4/4\pi\hbar^3 c$ (Ry $= 2.1799 \times 10^{-18}$ J $= 13.6$ eV). Sometimes the unit of energy is taken to be $2 \times$ Ry.

The atomic unit of pressure is $1\,\text{Ry}/a_0^3$ or $147.107 \times 10^{11}\,\text{Pa}$, i.e. $147.107\,\text{Mbar}$. (A value of $147.15\,\text{Mbar}$ will also be found in the literature.)

Another way of expressing the energy is by the equivalent temperature, that is, if the energy is kT per electron or atom, it is given as T K. (1 K is equivalent to $1.3807 \times 10^{-23}\,\text{J}$ or $0.8617 \times 10^{-4}\,\text{eV}$.)

It is convenient to have a value for the energy of a kilogramme mole of hydrogen which contains 6.022×10^{26} electrons. Thus 1 Ry per electron is equivalent to $1.3127 \times 10^9\,\text{J/kg mole}$, 1 eV per electron is equivalent to $9.648 \times 10^7\,\text{J/kg mole}$, and 1 K per electron is equivalent to $8.3146 \times 10^3\,\text{J/kg mole}$.

7.2 The free-electron gas (see Wilson, 1966)

Let us take as the simplest model of a metal N electrons in box of side L. The wave-functions of the electrons will be periodic with a spatial part, ψ, equal to

$$\exp\left[\frac{2\pi i}{L}(n_1 x, n_2 y, n_3 z)\right].$$

As the electrons are free, apart from being confined by the walls of the box, Schrödinger's equation reads

$$\frac{\hbar^2}{2m}\nabla^2 \psi + E_r \psi = 0,$$

so that E_r, the energy of the rth wave-function, is

$$\frac{h^2}{2mL^2}(n_1^2 + n_2^2 + n_3^2).$$

The total energy is obtained by summing E_r over the possible combinations of n_1, n_2, n_3. We assume that the total number of electrons is very large and so replace the sum by an integral. In fact the multiple integral can be reduced to a single integral

$$\sum (n_1^2 + n_2^2 + n_3^2) = \int_0^{v_0} v^{2/3}\,dv = \tfrac{3}{5}v_0^{5/3}.$$

where v_0 is related to N, the total number of electrons. In fact, N is equal to the number of combinations of (n_1, n_2, n_3) multiplied by 2 because two states of spin $\tfrac{1}{2}$ correspond to each ψ. Thus N is twice the volume of a sphere of radius $v^{1/3}$, or

$$N = \frac{8\pi}{3}v.$$

Consequently

$$E = \frac{3}{5} \frac{h^2}{2mL^2} \left(\frac{3}{8\pi} \right)^{2/3} N^{5/3}.$$

But the volume of the gas is V, and so, finally,

$$E = \frac{3}{5} \frac{h^2}{2m} N \left(\frac{3}{8\pi} \frac{N}{V} \right)^{2/3}.$$

The pressure, p, is $-\partial E/\partial V$, that is

$$p = \frac{1}{5} \frac{h^2}{m} \left(\frac{3}{8\pi} \right)^{2/3} \left(\frac{N}{V} \right)^{5/3}.$$

So far, the temperature of the gas has been ignored, as is to a large extent justified by the fact that the energy of 1 Ry per electron corresponds to a temperature of nearly 2×10^5 K. However, as the temperature rises, states higher than the lowest possible ones become occupied, as described by the Fermi function $f_0(E)$, equal to $\{1 + \exp{[(E - \mu)/kT]}\}^{-1}$, where μ is the chemical potential.

$f_0(E)$ is equal to the number of occupied states of energy E. If E is less than μ, $f_0(E)$ is almost unity, i.e. each state is occupied, but, as E increases above μ, $f_0(E)$ falls away much like a Maxwell–Boltzmann distribution (Figure 7.1). The fraction of electrons in states with $\mu > E$ is of the order of kT/μ, and it is the variation in their number with T that provides the electronic part of the specific heat of metals (only detectable at low temperatures when the lattice part, proportional to T^3, is negligible).

Let a degeneracy temperature, T_0, be defined as

$$T_0 = \frac{h^2}{2mk} \left(\frac{3}{8\pi} \frac{N}{V} \right)^{2/3}.$$

Figure 7.1. The Fermi–Dirac distribution.

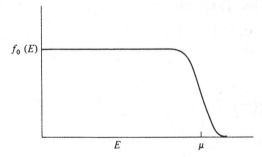

At temperature less than T_0 the gas of electrons will be highly degenerate and to a first approximation we may ignore the temperature in calculating the equation of state. Taking the density of metallic hydrogen to be $1000 \, \text{kg/m}^3$, and allowing one electron per atomic volume, it will be found that

$$N/V = 6 \times 10^{29}/\text{m}^3.$$

Now

$$h^2/2mk = 1.7 \times 10^{-15} \, \text{m}^2 \, \text{deg},$$

and so the degeneracy temperature for metallic hydrogen is about $2 \times 10^4 \, \text{K}$. The temperature within Jupiter is estimated to be about $10^3 \, \text{K}$ so that to a first approximation we may ignore it in calculating the equation of state of metallic hydrogen in Jupiter (and Saturn).

The internal energy at low temperature T is then

$$\tfrac{3}{5}NkT_0\left[1 + O\!\left(\frac{T^2}{T_0^2}\right)\right]$$

and the free energy F is

$$\tfrac{3}{5}NkT_0\left[1 - O\!\left(\frac{T^2}{T_0^2}\right)\right],$$

where the remainder $O(T^2/T_0^2)$ is the same for E and F to $O(T^2)$. Then

$$p = -\frac{\partial F}{\partial V} = \frac{2}{5}\frac{NkT_0}{V} + \cdots$$

$$= \left(\frac{3}{8\pi}\right)^{2/3}\frac{h^2}{5m}\left(\frac{N}{V}\right)^{5/3}\left[1 + O\!\left(\frac{T^2}{T_0^2}\right)\right]$$

and the entropy

$$S = -\frac{\partial F}{\partial T} = \frac{\pi^2 mk}{h^2}\left(\frac{8\pi}{3}\frac{V}{N}\right)^{2/3}NkT + \cdots$$

Our final results for the equation of state of the free-electron gas at a temperature much less then T_0 are that T is then

$$E = \tfrac{3}{5}N\frac{h^2}{2m}\left(\frac{3}{8\pi}\frac{N}{V}\right)^{2/3}$$

and

$$p = \left(\frac{3}{8\pi}\right)^{2/3}\frac{h^2}{5m}\left(\frac{N}{V}\right)^{5/3}$$

while S is negligible.

It is convenient to express these results in terms of the separation of the protons. We introduce the idea of the Wigner–Seitz sphere: let a sphere of radius r_s be drawn around a proton, such that it encloses a volume V/N. Then

$$\frac{4\pi}{3} r_s^3 = \frac{V}{N}$$

or

$$r_s = \left(\frac{3}{4\pi}\frac{V}{N}\right)^{1/3}$$

while

$$\frac{N}{V} = \left(\frac{3}{4\pi}\right)\frac{1}{r_s^3}.$$

Thus

$$E = \tfrac{2}{5}N\frac{h^2}{2m}\left(\frac{9}{32\pi^2}\right)^{3/2}\frac{1}{r_s^2} = \tfrac{3}{10}N\frac{\hbar^2}{mr_s^2}\left(\frac{9\pi}{4}\right)^{2/3},$$

while

$$p = \left(\frac{3}{8\pi}\right)^{2/3}\frac{h^2}{5m}\left(\frac{3}{4\pi}\right)^{5/3}\frac{1}{r_s^5}.$$

In the literature it is common to express these results in terms of atomic units, by writing r_s as a multiple of the Bohr radius. Similarly the energy per electron is commonly expressed as a multiple of the Rydberg.

In terms of the Rydberg and the Bohr radius

$$E = \frac{2.209}{r_s^2}\text{ Ry per electron.}$$

At this point three things must be emphasized. The first is that we have supposed that the gas of fermions is enclosed by a rigid box, like the molecules in a classical gas. The second follows from that: the energy of the electrons just calculated is entirely kinetic energy. Thirdly, the interactions between the electrons have been neglected. The model so limited is clearly an unrealistic model of a metal. It ignores the positive charges of the ions which actually hold the electrons together and thus although the free-electron model accounts well for the electrical and thermal properties of a metal it excludes *ab initio* the means by which the metal is held together so that it cannot account for the fact that the metal coheres at zero pressure nor can it predict the variation of density with pressure.

The effect of the lattice of protons may be calculated to first order on a classical basis by considering the protons to be embedded in a uniform background of negative charge of density ρ equal to $e(\frac{4}{3}\pi r_s^3)^{-1}$ (Figure 7.2). The justification for this approximation is akin to the justification of the Born–Oppenheimer approximation in molecules, namely that the motion of the protons may be ignored in comparison with that of the electrons because of their much greater mass. Accordingly, the electrons can be considered as a smeared out distribution of charge, whereas the protons must be treated as fixed charge points. The charge inside a sphere of radius r (less than r_s) about a proton is then equal to $+e$ on the proton together with the negative charge inside r, that is

$$q(r) = e - e(r/r_s)^3$$

The potential at r, $\phi(r)$, due to the charge within r, is $q(r)/r$, and the potential energy of the negative charge of density ρ lying between r and $r+\mathrm{d}r$ is

$$4\pi r^2 \rho \, \mathrm{d}r \phi(r)$$

and thus the potential energy of the charge inside the radius r_s is

$$-\int_0^{r_s} \frac{3e^2}{r_s^3}\left(r - \frac{r^4}{r_s^3}\right) \mathrm{d}r = -\frac{9e^2}{10 r_s}.$$

The total energy of the Fermi gas in a lattice of protons is accordingly

$$\frac{3}{10}\left(\frac{9\pi}{4}\right)^{2/3}\frac{\hbar^2}{mr_s^2} - \frac{9e^2}{10 r_s} = \frac{\hbar^2}{ma_0^2}\left[\frac{3}{10}\left(\frac{9\pi}{4}\right)^{2/3}\left(\frac{a_0}{r_s}\right)^2 - \frac{9}{10}\left(\frac{a_0}{r_s}\right)\right];$$

with r_s measured in units of the Bohr radius, a_0, the energy per electron is

$$\frac{2.21}{r_s^2} - \frac{1.8}{r_s} \quad \text{Rydberg.}$$

Figure 7.2. The distribution of charge in an elementary classical model of a metal.

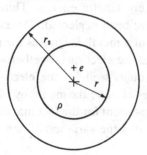

The energy has a minimum at $r_s = 2.45$; that is to say at a volume of $5.488 \text{ m}^3/\text{kg}$ mole, or a density of 182 kg/m^3. The pressure at a given volume is

$$-\frac{\partial E}{\partial V} = -\frac{1}{4\pi r_s^2}\frac{\partial E}{\partial r_s},$$

that is

$$\frac{1}{4\pi r_s^5}[4.42 - 1.8r_s],$$

in atomic units of pressure equal to 147.11×10^{11} Pa. The bulk modulus is an important property of planetary materials; it is given by

$$K = -V\frac{\partial p}{\partial V}$$

$$= -V\frac{dr_s}{dV}\frac{\partial p}{\partial r_s}$$

$$= \frac{1}{12\pi r_s^5}[22.1 - 7.2r_s]$$

again in atomic units of pressure.

Finally, values of the Gibbs free energy, G, are required for a discussion of the transition between metallic and non-metallic forms.

At zero temperature,

$$G = E + pV.$$

The behaviours of E, G and p as functions of r_s for this, the so-called 'Jellium' model of a metal are shown in Figure 7.3(a).

While the foregoing model shows how a stable metal may form, it is an inadequate description of a real metal. In the next section the quantum mechanical problem of electrons in a lattice of protons is discussed.

7.3   **Electrons in a lattice of protons** (see Pines and Nozieres, 1966)
        We consider a set of protons and an equal number of electrons and wish to calculate from quantum mechanical principles the state of lowest energy and the value of that energy for a given external pressure and given temperature. We recognize that this is a many-body problem which cannot be solved exactly and that some step-by-step approximate procedure will have to be adopted. Let us, however, write down formally a Hamiltonian for the system. It will be

$$H_e + H_p + H_i,$$

where H_e is the Hamiltonian for the electrons by themselves, H_p is that for the protons by themselves and H_i represents the interaction of protons and electrons.

Already there is an approximation implicit in this way of splitting up the Hamiltonian, namely that the velocities of the electrons are so much greater than those of the protons that we may suppose the motion of the lattice is not affected by the former. As was remarked above this is equivalent to the Born–Oppenheimer approximation in molecular physics; it is reasonably accurately satisfied there and, no doubt, also in metals.

Figure 7.3. (*a*) Comparison of internal energy, *E*, Gibbs free energy, *G*, and pressure, *p*, for the elementary classical model of metallic hydrogen without exchange forces ('Jellium') and for the Wigner–Seitz model. (*b*) Dependence of bulk modulus, *K*, upon pressure for the Wigner–Seitz model of metallic hydrogen.

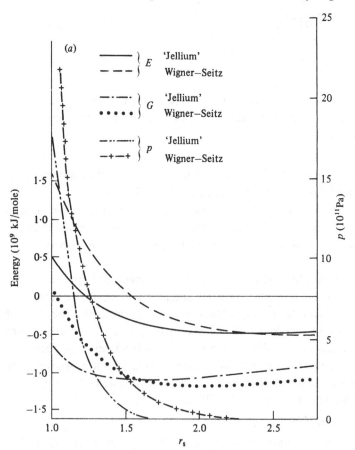

Accordingly H_e will be the Hamiltonian for a set of electrons in the field of the positive charges of a set of protons that we may assume to be fixed in their mean positions. In fact the protons are in motion whether they form a solid lattice or whether the metal is, as it may be, liquid, but, in the spirit of the Born–Oppenheimer approximation, we take the protons to be fixed in forming H_e. We must also arrange that the wave-function for all the electrons is anti-symmetrized and must explicitly recognize that electrons have spin $\frac{1}{2}$. The leading term of H_e will of course be that corresponding to the Fermi distribution, namely

$$\sum p^2/2m,$$

where p, equal to $\hbar k$, is the momentum of a simple harmonic wave-function of the form given in the previous section and having a wave-number k.

Next consider H_p; it contains two parts, the first the kinetic energy of the protons, which may be in a solid lattice or may move as in a liquid, and

Figure 7.3 *cont.*

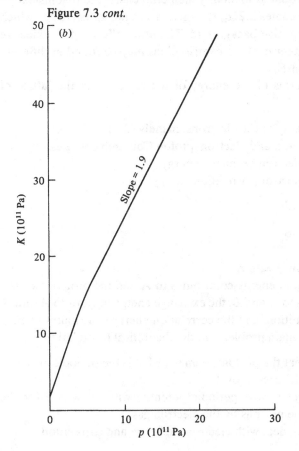

the potential energy of the Coulomb interaction of the protons. H_p will of course be different for a liquid or a solid and will be different for different types of lattice. Lastly, there is the term H_i which describes the interaction between electrons and protons. Again, in evaluating this term, we suppose that the protons are fixed in their mean positions. The obvious contribution to H_i is the Coulomb energy of the electrons in the field of the protons, but in evaluating it we have to take account of the fact that each electron is present in the field of all other electrons. In particular, the electrons move in a coherent way as a plasma so that it is not possible to build up H_i as the sum of terms of individual electrons each in the field of its own proton. It is to some extent a matter of choice which terms appear in H_e and which in H_i, for it is possible to consider the effect of the protons as to a first approximation to be that of a uniform background of positive charge which one takes as given in evaluating terms of H_e. One effect of the positive background is to modify the Fermi energy, so that instead of being $2.21/r_s^2$ it becomes $2.21\alpha/r_s^2$, where α is a numerical factor which accounts for the positive background. The other effect is the collective plasma energy. Either of these contributions may be found in different theories in H_e or in H_i.

The principal terms in the energy of the electrons in the lattice of protons are thus:

(a) the kinetic energy of the electrons, namely $(2.21/r_s^2)$,
(b) the proton–proton and electron–proton Coulomb energies,
(c) the electron–electron exchange energy,
(d) the electron–electron correlation energy.

The general expression is

$$\frac{2.21}{r_s^2} - \frac{A}{r_s} + B,$$

where B varies slowly with r_s.

The proton–proton energy contributes to A and the proton–electron energy contributes to A and B, the exchange energy to A, or to A and B, depending on definition, and the correlation energy contributes to B.

There are three main problems in the theoretical formulation:

(a) How do we treat the protons: namely, what is the periodic potential in which the electrons move?
(b) How do we set up the periodic potential with allowance for the screening of the protons by the electron gas?
(c) How should we deal with electron exchange and correlation?

Methods which have been developed to deal with the calculations will now be explained. It must be emphasized that any calculation of the properties of metallic hydrogen is approximate, since it is not possible to solve the many-body problem exactly. Two points may be made at the outset. Hydrogen is, we believe, the simplest metal in the sense that we know that the potential of the proton is exactly the Coulomb potential and thus it is not necessary to deal with a pseudopotential as with other ions (Chapter 4), but, on the other hand, we do not know from observation what lattice structure the solid metal may assume. The second point is that the results of calculations according to different procedures are in rather good agreement, especially at high pressures; in consequence, uncertainties in the behaviour of hydrogen at pressures between say 10^{10} and 10^{11} Pa or so arise much more from doubts about the properties of the molecular form than the metal.

In the earliest work on metallic hydrogen the interaction energy was calculated according to the method of Wigner and Seitz (1933, 1934). The space within the metal is divided into similar polyhedra conformably with the lattice structure such that there is one proton at the centre of each polygon. It is supposed that there is also just one electron in each polygon and the Hamiltonian is then

$$\left(-\frac{p^2}{2m} + V\right)\psi,$$

where p is the momentum of the electron and $V = -e^2/r$ is the potential of the electron in the field of the proton.

We require to find the lowest eigenvalue E corresponding to this Hamiltonian. It would be difficult to solve Schrödinger's equation for a polyhedron, so it is assumed that the polyhedron may be replaced by the sphere of equal volume, the Wigner–Seitz sphere of radius r_s.

The assumption made here is akin to St Venant's principle in elasticity, namely that the fields in the space between the polyhedron and the sphere have a minor effect on the eigenvalues. The boundary conditions for the spherical problem are taken to be the same as for the polyhedral problem. First ψ must be finite at the centre. Secondly the solution in one polyhedron must match that in the adjacent one. Now, by dealing with just one sphere, we of course ensure that the values of ψ are the same for all spheres, but we do not ensure that grad ψ is continuous. We therefore impose the condition that

$$\frac{\partial \psi}{\partial r} = 0$$

on the surface of the Wigner–Seitz sphere, corresponding to the physical condition that

$$\frac{\partial \psi}{\partial n} = 0$$

on the surface of any polyhedron.

The problem so posed has been solved in two ways, either analytically, explicit series developments of E_0 and ψ being obtained, or numerically.

Wigner and Huntington (1935) found for zero momentum of the electron

$$E_0 = -\frac{3}{r_s} - \frac{(1.36)^2}{20.02 - 2.65 r_s} - \cdots$$

by an analytical treatment, while Kronig, de Boer and Korringa (1946) used two analytical approaches which give effectively identical results, namely

$$E_0 = \frac{-3}{r_s - \frac{1}{30} r_s^2 + \cdots}.$$

Here, as before, E_0 is in Rydbergs and r_s in Bohr radii.

de Marcus (1958), in his work, made a numerical solution for E_0, and other authors (for example, Neece, Rogers and Hoover, 1971) have from time to time revised the Wigner–Seitz calculations; summaries of the results are given later in this chapter.

It would seem evident that a major defect of the Wigner–Seitz approach is that it ignores the presence of other electrons and protons. Thus, it is spoken of as the independent electron approach. At one time the problem of the effect of other electrons appeared to give considerable difficulty because, when the long range Coulomb interactions with all other electrons were allowed for, the calculated behaviour of simple metals appeared to agree less well with observations than if the electrons were treated as independent. The issue was resolved through studies of the shielding of the effects of distant electrons, from which it appears that the effective range of the electrostatic force between proton and electron is of the order of the distance between protons and thus it is quite a good approximation to ignore other electrons in working out the ground state potential energy of the electron in the field of the proton.

The effect of the screening of other electrons may be described by a dielectric function introduced by Lindhard (1954). If η is the ratio $k/2k_F$, where k_F is the Fermi wave-number, i.e. the value of k corresponding to the Fermi energy, then the dielectric function $\varepsilon(\eta)$ is

$$\varepsilon(\eta) = 1 + \frac{1}{2\pi} \left(\frac{4}{9\pi}\right)^{1/3} r_s g(\eta),$$

with

$$g(\eta) = \frac{2}{\eta^2}\left(\frac{1-\eta^2}{4\eta} \ln\left|\frac{\eta+1}{\eta-1}\right| + \frac{1}{2}\right).$$

Hammerberg and Ashcroft (1974) employ Lindhard's result in a somewhat different decomposition of the total Hamiltonian from that used in the Wigner–Seitz procedure. They add the part of the electron-electron interaction for which the wave-number is zero (stationary electrons) to the proton–proton interaction, so giving the Madelung energy of the lattice. That is of the form $-A_m/r_s$, where A_m has the values:

    simple cubic lattice: 1.760 122 Ry per electron
    face-centred cubic: 1.791 749 Ry per electron
    body-centred cubic: 1.791 861 Ry per electron.

The remaining part of the electron–electron interaction is then obtained with the Lindhard dielectric function, using the result

$$E = -\frac{1}{6\pi^2} \sum \eta^{-2} \frac{g(\eta)}{\varepsilon(\eta)}$$

where $\eta = 0$ (i.e. $k = 0$) is excluded from the summation since the corresponding contribution already enters the Madelung energy.

A third method of calculating the effect of the proton potential is by means of the combination of wave-functions corresponding to the solution of Schrödinger's equation about a single proton; the method is called the *linear combination of atomic orbitals* (LCAO) (Harris and Monkhorst, 1969). Suppose $\psi(r_i)$ to be the wave-function for an electron in the neighbourhood of an isolated proton with the position vector r_i; it is supposed that the wave-functions for electrons in the lattice are obtained by suitably adding similar wave-functions for electrons in the fields of all the protons. In order that the wave-function should have the periodicity of the lattice of protons it must be in the Bloch form. Ross and McMahon (1976) write the wave-function for the electron with momentum k as

$$\psi_k(r) = e^{k \cdot r} \sum_n \phi(r - R_n),$$

where ϕ is a wave-function of atomic form for the electron at r in the field of the proton at R_n; ϕ is taken to be a Slater function, $e^{-\alpha x}$, and α is varied to obtain the minimum energy.

The Thomas–Fermi–Dirac statistical method (see Chapter 4) is the fourth method to have been applied to the calculation of energies of electrons in the proton lattice (Salpeter and Zapolsky, 1967). On the face of it the method is inappropriate to hydrogen since it supposes that there are many electrons around each ion. Nonetheless, the electron density is found from Fermi–Dirac statistics, using an appropriate chemical potential.

Salpeter and Zapolsky (1967) include an additional energy corresponding to correlation between electrons, using the form found by Gell-Mann and Brueckner (1957) which, however, is appropriate to densities very much greater than those likely to be encountered in the planets. There is no inconsistency in Salpeter and Zapolsky's use of that form, for their aim was to obtain an asymptotic form of the equation of state at very high pressures; the Gell-Mann and Brueckner form is not, however, the correct one to use at planetary pressures. Finally, there is a fifth method of calculating the energy of the electron gas in a lattice, which is based on plane wave-functions and is known as the method of *augmented plane waves*.

All the foregoing methods have been used to calculate the equation of state of metallic hydrogen at planetary pressures and the results, as will be seen in section 7.5, are on the whole in close agreement with the approximation which includes just the kinetic energy and the exchange energy (the terms $2.21/r_s^2 - 0.916/r_s$). To this the correlation energy (next section) has to be added and, as will be seen, there is still some uncertainty in how to handle it. Further, the lattice structure also has a small effect upon the equation of state and the free energy. The small differences between the Madelung energies for simple cubic, face-centred cubic and body-centred cubic lattices in Hammerberg and Ashcroft's (1974) work have already been seen and Wigner and Huntington (1935) pointed out that a layered lattice would have the lowest energy and be the preferred form at low pressures.

The most extensive calculations on the properties of different lattices are those of Brovman and his colleagues (Brovman, Khagan and Kholas 1972*a*, *b*; Brovman, Khagan, Kholas and Pushkarev, 1973) whose results are set out later. Some of the structures will be unstable at very low pressures, but not at pressures of 10^{11} Pa or more (see also, Beck and Strauss, 1975).

Because the energy of electrons in a lattice cannot be calculated exactly, a perturbation method has always to be adopted, leading in effect to an expansion in powers of r_s. The leading terms correspond to the kinetic energy of a free–electron gas $(2.21/r_s^2)$ and to the potential energy of the exchange interaction of electrons with parallel spin $(-0.916/r_s)$. Everything else is grouped together as the correlation energy..

The correlation energy has been thoroughly discussed by Nozieres and Pines (1958) (see also Pines and Nozieres, 1966) who point out that metals, in particular metallic hydrogen, lie in an intermediate range of density. At very low densities $(r_s \gg 10)$ Wigner showed that the correlation energy could be put in the form

$$\frac{U}{r_s} + \frac{V}{r_s^{3/2}} + \frac{W}{r_s^2},$$

where U, V and W are constants, while, for very high densities $(r_s \ll 1)$, Gell-Mann and Brueckner (1957) obtained the expression

$$0.0622 \ln r_s - 0.096.$$

In metals r_s is of the order of 1, and Nozieres and Pines discuss the form of approximation that gives an appropriate interpolation between the high density and low density forms. They find that the correlation energy should be represented by

$$-0.115 + 0.031 \ln r_s.$$

More recently a very similar expression has been derived by Vashista and Singwi (1972).

7.4   Equation of state of molecular hydrogen and helium

It is much more difficult to make reliable estimates of the equation of state of molecular hydrogen than of that of the metallic form. As already mentioned (section 7.1) experimental results are restricted to pressures less than 2.3×10^9 Pa (Stewart, 1956; Anderson and Swenson, 1974). So far as theoretical calculations go, the interactions between H_2 molecules in a gas or liquid are much more difficult to describe than those between protons and electrons in the metal. At the same time, while molecular hydrogen is the stable form at low pressure, experimental studies of the equation of state have not so far been carried to sufficiently high pressures for the results to be extrapolated confidently to the pressures at which the transition to the metallic form may occur.

In early work it was supposed that the potential of the H_2 molecule was spherically symmetrical and a Lennard-Jones potential of the form

$$\left(-\frac{a}{r^6} + \frac{b}{r^{12}}\right)$$

was adopted. The constants a and b were found by fitting calculated isotherms to measured properties at low pressure and then used to extrapolate to high pressure. When that was done it was found that use of the Lennard-Jones potential enabled some properties to be reproduced, but gave rise to large discrepancies with others (see Etters, Danilowicz and England, 1975); such failures of theoretical calculations led de Marcus (1958) to extrapolate the experimental results of Stewart (1956), which extended to 2×10^9 Pa, by empirical methods to 3×10^{11} Pa.

If one considers two hydrogen molecules arbitrarily oriented (Figure 7.4) then it can readily be seen that the interaction between them can be expressed as a function of the four distances r_{13}, r_{14}, r_{23}, r_{24} between each pair of atoms and, in consequence, the effective potential will in general be anisotropic.

The forces between the molecules arise in two ways. First, there are the repulsive forces between pairs of atoms. Secondly, there are attractive forces which originate in the dipoles and quadrupoles induced in one molecule by the other. (The $1/r^6$ term in the Lennard-Jones potential accounts for induced dipole interactions.)

Trubitsyn (1966) took the repulsive potential between hydrogen atoms to have the form $a_1 e^{-br}$, where $a_1 = 2.17$ in atomic units of energy and $b = 1.81$ in atomic units of distance. The atomic units employed by Trubitsyn are

    distance: 5.29×10^{-11} m
    energy: 27.2 eV
    pressure: 294×10^{11} Pa.

On integrating over all orientations of the atoms in two molecules, the repulsive potential between molecules is found to be of the form

$$\Phi_1(R) = 4a_1\left(1.66 - \frac{1.30}{R}\right)e^{-bR}$$

Figure 7.4. Distances between atoms of two hydrogen molecules.

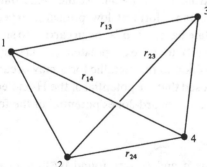

where $a_1 = 3.04$ and R is the distance between the centres of the molecules. Trubitsyn considers that, over the significant range of R, it is possible to write

$$\Phi_1(R) = a\,e^{-bR},$$

where $a = 5.6$ and $b = 1.81$.

The potential of the attractive van der Waals forces is taken to be

$$\Phi_2(R) = -cR^{-6}(1 + dR^{-2}),$$

with $c = 10.9$ and $d = 10.6$.

Trubitsyn adds the zero point energy of the lattice vibrations at zero temperature to the sum of the attractive and repulsive potentials and finds for the pressure

$$p = \tfrac{1}{3}ABV^{-2/3}\exp\left(-BV^{1/3}\right) - 2CV^{-3}(1 + 1.3DV^{-2/3}$$
$$+\,8.7 \times 10^{-2}V^{-4}),$$

where $A = 3a$, $B = 2^{1/2}b$, $C = 0.452c$, $D = 0.44d$, and

$$V = (R/2^{1/2})^3.$$

Trubitsyn finds that the results of Stewart (1956), already referred to, fit an expression of somewhat similar form, namely

$$p = \tfrac{1}{2}\alpha\beta V^{-2/3}\exp\left(-\beta V^{1/3}\right) - 2\delta V^{-3}$$

and shows that there is quite a close fit between his calculations and Stewart's results.

Trubitsyn gives the following expression for the bulk modulus, K, which is equal to $-V\,dp/dV$:

$$K = \tfrac{1}{9}AB\,V^{-2/3}(2 + BV^{1/3})\exp(-BV^{1/3})$$
$$-\,6CV^{-3}(1 + 1.6DV^{-2/3}).$$

He also considers the effect of higher temperatures. If molecular hydrogen behaves as a Debye solid, then the free energy, F, is equal to

$$E_0 + kT\{3\ln[1 - \exp\left(-\Theta/T\right)] - D(\Theta/T)\},$$

where Θ is the Debye temperature of the lattice and D is the Debye function equal to

$$\frac{3}{x^3}\int_0^x \frac{du}{e^u - 1}.$$

Thus the pressure at volume V and temperature T, which is $-(\partial F/\partial V)_T$, is equal to

$$p_0 + \frac{3kT}{V} \gamma D\left(\frac{\Theta}{T}\right),$$

where p_0 is the pressure at $T = 0$ and γ is $d \ln \Theta/d \ln T$. γ is Grüneisen's ratio.

Thus, with an estimate of γ, Trubitsyn writes

$$p(\rho, T) = 5.01 \times 10^{11} \rho^{2/3} \exp\left(-0.491\rho^{1/3}\right)$$
$$-2.34 \times 10^8 \rho^3 + 5.4 \times 10^6 TD(\Theta/T),$$

where now p is in Pa and ρ in kg/m^3.

The equation of state of unionized helium is somewhat easier to calculate than that of molecular hydrogen because the atoms are spherically symmetrical and there is no question of calculating averages over all orientations as with molecular hydrogen. The interatomic potential may be calculated from the wave-function of the ground state of helium, and Trubitsyn (1967) takes it to be

$$\Phi(R) = a \, e^{-bR} - (cR^{-6} + dR^{-8}) \exp\left(-fR^{-6}\right).$$

Here a cut-off factor, $\exp\left(-fR^{-6}\right)$ is applied to the normal van der Waals potential. The constants that Trubitsyn adopts are, in atomic units, $a = 14.5$, $b = 2.35$, $c = 1.47$, $d = 113$, $f = 160$.

The energy per atom at zero temperature is then found to be

$$A \exp\left(-BV^{1/3}\right) - CV^{-2}\left(1 + DV^{-2/3}\right) \exp\left(-FV^{-2}\right),$$

where $A = 6a$, $B = 2^{1/2}b$, $C = 3.61c$, $D = 6.703 \, d/c$ and $F = \frac{1}{2}f$.

The pressure and bulk modulus follow by differentiating the energy.

A somewhat similar calculation to that of Trubitsyn (1966) was performed for hydrogen by Neece, Rogers and Hoover (1971) who represented the repulsive forces by the exponential form

$$\varepsilon \exp\left(-r_i/r_0\right),$$

where $\varepsilon = 3.2e^2/a_0$ and $r_0 = 0.03$ nm and took the average of this potential over the four distances, r_i, between atoms of two molecules, as in Figure 7.4. They represented the attractions between induced dipoles and quadrupoles by the potential

$$\left[-10.9\left(\frac{a_0}{R}\right)^6 - 111\left(\frac{a_0}{R}\right)^8\right]\frac{e^2}{a_0},$$

where R is now the centre-to-centre separation of two molecules.

They found that their numerical results were close to

$$E = 5.645 \times 10^{10} \exp(5.05\rho^{-1/3}) - 9.425\rho^2,$$

where E is in J/kg and ρ is in kg/m^3.

Neece, Rogers and Hoover originally expressed their results in different units, namely

$$E = 1\,129\,000 \exp(-4.01\,V^{1/3}) - 1885\,V^{-2},$$

where E is in kbar cm^3/g mole H$_2$ and V is in cm^3/g mole H$_2$.

The pressure is then given by

$$p = -9.502 \times 10^{10}\rho^{2/3} \exp(-5.05\rho^{-1/3}) + 18.89\rho^3 \text{ Pa},$$

while G at zero temperature, namely $E + pV$, is

$$(5.465 - 9.502\rho^{-1/3}) \times 10^{10} \exp(5.05\rho^{-1/3}) + 9.455\rho^2 \text{ J/kg}.$$

Etters, Danilowicz and England (1975) carried out a systematic investigation of the effects of orientation of H$_2$ molecules one to another; they calculated the interactions between molecules by quantum mechanics from first principles and then took a spherical average over all orientations. Figure 7.5, reproduced from their paper, shows how the

Figure 7.5. Comparison of Lennard-Jones and spherically averaged potential for H$_2$. (After Etters, Danilowicz and England, 1975.)

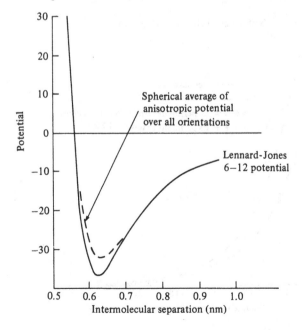

spherically averaged potential so obtained differs from a Lennard-Jones
(6–12) potential and it will be seen that, although the latter agrees quite
well with the former at small and large distances, there are appreciable
discrepancies around the minimum. Figure 7.6 shows how the calculated
results fit Stewart's observations at low pressures quite closely.

7.5   Numerical results and comparison of equations of state
7.5.1   *Metallic hydrogen*
The first approximation to the theory of metallic hydrogen given
at the end of section 7.2 is inadequate because the potential energy
$(-1.8/r_s)$ is calculated on a purely classical basis. The potential energy of
exchange for a fluid of free electrons in the ground state must be
included to give an adequate first order model of metallic hydrogen.

Figure 7.6. Comparison of theoretical and experimental variations
of pressure with volume for solid H_2. (After Etters, Danilowicz and
England, 1975.)

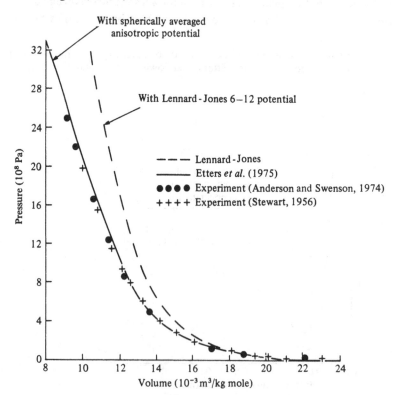

The theoretical form of the exchange energy is

$$-\frac{3}{2\pi}\left(\frac{9\pi}{4}\right)^{1/3} r_s^{-1} \text{ Ry per electron}$$

or $-0.916 r_s^{-1}$ Ry per electron.

Thus the total potential energy to first order is

$$-2.716 r_s^{-1} \text{ Ry per electron.}$$

The same result is found by the Wigner–Seitz calculation taken to first order. Kronig, de Boer and Korringa (1946), for example, find that the lowest eigenvalue for the solution of Schrödinger's equation for an electron and a proton in a Wigner–Seitz sphere is

$$\frac{-3}{r_s - \frac{1}{30} r_s^2}$$

or $-3/r_s + 0.100 + \cdots$

They take the exchange energy to be that calculated by Bardeen (1938), namely

$$\frac{0.284}{r_s} - \frac{0.58}{r_s + 5.1};$$

the total potential energy is therefore, as above, $-2.716 r_s^{-1}$ to first order.

When E is

$$\frac{2.21}{r_s^2} - \frac{2.716}{r_s},$$

the minimum value occurs at

$$r_s = 1.627,$$

corresponding to a density of 680 kg/m^3, which is thus the estimate of the density of metallic hydrogen at zero pressure.

The pressure is given, in atomic units, by

$$p = \frac{1}{4\pi}\left(\frac{4.42}{r_s^5} - \frac{3.62}{r_s}\right) \text{ Ry per electron,}$$

while the Gibbs free energy, G, is

$$\frac{3.68}{r_s^2} - \frac{3.62}{r_s} \text{ Ry per electron.}$$

Values of E, p and G according to the above Wigner–Seitz calculations are plotted in Figure 7.3(*a*) and show the effect of the exchange energy in the differences from 'Jellium'.

The dielectric function approach of Hammerberg and Ashcroft (1974) gives a similar result to first order except that explicit calculations for the Madelung energy of different lattices are used instead of the value $-1.8r_s^{-1}$ that comes from the simple 'jellium' calculation. The values, already given in section 7.3 are

     simple cubic lattice: $-1.760\,r_s^{-1}$ Ry per electron

     face-centred cubic: $-1.792\,r_s^{-1}$ Ry per electron

     body-centred cubic: $-1.792\,r_s^{-1}$ Ry per electron.

When more precise calculations are to be made, the higher terms of the potential energy, the correlation energy, must be included; one form for the correlation energy is that given by Nozieres and Pines (1958)

$$-0.115 + 0.031 \ln r_s,$$

while Ross and McMahon (1976) use the expression

$$-0.1303 + 0.0495 \ln r_s.$$

The various calculations of the properties of metallic hydrogen will now be compared in more detail. In the first place, some different calculations for the same model are compared, and then calculations for different models.

There are three points to beware of in comparing published calculations. First, a number of different systems of units have been employed. The most common is the atomic system (of which, as will have been noticed already, there are two versions), but other systems are used, and in the present comparisons SI units are used, namely, joules per kilogramme mole for energy, and pascals for pressure.

Secondly, different values are taken for zero energy. When the energies of metallic and molecular hydrogen are compared, the dissociation energy of molecular hydrogen and the ionization energy of atomic hydrogen must be taken into account; the energy of the metallic form that should be compared with the molecular form is

$$E_{\text{metal}} + \tfrac{1}{2}E_D + E_I,$$

where E_D is the dissociation energy of one molecule and E_I is the ionization energy of one atom:

$$E_D = 0.329 \text{ Ry/molecule} = 0.216 \times 10^9 \text{ J/kg mole}$$

$$E_I = 1 \text{ Ry/atom} = 1.3128 \times 10^9 \text{ J/kg mole}.$$

Thirdly, some authors calculate the zero point energy of the proton lattice while others do not.

Table 7.1(*a*) gives a comparison of two calculations by the Wigner–Seitz method, one by de Marcus (1958) and the other by Ross and

Table 7.1. *Metallic hydrogen at* $T = 0$
(a) Calculations by the Wigner–Seitz method

r_s (bohr)	ρ (kg/m³)	E (10⁹ J/kg)			G (10⁹ J/kg)			p (10¹¹ Pa)		
		1	2	3	1	2	3	1	2	3
1.70	540	−1.410	−1.369	−1.246	−1.409	−1.359	−1.289	0.04	0.05	−0.18
1.39	1000	−1.374	−1.312	−1.205	−1.226	−1.114	−1.069	1.62	1.90	+1.38
1.34	1111	−1.357	−1.287	−1.192	−1.167	−1.043	−1.015	2.27	2.60	2.03
1.29	1250	−1.334	−1.256	−1.159	−1.092	−0.957	−0.932	3.20	3.62	3.14
1.23	1430	−1.296	−1.210	−1.117	−0.987	−0.831	−0.811	4.57	5.40	4.41
1.17	1667	−1.242	−1.145	−1.066	−0.849	−0.677	−0.729	6.82	7.40	6.47
1.10	2000	−1.122	−1.048	−0.979	−0.653	−0.453	−0.470	10.51	11.96	10.33

1: Ross and McMahon (1976).
2: de Marcus (1958).
3: Carr (1962).
The values from Ross and McMahon (1976) include correlation and zero point terms as in de Marcus (1958); those from Carr (1962) include correlation but not zero point energy. Different expressions were used for the correlation energy in the three calculations.
The values taken from de Marcus have been reduced by the ionization of a single atom (1.3128×10^9 J/kg).

Table 7.1. continued
(b) Calculations by the dielectric function method

r_s (bohr)	ρ (kg/m³)	E (10⁹ J/kg) 1	E 2	G (10⁹ J/kg) 1	G 2	p (10¹¹ Pa) 1	p 2
1.70	540	-1.374		-1.416		-0.23	
1.39	1000	-1.351	-1.339	-1.216	-1.204	1.36	1.36
1.34	1111	-1.337	-1.321	-1.155	-1.132	1.98	2.08
1.29	1250	-1.311	-1.297	-1.080	-1.057	2.89	3.00
1.23	1430	-1.276	-1.258	-0.979	-0.946	4.25	4.49
1.17	1667	-1.224	-1.197	-0.838	-0.810	6.49	6.65
1.10	2000	-1.143	-1.122	-0.659	-0.637	10.06	10.33

1: Ross and McMahon (1976).
2: Hammerberg and Ashcroft (1974).
The values from Ross and McMahan (1976) include the correlation but not the zero point terms, as in Hammerberg and Ashcroft (1974).
The Hammerberg and Ashcroft (1974) values are for the face-centred cubic lattice to correspond with those of Ross and McMahan (1976).

McMahon (1976) using the formula of Neece, Rogers and Hoover (1971). There is some doubt about the basis of the comparison, for it is clear that de Marcus must have added a constant to the energy and Gibbs free energy and the best agreement between energies is obtained if that constant is taken to be the ionization energy, whereas the Gibbs free energies agree best if the constant is taken to include half the dissociation energy as well. As explained in the table, de Marcus's values have been reduced by the ionization energy.

The calculations of Carr (1962), which were the basis of Trubitsyn's (1966) work, are somewhat similar to Wigner–Seitz calculations and are accordingly included in Table 7.1(a). His results correspond to the formula:

$$E = \frac{2.209}{r_s^2} - \frac{2.708}{r_s} - 0.0905 - 0.018 r_s \text{ Ry per electron,}$$

with a small difference of $0.000\,14\ r_s^{-1}$ Ry per electron between body-centred cubic and face-centred cubic lattices. The correlation energy is not the same as that used by Ross and McMahon (1976).

Table 7.1(*b*) gives a comparison of two calculations by the dielectric function method, one by Ross and McMahon (1976) and the other by Hammerberg and Ashcroft (1974). It should be noted that Hammerberg and Ashcroft (1974) tabulate E and p as functions of r_s, from which it is a simple calculation to obtain them as functions of density, whereas they tabulate G as a function of pressure, entailing a slightly more involved calculation to obtain it as a function of density.

Table 7.2. *Calculations of properties of metallic hydrogen. Maximum differences between different calculations by the same method*

Wigner–Seitz method	A	B
$\Delta E_{max}(10^9 \text{ J/kg})$	0.179	0.086
$\Delta G_{max}(10^9 \text{ J/kg})$	0.200	0.200
$\Delta p_{max}(10^{11} \text{ Pa})$	1.63	1.45
Dielectric function method		
$\Delta E_{max}(10^9 \text{ J/kg})$	0.027	
$\Delta G_{max}(10^9 \text{ J/kg})$	0.033	
$\Delta p_{max}(10^{11} \text{ Pa})$	0.27	

A: including Carr's (1962) calculations.
B: excluding Carr's calculations.

The greatest differences between the various calculations over the range of density shown in Tables 7.1(*a*) and 7.1(*b*) are listed in Table 7.2. The spread of energies attains about 20 per cent for the Wigner–Seitz calculations but only about 3 per cent for the dielectric function calculations and the spread of pressure attains about 15 per cent for the Wigner–Seitz, but again 3 per cent for the dielectric function calculations.

Ross and McMahon (1976) compared the results of four methods of calculation, and their results, in SI units, are given in Table 7.3. It will be

Table 7.3. *Calculations of the properties of metallic hydrogen by different methods* (after Ross and McMahan, 1976)
(*a*) Energies

r_s (bohr)	ρ (kg/m³)	Energy (10⁹ J/kg) WS	DF	LCAO	APW	Δ (per cent)
1.70	540	−1.390	−1.355	−1.340	−1.388	3.7
1.39	1000	−1.348	−1.325	−1.306	−1.339	3.2
1.34	1110	−1.330	−1.308	−1.288	−1.319	3.2
1.29	1250	−1.304	−1.281	−1.260	−1.289	3.4
1.23	1430	−1.264	−1.244	−1.223	−1.250	3.3
1.17	1667	−1.208	−1.189	−1.168	−1.190	3.4
1.10	2000	−1.122	−1.105	−1.084	−1.101	3.4

Values include correlation energy and zero point energy.
The greatest differences are between WS and LCAO.

(*b*) Gibbs free energies

r_s (bohr)	ρ (kg/m³)	Gibbs Free Energy (10⁹ J/kg) WS	DF	LCAO	APW	Δ (per cent)
1.70	540	−1.380	−1.387	−1.365	−1.386	1.5
1.39	1000	−1.187	−1.176	−1.153	−1.162	2.9
1.34	1110	−1.125	−1.113	−1.090	−1.099	3.2
1.29	1250	−1.048	−1.036	−1.011	−1.017	3.6
1.23	1430	−0.948	−0.935	−0.916	−0.914	3.7
1.17	1667	−0.798	−0.786	−0.764	−0.763	4.5
1.10	2000	−0.597	−0.597	−0.559	−0.553	6.2

The greatest differences are mostly between WS and LCAO and the magnitude is nearly constant at about 0.035×10^9 J/kg.

(c) Pressures

r_s (bohr)	ρ (kg/m³)	Pressure (10¹¹ Pa)				Δ (per cent)
		WS	DF	LCAO	APW	
1.70	540	+0.05	−0.18	−0.14	−0.003	>100
1.39	1000	1.69	+1.49	+1.54	+1.76	17
1.34	1110	2.27	2.14	2.19	2.48	15
1.29	1250	3.20	3.08	3.13	3.41	10
1.23	1430	4.57	4.48	4.53	4.85	8
1.17	1667	6.82	6.69	6.74	7.13	6
1.10	2000	10.45	10.44	10.49	10.97	5

The greatest differences are mostly between DF and APW and lie between 0.23 and 0.53×10^{11} Pa.
WS: Wigner–Seitz.
DF: dielectric function.
LCAO: linear combination of atomic orbitals.
APW: augmented plane wave.
Δ: range of E, G or p.

seen that the spread of energies is less than that in Table 7.1(a) for the different Wigner–Seitz calculations, but the spread of pressure is greater. The results listed in Table 7.3 are shown graphically in Figures 7.7–7.9.

7.5.2 *Molecular hydrogen*

It seems that the only calculations for molecular hydrogen that should be considered seriously are those in which spherical averages have been taken over all relative orientations of pairs of molecules, namely the calculations of Trubitsyn (1966), Neece, Rogers and Hoover (1971) and Etters, Danilowicz and England (1975).

Table 7.4 gives values of energy, pressure and Gibbs free energy for values of the density from 100 to 1000 kg/m³ as derived from the calculations of Neece, Rogers and Hoover (1971) and Etters, Danilowicz and England (1975). The latter authors published numerical tables of their results, from which the values given in Table 7.4 are derived by interpolation and by conversion of units, while the former authors expressed their numerical calculations in the form of an interpolation formula equivalent to

$$E = 5.645 \times 10^{10} \exp\left(-50.52\rho^{-1/3}\right) - 4.712 \times 10^{-4}\rho^2,$$

where E is in J/kg and ρ is in kg/m³.

It follows that the pressure in pascals is given by

$$p = 9.506 \times 10^{11} \rho^{2/3} \exp(-50.52\rho^{-1/3}) - 9.425 \times 10^{-4} \rho^3.$$

There are unfortunately inconsistencies in Trubitsyn's (1966) published results which prevent the tabulation of any consistent values derived from his work. He himself gave his results in three forms. His expression for the internal energy at zero temperature is

$$A \exp(-BV^{1/3}) - CV^{-2}(1 + DV^{-2/3}) \text{ per atom},$$

where the volume is in atomic units per atom. A is equal to 16.8 atomic units of 27.2 eV and b is 1.81.

On conversion to SI units, the expression for the internal energy becomes

$$4.4089 \times 10^{10} \exp(-57.45\rho^{-1/3}) - 0.0165\rho^2 \text{ J/kg},$$

Figure 7.7. Comparison of internal energy, E, and Gibbs free energy, G, of metallic hydrogen for different calculations.
1: Carr (1962), Wigner–Seitz;
2: de Marcus (1958), Wigner–Seitz;
3: Hammerberg and Ashcroft (1974), dielectric function;
4: Ross and McMahan (1976), dielectric function;
5: Ross and McMahan (1976), Wigner–Seitz.

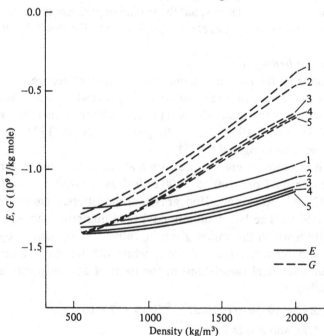

and the equivalent expression for the pressure would be

$$8.443 \times 10^{11} \rho^{2/3} \exp(-57.45 \rho^{-1/3}) - 0.0335 \rho^3 \text{ Pa.}$$

Trubitsyn himself gives the following expression for the pressure at zero temperature:

$$p = 5.01 \times 10^8 \rho^{2/3} \exp(-4.91 \rho^{-1/3}) - 2.34 \times 10^6 \rho^3,$$

where p is in atmospheres and ρ in g/cm^3.

In SI units the equivalent expression is

$$p = 5.08 \times 10^{11} \rho^{2/3} \exp(-49.1 \rho^{-1/3}) - 237 \rho^3,$$

which is evidently different from the expression derived from the energy above.

Neither formula agrees with Trubitsyn's tabulated values. Thus he gives a table (Trubitsyn, 1971) in which the molecular hydrogen values

Figure 7.8. Dependence of pressure on density of metallic hydrogen for different calculations. Key as for Figure 7.7.

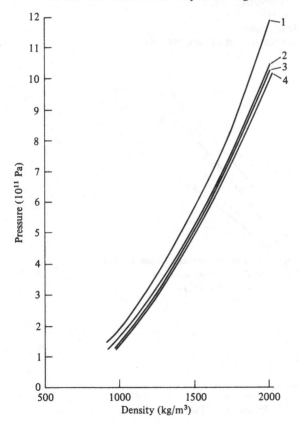

are taken from the 1966 paper, according to which the pressure at a density of $850 \, \text{kg/m}^3$ is 2.0×10^{11} Pa. His own formula for the pressure gives 1.21×10^{11} Pa, whilst that derived above from his expression for the energy leads to 1.76×10^{11} Pa.

The energies and pressures derived from any of Trubitsyn's forms are less than those of the other two authors.

7.5.3 *The transition pressure*

The transition from the molecular to metallic form of hydrogen, or vice versa, will take place at the pressure at which the Gibbs free energies of the two forms are the same. The results of sections 7.5.1 and 7.5.2 are therefore now presented in a slightly different way, namely as graphs showing the dependence of Gibbs free energy on pressure (Figure 7.9).

Figure 7.9. Gibbs free energy as a function of pressure for metallic (1–4) and molecular hydrogen (*a*, *b*).
1: Wigner–Seitz;
2: dielectric functions;
3: linear combination of atomic orbitals;
4: augmented plane wave.

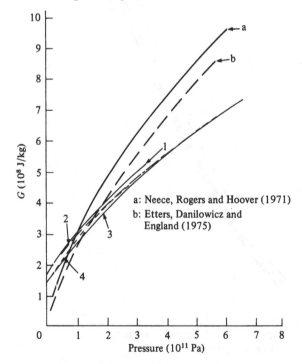

It will be remembered that the energies for the metallic form were calculated from a state in which the protons and electrons were widely separated, whereas those for the molecular form were calculated from a state in which the molecules were widely separated. To make the two sets of energies comparable, the dissociation energy of the molecule and the ionization energy of the atom must be added to the internal energies and Gibbs free energy of the metallic form as calculated above. If E_0 is the internal energy of the metallic form calculated from the state of widely separated protons and electrons, the energy to be compared with that of the molecular form is

$$E = \tfrac{1}{2}E_D + E_I + E_0 \text{ per atom,}$$

where $\tfrac{1}{2}E_D + E_I$ is equal to 1.528×10^9 J/kg.

It will be seen from Figure 7.9 that the molecular and metallic curves of G against p cross at a relatively small angle so that the estimated transition pressure is rather strongly dependent on errors in the calculations. According to which calculations are preferred, the estimated transition pressure may range from 0.75×10^{11} Pa to 2.3×10^{11} Pa. Any of Trubitsyn's calculations for molecular hydrogen gives lower energies and pressures at a given density and a lower Gibbs free energy at a given pressure, and consequently the transition pressures estimated from his results would be greater than those shown in Figure 7.9 by something of the order of 10^{11} Pa. Trubitsyn's (1966) estimate, using his own molecular calculations and Carr's (1962) metallic calculations, is 4.6×10^{11} Pa.

Table 7.4. *Calculations of the properties of molecular hydrogen*

ρ (kg/m^3)	E (J/kg) 1	E (J/kg) 2	G (J/kg) 1	G (J/kg) 2	p (10^8 Pa) 1	p (10^8 Pa) 2
100	1.1×10^6	-3.3×10^5	4.9×10^6	1.2×10^5	3.8	0.5
200	1.0×10^7	2.5×10^6	3.9×10^7	1.5×10^7	57.5	22.5
300	3.0×10^7	1.2×10^7	1.05×10^8	5.0×10^7	2.2×10^2	1.2×10^2
400	5.9×10^7	2.7×10^7	1.95×10^8	1.0×10^8	5.4×10^2	3.1×10^2
500	9.7×10^7	5.1×10^7	3.0×10^8	1.9×10^8	1.0×10^3	7.0×10^2
600	1.4×10^8	8.7×10^7	4.4×10^8	3.2×10^8	1.7×10^3	1.4×10^3
700	1.9×10^8	1.3×10^8	5.5×10^8	4.4×10^8	2.5×10^3	2.3×10^3
800	2.4×10^8	—	6.8×10^8	—	3.6×10^3	—
900	3.0×10^8	—	8.3×10^8	—	4.7×10^3	—
1000	3.6×10^8	2.9×10^8	9.7×10^8	8.5×10^8	6.1×10^3	5.6×10^3

1: From the formula of Neece, Rogers and Hoover (1971).
2: Interpolated from the table of Etters, Danilowicz and England (1975).

All the foregoing calculations for metallic as well as molecular hydrogen have been made for zero temperature. At finite temperatures the energy will contain a term corresponding to the lattice vibrations, which may be estimated from Debye theory. Trubitsyn shows that the effect of temperature is small compared with the uncertainties in the calculation of the transition pressure at zero temperature.

It is not clear whether the transition from the molecular to the metallic form of hydrogen has been observed. Grigorev *et al.* (1972) as already mentioned, considered that their observations implied a transition at a pressure of about 2.8×10^{11} Pa. More recently, Vereshchagin, Yakolev and Timofeev (1975) claim to have found the transition at 10^{11} Pa, but, as will be seen below, some doubt must be felt about this result.

7.6 Melting and the phase diagram of hydrogen

The following phases of hydrogen are known or postulated:

(*a*) in the molecular form: gas, liquid and solid;
(*b*) in the metallic form: liquid and solid.

Experimental results for molecular hydrogen are known at low pressures and low temperatures, where it is found that the melting temperature T_m depends on the melting pressure p_m according to the rule

$$T_m = T_m^0 \left(1 + \frac{p_m}{a}\right)^{1/c};$$

$T_m^0 = 14.155$ K, $a = 2.742 \times 10^7$ Pa, and $c = 1.747$ (Trubitsyn, 1971).

At higher pressure, the melting temperature is supposed to be found from Lindemann's rule, according to which a solid melts when the thermal vibrations of the lattice attain an amplitude comparable with the lattice spacing. The rule leads to the following expression:

$$\frac{T_m}{T_m^0} = \left(\frac{\Theta}{\Theta_0}\right)^2 \left(\frac{V}{V_0}\right)^{2/3},$$

where Θ is the Debye temperature and V the molecular volume at the pressure p corresponding to the temperature T_m, and Θ_0 and V_0 are the Debye temperature and molecular volume at zero pressure.

The Lindemann rule may similarly be used to estimate the melting temperature of metallic hydrogen. Neither Θ_0 nor T_m^0 is known for metallic hydrogen, but Trubitsyn (1971) assumed that metallic hydrogen is similar to the alkali metals and so found

$$T_m = 1.8 \times 10^{-5} \Theta^2 V^{2/3} \text{ K},$$

where V is in atomic units.

On substituting his calculated value of Θ, he obtained

$$T_m = 4 \times 10^2 \rho^{1/3} \, \text{K}$$

if ρ is in kg/m^3.

Leung, March and Motz (1976) adopt a different rule for extrapolating the melting temperature of molecular hydrogen, supposing that the melting pressure is proportional to $T_m^{1.78}$, a result obtained from work of Mills and Grilly (1956) and Woolley, Scott and Brickewedde (1948). They use Lindemann's rule to obtain the melting curve of metallic hydrogen. The melting relation for metallic hydrogen has also been derived by Stephenson and Ashcroft (1974), again by the Lindemann rule; they found a melting temperature of 3×10^4 K at a density of 10 000 kg/m^3, to be compared with a value of about 9000 K that would be predicted by Trubitsyn's (1971) rule.

The line between the molecular and metallic solid probably lies at lower pressures as the temperature rises; the evidence for this prediction comes from a criterion for the metal–insulator transition given by Herzfield (1927) and on experimental observations of Vereshchagin (1973) on the transition in carbon.

The phase diagram proposed by Leung *et al.* (1976) is reproduced in Figure 7.10 and that of Trubitsyn (1971) is very similar. Leung *et al.* point out that it is possible to join up the molecular and metallic liquid lines and then the solid metallic–molecular liquid could meet them at the point of intersection, but the liquid metallic–molecular line could not then pass through that same point. However the nature of these intersections is at present uncertain and is left so in the diagram.

Figure 7.10. Sketch of phase diagram of hydrogen. (After Leung, March and Motz, 1976.)

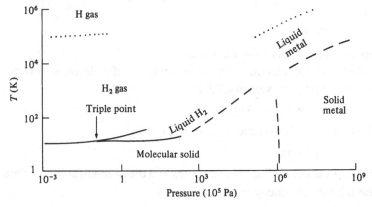

7.7 Transport properties

A number of important questions about the interiors of Jupiter and Saturn turn on a knowledge of the electrical and thermal conductivity of metallic hydrogen. If the magnetic field of Jupiter is to be generated by dynamo action in an electrically conducting fluid (as is generally postulated) then the electrical conductivity must be high and also the temperature must be high enough for metallic hydrogen to be liquid; to assess the likelihood of metallic hydrogen being liquid requires some ideas about the internal temperature as well as about the phase diagram, and, hence, some estimate of the thermal conductivity.

The theory of transport properties of simple metals is well understood (for an elementary account see, for example, Kittel, 1968). The momentum, p, of a free electron is $\hbar k$. In an electric field, E, the force exerted on an electron is eE, and the rate of change of momentum is consequently

$$\frac{\mathrm{d}p}{\mathrm{d}t} = -eE$$

or

$$\delta k = -\frac{eE\delta t}{\hbar}.$$

If electrons moved without collisions δt would increase without limit and so would δk: the metal would be a superconductor. But, in ordinary metals, electrons collide with the ions of the lattice. If the average time between collisions is τ and if the collisions destroy all coherence between motion before and after collision, the time for which the electric field, E, acts will be in effect τ, and thus the average change of k is

$$-eE\tau/\hbar.$$

The change of velocity, δv is

$$\hbar \delta k/m,$$

that is

$$-eE\tau/m,$$

where m is the mass of the electron.

If the number of electrons per unit volume is n, the electrical current, j, corresponding to the velocity, δv, is

$$j = ne\, \delta v = -ne^2 E\tau/m.$$

Thus, the specific conductivity, σ, is given by

$$\sigma = ne^2\tau/m.$$

In this expression, n, e and m are known and the problem is to calculate τ.

The thermal conductivity, K, is equal to

$$\tfrac{1}{3}C\,\delta vl,$$

where C is the heat capacity of the electrons (the phonon contribution is ignored), δv is the drift velocity of the electrons and l is the mean free path of the electrons.

Now the theory of the Fermi gas shows that

$$C = \tfrac{1}{2}\pi^2 mkT/T_0$$

where T_0 is the degeneracy temperature.

l is given by τv_{F}, where v_{F} is the velocity of electrons at the edge of the Fermi surface. It then follows that the electron conductivity, K_{el}, equals

$$\frac{\pi^2 n k^2 T\tau}{3m}.$$

The expression for K_{el} and σ combine to give the Wiedmann–Franz law, namely

$$\frac{K_{\mathrm{el}}}{\sigma} = \frac{\pi^2}{3}\left(\frac{k}{e}\right)^2 T,$$

a result which is probably reliable for metallic hydrogen.

The most complete treatment of collisions of electrons with the hydrogen lattice is that of Stephenson and Ashcroft (1974). They estimated the melting temperature of metallic hydrogen from the Lindemann criterion (see below, section 7.8) to be about 3×10^4 K at a density of 10 000 kg/m^3 and, using that value, found that the electrical conductivity at 16 000 K would be 2×10^{17} e.s.u. or 2×10^7 S/m. The Wiedemann–Franz law then led to values of the thermal conductivity which, for a specific model of Jupiter, ranged from 9×10^3 W/m deg at the centre to 10^3 W/m deg at the boundary of the metallic part. The values of the electrical conductivity obtained by Stephenson and Ashcroft (1974) are higher than those of Hubbard and Lampe (1969). Other calculations have been made by Hubbard (1970) and earlier by de Marcus (1958).

7.8 Is metallic hydrogen a superconductor?

The alkali metals, which are presumably the closest analogues of metallic hydrogen, are not found to be superconducting at the lowest temperatures at which experiments have been done. Ashcroft (1968) has however pointed out that metallic hydrogen may differ significantly from the alkali metals and could be a superconductor at quite high temperatures.

The temperature T_c at which a metal becomes superconducting is, according to the Bardeen–Cooper–Schrieffer theory, given by

$$T_c = 0.85\Theta \exp\left(-1/N_0 V\right),$$

where Θ is the Debye temperature and $N_0 V$ is the non-dimensional product of the density of (Fermi) states with the lattice interaction. The product $N_0 V$ depends on the velocity of sound in the metal. Ashcroft estimates that Θ is about 3.5×10^3 K, much greater than the values for the alkali metals which are about 100 K. He further estimates that the product $N_0 V$ is not less than 0.25. Thus T_c is not less than 60 K and, because of the strong dependence of T_c on $N_0 V$, it could be much greater.

The dependence of the transition to the superconducting state on pressure has been summarized by Bowen (1967).

7.9   Conclusion

It has been my aim in this chapter to give a general idea of the state of theoretical calculations on the properties of hydrogen and helium at pressures such as are attained in the major planets. I have also tried to summarize and compare numerical calculations where possible. Within recent years, a great deal of interest has been taken in the properties of metallic hydrogen, and various methods have been employed to calculate them, but it is clear that the spread of the calculations made by different authors is not great. Calculations on molecular hydrogen are, by that criterion, less successful, and the uncertainties of estimates of the pressure of the transition from the molecular to the metallic form depend primarily on the molecular calculations; at the same time, because the equations of state of the metallic and molecular forms are not so very different, the uncertainty in the transition pressure does not greatly affect the construction of dynamical models of the major planets.

Nothing has been said about the effect of temperature upon the transition pressure, save for a study by Trubitsyn (1966). The difficulty lies in the fact that the difference of the internal energies of the molecular and metallic forms is the small difference between quantities having uncertainties comparable with the difference. Thus the latent heat of the transition, needed in the Clausius–Clapeyron equation, is very uncertain.

The melting temperature of the metallic form is also a quantity which is very uncertain. Here the difficulty is that there is no reliable theory of melting that can be applied to even so apparently simple a system as metallic hydrogen. All current theories are to some extent empirical and consequently unreliable at the conditions in the major planets. One of the principal needs in planetary physics is for a reliable theoretical account of melting.

8

Jupiter and Saturn, Uranus and Neptune

8.1   Introduction

The major planets are distinguished from the terrestrial planets most obviously by their great sizes, their great distances from the Sun (Table 8.1) and their low densities. It has already been seen (Chapter 1) that their low densities imply that Jupiter and Saturn can only be formed in the main of hydrogen and helium; the mean densities of Uranus and Neptune are greater than those of Jupiter and Saturn (Table 8.8) while the central pressures are less and, accordingly, the mean atomic weights of Uranus and Neptune must be somewhat greater than that of a mixture of hydrogen and helium.

A further distinctive feature of all the major planets is that they spin much faster than any of the terrestrial planets. The spin periods themselves are of the order of a day, but the factor m which expresses the spin angular acceleration at the surface in terms of surface gravity is much greater than in the terrestrial planets (Table 8.8). It will be recalled that m is equal to $a^3\omega^2/GM$ and, because the mean densities of the major planets are less than those of the terrestrial planets, the factor a^3/M is greater for the major planets. Thus m ranges from about 2×10^{-2} to 1.6×10^{-1} for the major planets, Saturn having the greatest value, whereas the greatest value for the terrestrial planets is 4.6×10^{-3} for

Table 8.1. *Some properties of the major planets*

	Jupiter	Saturn	Uranus	Neptune
Distance from Sun (AU)	5.20	9.55	19.2	30.1
Greatest angular diameter seen from the Earth (arc sec)	46.9	19.5	3.6	2.1
Number of satellites	13	9	5	2

Mars. Further, the material of the major planets, is more compressible than that of the terrestrial planets, for the bulk modulus tends to $2p$ at high pressures (p is the pressure) instead of to $3p$. Accordingly, and because, in addition, the central pressures are greater in the major planets, the central compression is greater and thus the dimensionless moment of inertia is less in the major planets than in the terrestrial planets. Darwin's formula relating J_2, m and C/Ma^2 for a planet in hydrostatic equilibrium then shows that J_2 will be much greater for the major planets than for the terrestrial planets, and, consequently, that the major planets will be more flattened at the poles. These expectations are borne out, indeed the polar flattenings of Jupiter and Saturn are so great ($\frac{1}{16}$ and $\frac{1}{10}$ respectively) that they may readily be seen through a telescope. Uranus and Neptune are less flattened, about $\frac{1}{33}$, but, even so, much more than any of the terrestrial planets. In summary, the major planets are larger and softer and spin faster than the terrestrial planets and so are more flattened.

Information about the major planets is, however, much more difficult to obtain because they are so much more distant from the Earth than any of the terrestrial planets. Radar observations have in consequence not been rewarding, and it is only quite recently that two space craft, Pioneers 10 and 11, have passed by Jupiter and Saturn. Tracking of Pioneers 10 and 11 has yielded valuable information about the masses and gravity fields of Jupiter and Saturn, but both have many well-observed satellites (Table 8.1), and from their motions the masses and J_2 and J_4 had been reasonably well determined. The other major planets likewise have satellites, the motions of which have been used to estimate the masses and values of J_2, but Uranus presents an interesting difficulty. The polar axis of Uranus lies almost in the orbital plane, itself close to the ecliptic, while the planes of the satellite orbits, like those of the satellites of other planets, are close to the ecliptic. Thus the satellites are nearly polar satellites instead of being nearly equatorial as with other planets. The orbits are also nearly circular; in consequence, the motions of the nodes and periastra are difficult to determine. The sizes and geometrical flattenings present problems. The major planets are small objects when seen through a telescope (Table 8.1), and the errors of visual observations, especially of Uranus and Neptune, are rather large; some additional estimates of size and flattening come from occasional occultations of stars. There are likewise considerable difficulties in observing spins of Uranus and Neptune. Furthermore, the outer parts of the major planets are not solid but fluid, so that there is no uniquely defined surface;

rather what is seen in the telescope is a surface of particular optical depth in the fluid, while, at the same time, the surface is mobile, and so there are some problems in defining the rotation of the planet as a whole. Jupiter and Saturn have magnetic fields. The significance of this fact for the internal structure is elaborated in Chapter 9, but it will suffice to mention here that on the basis of dynamo theory it implies that part of the planet is fluid and electrically conducting, as is implied also by the study of metallic hydrogen and helium in Chapter 7. This chapter is concerned with the dynamical properties of the major planets and with the internal structures entailed by them. The dynamical properties themselves are discussed in the next section.

So far as is known, the major planets are in hydrostatic equilibrium. However, the theory of planets in hydrostatic equilibrium, developed in Chapter 3, is inadequate for a study of the major planets. It is a first order theory, sufficient for planets not greatly flattened nor with definite departures from hydrostatic equilibrium, and it is not good enough for highly flattened planets. Therefore, in section 8.3 the theory of a planet in hydrostatic equilibrium will be carried to an order of approximation sufficient for highly flattened planets.

Jupiter and Saturn, as already pointed out, are composed essentially of hydrogen and helium, and models of their internal structures will be considered in section 8.4. Such models are inapplicable to Uranus and Neptune with their higher densities and the constitutions of those two planets are separately discussed in section 8.5.

8.2 Dynamical properties

Summaries of earlier estimates of the radii, masses and values of J_2 of the major planets will be found in the articles by Kovalevsky (1970) and Dollfus (1970*b*) in the compilative work on surfaces and interiors of planets and satellites edited by Dollfus (1970*a*). A somewhat more recent review of the properties of the major planets and their satellites has been provided by Newburn and Gulkis (1973).

The spin period of Jupiter as derived from the motions of surface features in the equatorial zone is 9 h 50 m 30 s (Allen, 1963), and the corresponding period in high latitudes is greater. The period of rotation of the magnetic field of Jupiter (or more strictly, of the magnetosphere) has been derived from extended observations of features of radio emission (Carr, 1971) and the value is still longer: 9 h 55 m 29.75 s. If it is assumed that the magnetic field rotates with the main body of the planet, this is the period to take in calculations of the dynamics of Jupiter. As on the Earth, it appears that winds in the upper atmosphere rotate more rapidly than the body of the planet.

The best estimates of the radius of Jupiter are obtained from two groups of observations. Dollfus (1970*b*, *c*) has devoted a great deal of effort to the measurement of the diameters of the planets using an optical double micrometer on large telescopes; his result for Jupiter (Dollfus 1970*b*, *c*; Table 8.2) is:

    equatorial radius: 70 850 ± 100 km

    polar radius: 66 550 ± 100 km,

values which entail a polar flattening of 0.060 69 ± 0.000 12, or 1/16.48. A more recent value has been derived from observations of the occultation of the star β-Scorpii by Jupiter; the value for the equatorial radius is 71 400 ± 100 km. Dollfus's measurement is to the top of the cloud layer over Jupiter and the occultation measurement no doubt refers to a somewhat similar, though not necessarily identical, surface.

The flattening originally obtained from the occultation was 0.060 ± 0.001, which is consistent with the observations of Dollfus. However, as a result of a study of the orbit of Io, Hubbard (1977) has obtained a new position for the centre of mass of Jupiter and that in turn entails a revision of the flattening calculated from the occultation, leading to a value of 0.063 ± 0.001. It will be seen below that this is in much better accord with the gravity data than the value of 0.0607. It should be noted that the radii of Jupiter and Saturn refer to some rather ill-defined layer in the atmosphere.

The mass of Jupiter given by Kovalevsky (1970) is equivalent to $(1.899\ 67 \pm 0.000\ 07) \times 10^{27}$ kg. The data of the estimate are observations of the sizes and periods of satellites of Jupiter and perturbations of the orbits of the asteroids. It should perhaps be noted that the mass obtained by such methods is expressed as a fraction of the mass of the Sun, and Kovalevsky gives the mass of Jupiter as $1/(1047.38 \pm 0.04)$ times the solar mass. When, on the other hand, the mass is derived from the tracking of space craft, the result is often expressed as the product, GM, of the constant of gravitation and the mass. Now the value of GM for the Sun is $1.327\ 125 \times 10^{20}\ \text{m}^3/\text{s}^2$ and so Kovalevsky's result is equivalent to a value of GM for Jupiter equal to $1.267\ 08 \times 10^{17}\ \text{m}^3/\text{s}^2$. The value of G is taken throughout this book to be $6.67 \times 10^{-11}\ \text{m}^3/\text{s}^2$ kg, and the mass of Jupiter in kilogrammes corresponding to Kovalevsky's result is as given above. The mass of Jupiter has also been obtained from analyses of the orbit of Pioneer 10 as tracked by the variation of the Doppler shift of S-band radio signals from the space craft. As may be seen

Table 8.2. *Dynamical properties of Jupiter*

Period of rotation			
9 h 50 m 30 s			Allen (1963)
9 h 55 m 29.75 s			Carr (1971)

Radius

Equatorial radius (km)	Polar radius (km)	
71 400 ± 100		Hubbard and van Flandern (1972)
70 850 ± 100	66 550 ± 100	Dollfus (1970*b*, *c*)

Geometrical flattening, f

0.060 69 ± 0.000 12 (1/f = 16.48)	Dollfus (1970*b*, *c*)
0.063 ± 0.001	Hubbard (1977)

Mass

Satellites and asteroids

GM (m^3/s^2)	M (kg)	
$\left\{\begin{matrix}1.267\ 08\\ \pm 0.000\ 05\end{matrix}\right\} \times 10^{17}$	$\left\{\begin{matrix}1.899\ 67\\ \pm 0.000\ 07\end{matrix}\right\} \times 10^{27}$	Kovalevsky (1970)
Pioneer 10		
$\left\{\begin{matrix}1.267\ 136\\ \pm 0.000.025\end{matrix}\right\} \times 10^{17}$	$\left\{\begin{matrix}1.899\ 754\\ \pm 0.000\ 037\end{matrix}\right\} \times 10^{27}$	Anderson, Null and Wong (1974)

Gravitational potential

J_2	J_4	J_6	
Satellites			
$\left\{\begin{matrix}1.4710\\ \pm 0.0015\end{matrix}\right\} \times 10^{-2}$	$\left\{\begin{matrix}-6.7\\ \pm 3.7\end{matrix}\right\} \times 10^{-4}$		Kovalevsky (1970)
Pioneer 10			
$\left\{\begin{matrix}1.4720\\ \pm 0.0040\end{matrix}\right\} \times 10^{-2}$	$\left\{\begin{matrix}-6.5\\ \pm 1.5\end{matrix}\right\} \times 10^{-4}$		Anderson, Null and Wong (1974)
Pioneer 11			
$\left\{\begin{matrix}1.4750\\ \pm 0.0050\end{matrix}\right\} \times 10^{-2}$	$\left\{\begin{matrix}-6.5\\ \pm 1.5\end{matrix}\right\} \times 10^{-4}$	$\left\{\begin{matrix}5\\ \pm 6\end{matrix}\right\} \times 10^{-5}$	Null, Anderson and Wong (1975)

J_3	
$< 1.5 \times 10^{-4}$	Anderson, Null and Wong (1974)
$(1 \pm 4) \times 10^{-5}$	Null, Anderson and Wong (1975)
$C_{22}, S_{22} < 10^{-6}$	Anderson, Null and Wong (1974)

$$\frac{B - A}{Ma^2} = 4(C_{22}^2 + S_{22}^2)^{1/2} < 4 \times 10^{-6}$$

from Table 8.2, the estimates of the mass of Jupiter obtained by classical celestial mechanics and from the space craft track agree excellently.

The coefficients of the spherical harmonics in the gravitational potential have likewise been found both from the perturbations of the orbits of the natural satellites and from the tracking of space craft. J_2 was obtained by de Sitter (1931) from the motions of the Galilean satellites, while van Woerkom (1950) obtained a relation between J_2 and J_4 from which Kovalevsky (1970) derived the values of J_2 and J_4 given in Table 8.2. Values of J_2 and J_4 have been estimated from the tracks of Pioneers 10 and 11 and J_6 from Pioneer 11; it will be seen from Table 8.2 that there is excellent agreement between the classical and space craft results.

Some idea of possible departures of Jupiter from the hydrostatic state can be obtained from other harmonic coefficients that were investigated in the analysis of the Pioneer tracks. It will be seen from Table 8.2 that only upper limits can be set to values of J_3, C_{22} and S_{22}; the limits are small, and it may be said that no non-hydrostatic terms in the gravitational potential have been detected.

The radii of Saturn, Uranus and Neptune have been determined by Dollfus (1970b, c) by the optical double micrometer. The equatorial and polar radii of Saturn have been found from observations from the Pioneer 11 space craft (Gehrels *et al.*, 1980); the equatorial radius agrees well with Dollfus's (1970b, c) value, but the polar radius is rather different. Uranus has been observed at high resolution with the stratoscope telescope in orbit about the Earth – making observations outside the Earth's atmosphere permits much greater resolution than observations from the ground (Danielson, Tomasko and Savage, 1972) – and occultations of stars by Neptune and Uranus have been studied (Taylor, 1968; Kovalevsky and Link, 1969; Freeman and Lyngå, 1970; Elliot, Dunham and Mink, 1979; Table 8.3).

The masses of Saturn, Uranus and Neptune have all been estimated by Klepcynski, Seidelman and Duncombe (1971) in their general review of the masses of the planets. (See also, Duncombe, Klepcynski and Seidelman, 1971; Ash, Shapiro and Smith, 1971.) The mass of Saturn has been estimated subsequently by Garcia (1972) and Sinclair (1976) each of whom has analysed the motions of the closer satellites.

The values of J_2 and J_4 for Saturn have been quite well established from the perturbations of orbits of the satellites (Jeffreys, 1954; Sinclair, 1976) although Garcia (1972) obtained discrepant results. J_2 and J_4 have also been determined from Doppler tracking of Pioneer 11 (Anderson *et al.*, 1980); the values agree well with those found from natural satellites.

When it comes to comparing the geometrical (or optical) polar flattening for Jupiter or Saturn with that of an equipotential surface, as calculated from the harmonic coefficients of the gravity field, the first order relation between f, m and J_2 is no longer adequate. That relation is based on the assumption that the equipotential surface is an ellipsoid of revolution with its meridional sections being ellipses, and on that assumption the relations to the second order in f, m and J_2 are

$$J_2 = \tfrac{2}{3}[f(1 - \tfrac{1}{2}f) - \tfrac{1}{3}m(1 - \tfrac{9}{7}f)]$$
$$J_4 = -\tfrac{4}{35}f(7f - 5m)$$

(Cook, 1959, but note that the m used in that paper is $(1-f)a^3\omega^2/GM$ and not $a^3\omega^2/GM$ as here).

Let us first check whether the equipotential surfaces are spheroids. We calculate f from J_2 alone and find, for Jupiter, $f = 0.065\,89$ and, for

Table 8.3. *Dynamical properties of Saturn*

Period of rotation		
10 h 14 m		Allen (1963)
Radius		
Equatorial (km)	*Polar (km)*	
60 000 ± 240	53 450 ± 240	Dollfus (1970b, c)
60 000 ± 500	54 720 ± 500	Gehrels et al. (1980)
Geometrical flattening, f		
0.10917 ± 0.004		Dollfus (1970b, c)
0.088 ± 0.006		Gehrels *et al.* (1980)
Mass		
GM (m³/s²)	M (kg)	Dollfus (1970b, c)
$\left\{\begin{matrix} 3.7938 \\ \pm 0.0005 \end{matrix}\right\} \times 10^{16}$	5.6879×10^{26}	Klepcynski, Seidelman and Duncombe (1971)
3.7918×10^{16}	5.6848×10^{26}	Sinclair (1976)
$\left\{\begin{matrix} 3.7902 \\ \pm 0.002 \end{matrix}\right\} \times 10^{16}$	$\left\{\begin{matrix} 5.6824 \\ \pm 0.003 \end{matrix}\right\} \times 10^{26}$	Garcia (1972)
Gravitational potential		
J_2	J_4	
Satellites		
$\left\{\begin{matrix} 0.016\,67 \\ \pm 0.000\,03 \end{matrix}\right\}$	$\left\{\begin{matrix} -0.001\,03 \\ \pm 0.000\,07 \end{matrix}\right\}$	Jeffreys (1954)
0.0165	$\left\{\begin{matrix} -0.0010 \\ \pm 0.0001 \end{matrix}\right\}$	Sinclair (1976)
Pioneer 11		
$\left\{\begin{matrix} 0.016\,46 \\ \pm 0.000\,05 \end{matrix}\right\}$	$\left\{\begin{matrix} -0.000\,99 \\ \pm 0.000\,08 \end{matrix}\right\}$	Anderson *et al.* (1980)

Saturn, $f = 0.0797$. The value for Saturn is less than either optical flattening, and that for Jupiter is greater. Now we calculate J_4 from m, and the above values of f, with the following results. For Jupiter $J_4 = -0.000\,15$, and for Saturn $J_4 = +0.031$. The calculated values are far from the observed values and, for Saturn, even the sign is different. We conclude that the equipotential surfaces of Jupiter and Saturn are not ellipsoids of revolution.

Let us then suppose that the equipotential surface has the form

$$r = a[1 - e_2 - e_4 + e_2 P_2(\cos\theta) + e_4 P_4(\cos\theta)].$$

e_4 is $27e_2^2/35$ for an ellipsoid of revolution. The polar flattening of such a surface is

$$\tfrac{3}{2}e_2 + \tfrac{5}{8}e_4,$$

and the value of J_4 will differ from that for an ellipsoid. Consider only the changes Δe_4 due to departures from the elliptical section. Then

$$\Delta e_4 = -\Delta J_4,$$

where ΔJ_4 is the difference between the observed J_4 and that corresponding to an ellipsoid of revolution with the same J_2. The change in flattening, Δf, is then $-\tfrac{5}{8}\Delta J_4$.

The values are

Jupiter: ΔJ_4: -5×10^{-4}; Δf: 3.1×10^{-4}; $f + \Delta f$: 0.0662;
Saturn: ΔJ_4: -0.032; Δf: 0.02; $f + \Delta f$: 0.10.

The revised values are still discordant with the geometrical flattenings. The geometrical flattening of Jupiter seems to be less than is entailed by the gravity field, while the two values for Saturn lie one either side of the dynamical estimate.

To summarize, the dynamical properties of Jupiter and Saturn are reasonably well established, but the gravitational fields appear not to be those of ellipsoids of revolution, although some of the discrepancies may arise from uncertainties in the spin rates of the planets or in the positions of their centres of mass.

Properties of Uranus and Neptune are much less well known. Not only are the values of J_2 poorly known, and those of J_4 unknown, but there are large discrepancies between observations of the spin rates. The spin periods of both planets have recently been determined by two methods, one involving measurements of the Doppler shifts of Fraunhofer lines in sunlight reflected from the disc of the planet, and the other depending on periodic variations in the intensity of the reflected sunlight. Two spectroscopic determinations of the period of Uranus (Hayes and

Belton, 1977; Trafton, 1977) give values close to 24 h (Table 8.4), whereas two others (Brown and Goody, 1977; Trauger, Roesler and Münch, 1978) give values about 14.5 h. Similarly, Hayes and Belton (1977) found a value of 22 h for the rotational period of Neptune using the spectroscopic method, whereas Cruickshank (1978) and Slavsky and Smith (1977), both of whom employed the photometric method, obtained values in the neighbourhood of 18.5 h (Table 8.5). Smith and Slavsky (1977) estimated that the photometric period of Uranus was close to 24 h.

Neither for Uranus nor for Neptune can J_4 be found and the values of J_2 and the geometrical flattening are poorly known. The difficulties in obtaining J_2 from the motions of the satellites of Uranus have already been mentioned, but two values have been obtained from the orbits of the satellites Miranda (Whitaker and Greenberg, 1973) and Ariel and Umbriel (Dunham, 1971). The mean motions of the three satellites are approximately commensurate according to the relation

$$n_M - 3n_A + 2n_U = 0,$$

where n_M, n_A and n_U are the mean motions (angular velocities) of Miranda, Ariel and Umbriel respectively.

Table 8.4. *Dynamical properties of Uranus*

Spin period (h)	24 ± 3	Hayes and Belton (1977)
	23^{+5}_{-2}	Trafton (1977)
	15.57 ± 0.8	Brown and Goody (1977)
	13.0 ± 1.3	Trauger, Roesler and Münch (1978)
	23.92	Smith and Slavsky (1979)
Equatorial radius (km)	$25\,400 \pm 280$	Dollfus (1970c)
	$25\,900 \pm 300$	Danielson, Tomasko and Savage (1972)
	$25\,700 \pm 200$	Elliot et al. (1979)
Geometrical flattening, f	0.028 ± 0.01	Dollfus (1970c)
	0.01 ± 0.01	Danielson, Tomasko and Savage (1972)
	0.033 ± 0.007	Elliot et al. (1979)
J_2	0.005	Whitaker and Greenberg (1973)
	0.012	Dunham (1971)
	0.00343	Nicholson et al. (1978)
Mass (kg)	8.727×10^{25}	Klepcynski, Seidelman and Duncombe (1971)

Whitaker and Greenberg (1973) found that the motions of the node and pericentre of Miranda are not equal and opposite as would be expected from the secular theory of perturbation by the oblateness of Uranus, and, although they estimated J_2 from those motions, and found that Dunham's (1971) value of J_2 is inconsistent with the observed motions, they also point out that the commensurability of the mean motions probably means that the secular theory is inadequate. On the other hand, Miranda, being closest to Uranus, is most perturbed by the oblateness of Uranus.

Uranus has recently been found to have a system of rings of dust about it (Elliot *et al.*, 1978) and the fifth ring outwards (labelled ε) has been shown by a study of occultations to have an elliptical form which is precessing about Uranus. J_2 has been estimated from the apsidal precession (Nicholson *et al.*, 1978).

There is just one estimate of the value of J_2 (0.0050 ± 0.0004) for Neptune, derived from an analysis of the orbit of Triton (Gill and Gault, 1968; Kovalevsky 1970).

For each of the planets Uranus and Neptune there are at least two estimates of the ratio, m, and two of the geometrical flattening, f. There are three estimates of J_2 for Uranus and one for Neptune. Combinations of the estimates of m and J_2 lead to additional estimates of f according to hydrostatic theory, and some attempt will now be made to discriminate between the many estimates of f on the basis of physical arguments. In view of the wide range of the estimates of m and f, only first order theory

Table 8.5. *Dynamical properties of Neptune*

Spin period (h)	22 ± 4	Hayes and Belton (1977)
	18.17 ± 0.003	Cruickshank (1978)
	or 19.58 ± 0.003	
	18.43 ± 0.05	Slavsky and Smith (1977)
Equatorial radius (km)	$24\,300 \pm 450$	Dollfus (1970*c*)
	$24\,753 \pm 59$	Freeman and Lyngå (1970)
	$25\,225 \pm 60$	Kovalevsky and Link (1969)
Geometrical	0.0259 ± 0.0051	Freeman and Lyngå (1970)
flattening, f	0.021 ± 0.004	Kovalevsky and Link (1969)
J_2	0.0050 ± 0.0004	Kovalevsky (1970) and Gill and Gault (1968)
Mass (kg)	1.0296×10^{26}	Klepcynski, Seidelman and Duncombe (1971)

will be used, so that f is calculated from the formula (see Chapter 3)

$$f = \tfrac{3}{2}J_2 + \tfrac{1}{2}m.$$

Now, according to Darwin's first order theory of a planet in hydrostatic equilibrium, the dimensionless moment of inertia is given by

$$\frac{C}{Ma^2} = \frac{2}{3}\left[1 - \frac{2}{5}\left(\frac{5}{2}\frac{m}{f}-1\right)^{1/2}\right].$$

Physical restrictions on the possible values of C/Ma^2 entail corresponding restrictions on the possible values of m/f. The ratio C/Ma^2 must be real; accordingly m/f must exceed 0.4. Also C/Ma^2 must be less than 0.4 if the density is to increase towards the centre of the planet; m/f must therefore exceed 0.8. On the other hand, C/Ma^2 cannot be less than zero; m/f must therefore be less than 2.9.

The values of f and C/Ma^2 for Uranus, calculated for two values of the spin period and m, are collected in Table 8.6. Certain combinations are seen to be excluded by the criteria just set out, but there remains a wide range of possible values. Dunham's (1971) high value of J_2 (0.012) should perhaps be discounted in the light of the later findings of Whitaker and Greenberg (1973). There then remain two quite different possibilities, according to which value of the spin period is adopted. If the period is taken to be about 14.5 h, the two remaining values of J_2 and the geometrical flattenings of 0.028 and 0.033 are reasonably consistent, with m/f about 1.5 and C/Ma^2 about 0.22. On the other hand, when the period is taken to be 24 h, the geometrical flattenings of 0.028 and 0.033

Table 8.6. *Values of f and C/Ma^2 for Uranus*

Spin period: 24 h m: 0.0153						
J_2				0.005	0.012	0.00343
f	0.033	0.028	0.01	0.0152	0.0256	0.0128
m/f	0.464	0.546	1.53	1.0	0.60	1.20
C/Ma^2	0.56	0.5	0.22	0.34	0.4	0.29
Spin period: 14.5 h m: 0.0420						
J_2				0.005	0.012	0.00343
f	0.033	0.028	0.01	0.0285	0.39	0.0262
m/f	1.273	1.5	4.2	1.47	1.08	1.60
C/Ma^2	0.27	0.22	0	0.23	0.32	0.20

Adopted values of a: 25 650 km,
mass: 8.727×10^{25} kg.

are impossible, and there is no consistency between the various values of m/f and C/Ma^2.

The greater geometrical flattening of Neptune (0.0259) is impossible on the basis of either spin period, and the lesser one on the basis of the longer period. With the shorter period, however, the flattening of 0.021 can be roughly reconciled with the value of J_2.

These comments on the consistency of different estimates do not, of course, prove that the respective estimates are correct. A different line of argument suggests that the smaller values of C/Ma^2, to which the comments on Uranus appear to tend, may be unlikely. Uranus and Neptune have mean densities and central pressures which lie between those of the Earth, on the one hand, and of Jupiter and Saturn, on the other, the central pressures in effect probably being rather close to that in the Earth. The central compressions might therefore be expected to be closer to that in the Earth than in Jupiter or Saturn, and thus C/Ma^2 would be expected to be greater than the Jovian value (0.25) but less than that for the Earth (0.33). On this argument, the low values for Uranus in Table 8.6 would be unlikely, while the Neptunian value of just over 0.3 in Table 8.7 would be reasonable. However, the argument is none too secure, for, applied to Jupiter and Saturn, it would lead to the conclusion that C/Ma^2 should be less for Jupiter than for Saturn, whereas the reverse is the case. It is tempting to think that Uranus and Neptune, which have almost identical sizes, should be similar in other respects as well, but the masses and mean densities are substantially different, and it may well

Table 8.7. *Values of f and C/Ma^2 for Neptune·*

Spin period: 22 h			
m: 0.014			
J_2			0.0050
f	0.0259	0.021	0.0145
m/f	0.540	0.667	0.97
C/Ma^2	0.5	0.45	0.348
Spin period: 18.5 h			
m: 0.020			
J_2			0.0050
f	0.0259	0.021	0.0175
m/f	0.772	0.95	1.142
C/Ma^2	0.41	0.353	0.303

Adopted values of a: 24800 km,
mass: 1.0296×10^{26} kg.

be that so also are the flattenings and the values of C/Ma^2. It can only be concluded that large uncertainties remain in the dynamical properties of Uranus and Neptune and that no generally consistent set of data exists for either.

The dynamical properties of the major planets are summarized in Table 8.8, various alternative values being given for Uranus and Neptune where appropriate.

8.3 Theory of a rapidly spinning planet in hydrostatic equilibrium

The theory of a rotating planet in hydrostatic equilibrium given in Chapter 3 was confined to quantities of the order of J_2, and was adequate for the purposes for which it was used in discussing the internal states of the terrestrial planets. The uncertainties introduced by neglecting terms of order J_2^2 are probably not important in relation to the error committed in using Darwin's formula for C/Ma^2 when the planet is not in hydrostatic equilibrium. When, however, we turn to the major planets, the situation changes, for the values of m and J_2 are much greater and their squares must be taken into account in comparing the geometrical

Table 8.8. *Summary of the dynamical properties of the major planets*

	Jupiter	Saturn	Uranus	Neptune
Spin angular velocity (ω) (rad/s)	1.7734×10^{-4}	1.7055×10^{-4}	7.721×10^{-5} 1.204×10^{-4}	7.93×10^{-5} 9.43×10^{-5}
Equatorial radius (km)	71 200	60 000	25 650	24 800
Polar radius (km)	66 710	54 000	25 260	24 440
Mass (kg)	1.8997×10^{27}	5.685×10^{26}	8.727×10^{25}	1.0296×10^{26}
J_2	1.472×10^{-2}	1.65×10^{-2}	0.005 0.00343	0.0050
J_4	-6.5×10^{-4}	-1.0×10^{-3}		
m	0.08809	0.1657	0.0153 0.0420	0.014 0.020
Polar flattening geometrical	0.063	0.108	0.028 0.01	0.021
dynamical[a]	0.065 89	0.08		
C/Ma^{2}[b]	0.252	0.212		
Mean density (kg/m^3)	1337.7	705.3	1254	1635

[a] for an ellipsoid of revolution.
[b] using dynamical flattening.

and dynamical flattenings of Jupiter and Saturn. Furthermore, values of J_4 as well as J_2 are available. Use of the Darwin formula (a first order expression in any case) to calculate C/Ma^2 neglects the information available in J_4 and in any higher harmonic coefficients that may eventually be determined for those planets. In addition, with values of J_4 and J_6 available, other integrals of the density besides the mass and moment of inertia may be calculated. Thus, it is necessary to carry the theory of the gravity field of a spinning planet to higher order than it was in Chapter 3 to make use of the extra data available for the major planets. The theory presented in Chapter 3 has a further defect in its application to the major planets. The density is supposed known as a function of position (not strictly of radius, because the density is constant on equipotential surfaces which are not spherical), but, in the usual models of the major planets, the density is supposed known as a function of pressure, so that a theory of the internal gravity field which employs the density in that form is desirable.

Let us first recall the principles of the theory of Clairaut, Darwin and Callandreau as set out in Chapter 3. The density is supposed known as a function of position, and the potential (including the rotational contribution) is calculated at points within the planet. The condition is then imposed that the potential on a surface of constant density should be constant, and from that the form of each such surface is derived. It was seen that the condition led to an integral equation from which a differential equation for the polar flattening could be derived and also that it was possible to derive a relation (Darwin's) between C/Ma^2, f and m. When m is small, the surfaces of constant density and potential are nearly spherical, and it is natural to express their equations in the form of a series of even zonal spherical harmonics, other spherical harmonics being excluded by axial symmetry and by symmetry about the equatorial plane. The differential equation given by the first order theory is for the coefficient of the zonal harmonic of second degree, and it is tacitly assumed that the coefficients of all the terms of higher degree are negligible.

Let the radius vector of a surface of constant density be given by the expression

$$r = a' \left(1 + \sum_n e'_n S_n \right)$$

where S_n is a zonal harmonic of even degree and where the (small) coefficient e'_n is a function of a'. Let the density on this surface be ρ'.

According to the results obtained in Chapter 3, the potential of the whole body on the surface for which $a' = a_1$ is

$$\frac{-4\pi G}{3}\left[\frac{1-\sum_n e_{n1}S_n}{a_1}\int_0^{a_1} 3\rho' a'^2 \, da'\right.$$

$$+\sum_n \frac{3S_n}{2n+1}\frac{1}{a_1^{n+1}}\int_0^{a_1}\rho' \, d(a'^{n+3}e_n')$$

$$\left.+\sum_n \frac{3S_n}{2n+1}a_1^n\int_{a_1}^a \rho' \, d\left(\frac{e_n'}{a'^{n-2}}\right)\right]$$

$$-\frac{4\pi G}{3}\int_{a_1}^a \rho' 3a'^2 \, da' - \tfrac{1}{3}r^2\varpi^2(S_2-1),$$

the last term being the rotational potential. The variable a is the surface value of a', and e_{n1} corresponds to a_1.

Each harmonic term separately must be zero, and thus, equating coefficients of each harmonic to zero, we have

$$\frac{-e_n}{a_1}\int_0^{a_1}\rho' a'^2 \, da' + \frac{1}{2n+1}\frac{1}{a_1^{n+1}}\int_0^{a_1}\rho' \, d(a'^{n+3}e_{n1})$$

$$+\frac{a_1^n}{2n+1}\int_{a_1}^a \rho' \, d\left(\frac{e_n'}{a'^{n-2}}\right) = 0$$

for $n \neq 2$, and

$$\frac{-r^2\varpi^2}{4\pi G}$$

for $n = 2$.

In proceeding to the second order, Darwin (1899) took any surface to be of the form

$$r = a'[1-\tfrac{1}{3}h+\tfrac{167}{210}h^2+\tfrac{3}{35}e_2-\tfrac{2}{3}h(1+2h)P_2(\cos\theta)$$
$$+\tfrac{8}{35}(\tfrac{3}{2}h^2-e_2)P_4(\cos\theta)].$$

The rather complex form is of eventual algebraic convenience.

h is related to the first order flattening f by

$$h = f - \tfrac{25}{14}f - \tfrac{1}{7}e_2.$$

Notice that f is equal to $(a-b)/a$ and is not the same as e_1 in the earlier expression.

The expression for the internal potential may then be written in the form

$$-\frac{3V}{4\pi G} = \frac{S_0}{r} - \frac{2}{5}\frac{S_2 P_2(\cos\theta)}{r^3} + \frac{12}{35}\frac{S_4 P_4(\cos\theta)}{r^5}$$

$$-\tfrac{2}{5}r^2 T_2 P_2(\cos\theta) - \tfrac{8}{105}r^4 T_4 P_4(\cos\theta)$$

$$+\frac{\varpi^2 r^2}{4\pi}\{1 - P_2(\cos\theta)\},$$

where the coefficients of the zonal harmonics are the following integrals:

$$S_0 = \int_0^a \rho'\, \mathrm{d}[a'^3(1 - h - \tfrac{25}{14}h^2 + \tfrac{9}{35}e_2)],$$

$$S_2 = \int_0^a \rho'\, \mathrm{d}[a'^5(h + \tfrac{2}{7}h^2)],$$

$$S_4 = \int_0^a \rho'\, \mathrm{d}[a'^7(h^2 - \tfrac{2}{9}e_2)],$$

$$T_2 = \int_{a_1}^a \rho'\, \mathrm{d}[h + \tfrac{17}{7}h^2],$$

$$T_4 = \int_{a_1}^a \rho'\, \mathrm{d}[e_2/a'^2].$$

In these expressions a quantity $\mathrm{d}[X]$ stands for $(\mathrm{d}X/\mathrm{d}a')\,\mathrm{d}a'$.

The conditions for the potential to be constant on a surface of constant density are obtained by substituting for r in the expression for the potential and equating the coefficients of P_2 and P_4 each to zero. Then

$$\frac{S_0}{a}(h + \tfrac{20}{7}h^2) - \frac{3}{5}\frac{S_2}{a^3}(1 + \tfrac{11}{4}h) - \tfrac{2}{5}a^2 T_2(1 - \tfrac{22}{21}h)$$

$$+\frac{3\varpi^2 a^2}{8\pi}(1 + \tfrac{2}{7}h) = 0,$$

$$\frac{S_0}{a}(e_2 - \tfrac{1}{2}h^2) - \frac{9}{5}\frac{S_2 h}{a^3} + \frac{3}{2}\frac{S_4}{a^3} + \tfrac{6}{5}a^2 T_2 h - \tfrac{1}{3}a^4 T_4 + \frac{3\varpi^2}{4\pi}a^2 h = 0.$$

These are integral equations for h and e_2 because $S_0 \ldots, T_2 \ldots$ involve h and e_2 under the integral sign.

Darwin (1899) reduced the integral equations to the following differential equations:

$$\frac{\mathrm{d}^2 h}{\mathrm{d}a^2} + \frac{6\rho}{\rho_1 a}\frac{\mathrm{d}h}{\mathrm{d}a} - \left(1 - \frac{\rho}{\rho_1}\right)\left[\frac{6h}{a^2} - \frac{14h^2}{a^2} - \frac{66}{7}\frac{h}{a}\frac{\mathrm{d}h}{\mathrm{d}a}\right.$$

$$\left. -\frac{40}{7}\left(\frac{\mathrm{d}h}{\mathrm{d}a}\right)^2 + 4m\left(\frac{h}{a^2} + \frac{1}{a}\frac{\mathrm{d}h}{\mathrm{d}a}\right)\right] = 0,$$

and

$$\frac{d^2 e_2}{da^2} + \frac{6\rho}{\rho_1 a} \frac{de_2}{da} - \left(20 - \frac{6\rho}{\rho_1}\right) \frac{e_2}{a^2} + 12\left(1 - \frac{\rho}{\rho_1}\right) \frac{h^2}{a^2}$$

$$- \left(4 - \frac{18\rho}{\rho_1}\right) \frac{h}{a} \frac{dh}{da} - \left(1 - \frac{9\rho}{\rho_1}\right) \left(\frac{dh}{da}\right)^2 = 0.$$

The procedure would be to solve the first equation for h and then, after substituting h so found in the second equation, to solve that for e_2.

ρ_1 is the mean density out to a radius r, defined by

$$\int_0^r \rho' a'^2 \, da' = \tfrac{1}{3}\rho_1 r^2;$$

like ρ itself, it is a function of radius.

When, as is in general the case, the equations have to be solved numerically, there is no particular advantage in working with either the differential or the integral form, and in the latest development of the theory by Zharkhov and Trubitsyn (1969, 1975*b*) it is the integral form which has been chosen.

If, as earlier, we take the radius in the form

$$r = a\{1 + a_0 + e_2 P_2 + e_4 P_4 + e_6 P_6\},$$

where a is the equatorial radius, the integrals $S_0, \ldots, T_2, \ldots$ take on slightly different forms:

$$S_0 = \int_0^a \rho(a) \, d[a^3\{1 + 3a_0 + (3a_0^2 + \tfrac{3}{5}e_2^2)$$

$$+ (a_0^3 + \tfrac{3}{5}a_0 e_2^2 + \tfrac{2}{35}e_2^3)\}],$$

$$S_2 = -\tfrac{3}{2}\int_0^a \rho(a) \, d[a^5\{e_2 + (4a_0 e_2 + \tfrac{4}{7}e_2^2)$$

$$+ (6a_0^2 e_2 + \tfrac{12}{7}a_0 e_2^2 + \tfrac{6}{7}e_2^3 + \tfrac{8}{7}e_2 e_4)\}],$$

$$S_4 = \tfrac{35}{36}\int_0^a \rho(a) \, d[a^7\{(\tfrac{54}{35}e_2^2 + e_4)$$

$$+ (\tfrac{54}{7}a_0 e_2^2 + 6a_0 e_4 + \tfrac{108}{77}e_2^3 + \tfrac{120}{77}e_2 e_4)\}],$$

$$S_6 = -\tfrac{63}{104}\int_0^a \rho(a) \, d[a^9(\tfrac{24}{11}e_2^3 + \tfrac{40}{11}e_2 e_4 + e_6)],$$

$$T_0 = \int_a^{a_1} \rho(a) \, d[a^2(1 + 2a_0 + a_0^2 + \tfrac{1}{5}e_2^2)],$$

$$T_2 = -\tfrac{3}{2}\int_a^{a_1} \rho(a) \, d[e_2 - a_0 e_2 + \tfrac{1}{7}e_2^2 + a_0^2 e_2 + \tfrac{2}{7}a_0 e_2^2 + \tfrac{1}{7}e_2^3 - \tfrac{2}{7}e_2 e_4],$$

$$T_4 = -\tfrac{35}{8} \int_a^{a_1} \rho(a) \, \mathrm{d}[a^{-2}(-\tfrac{27}{35}e_2^2 + e_4 + \tfrac{108}{5}a_0e_2^2$$
$$- 3a_0e_4 + \tfrac{206}{385}e_2^3 - \tfrac{60}{77}e_2e_4)],$$

$$T_6 = -\tfrac{231}{80} \int_a^{a_1} \rho(a) \, \mathrm{d}[a^{-4}(\tfrac{90}{77}e_2^3 - \tfrac{25}{11}e_2e_4 + e_6)].$$

There then follow three integral equations for e_2, e_4 and e_6 (a_0, the constant part, is a linear combination of e_2, e_4 and e_6).

Zharkhov and Trubitsyn also show how to express the integrals and the integral equations, not only in terms of the equatorial radii of equipotential surfaces, but also in terms of the polar radii or the mean radii (de Sitter, 1924).

The theory of a rotating body in hydrostatic equilibrium given the density as a function of pressure, was developed for stars by James (1964) and later by Ostriker and Mark (1968) and has been applied to the problems of rapidly rotating planets by Hubbard, Slattery and de Vito (1975). If the density is known as a function of pressure, then equally, the pressure is known as a function of density, $p(\rho)$. The equation of hydrostatic equilibrium is then

$$\rho^{-1} \, \mathrm{d}p = \mathrm{d}V.$$

Let V be the potential within the body and V_s the value at the surface. Also let

$$Z(\rho) = V - V_s.$$

Evidently

$$Z(\rho) = \int_{\rho_s}^{\rho} \frac{1}{\rho'} \frac{\mathrm{d}p}{\mathrm{d}\rho'} \, \mathrm{d}\rho'.$$

$Z(\rho)$ incorporates the equation of state. It may be supposed that ρ_s is zero. Thus

$$V = V_s + Z(\rho),$$

a relation between potential and density, the latter being given in terms of pressure.

But V may also be calculated from the density as a function of position. Let $\rho(r')$ be the density at radius vector r'. The gravitational potential at the radius vector r is

$$G \int \mathrm{d}\tau \frac{\rho(r')}{|r - r'|},$$

where $\mathrm{d}\tau$ is the element of volume and the integral extends throughout the whole body.

If Q is the potential of rotation,

$$V = Q + G \int d\tau \frac{\rho(r')}{|r - r'|}.$$

This is a second expression for the potential, and, by equating it to the first, we obtain an equation from which the form of an equipotential surface and the coefficients J_2, J_4, J_6 of the external gravity field may be calculated. The method of solution adopted by Hubbard, Slattery and de Vito is similar to that of Clairaut and Darwin, in that all quantities are expanded in series of spherical harmonics: the two expressions for V are written as series of spherical harmonics, and the coefficients of a given degree are equated. To effect the expansions, Hubbard, Slattery and de Vito define projection operators, T_l, for the lth zonal harmonic, which form the lth harmonic component of a function $\phi(r, \mu)$ say:

$$T_l\phi(r, \mu) = \tfrac{1}{2}(2l + 1)P_l(\mu) \int_{-1}^{+1} d\mu' \, \phi(r, \mu')P_l(\mu'),$$

(μ is the cosine of the colatitude.)

This is just the usual expression for a term in an expansion of an arbitrary function of r and μ in zonal harmonics.

It is assumed that the density at a given radius may be expanded in even zonal harmonics:

$$\rho(r, \mu) = \sum_{l=0} \rho_{2l}P_{2l}(\mu).$$

ρ_0 is of course a function of r; if all other coefficients were zero, the distribution of density would be spherically symmetrical as in a non-rotating planet. The rotational potential is $\tfrac{1}{2}r^2\omega^2 \sin^2 \theta$, and its projections are

$$T_0Q = \tfrac{1}{3}r^2\omega^2,$$
$$T_2Q = -\tfrac{1}{3}r^2\omega^2 P_2(\mu),$$
$$T_{2l}Q = 0, \qquad (l > 1).$$

Formally we wish to equate all the projections of the two expressions for the potential, i.e.

$$T_{2l}(V_s + Z) = T_{2l}\left(Q + G \int \frac{d\tau \rho'}{|r - r'|}\right),$$

for $l = 0, 1, 2 \ldots$.

Q appears only in the first two projections and it induces the second harmonic distortion in surfaces of constant potential. Its effects are, however, not confined to the first two projections, because the expansions in the higher projections involve terms such as P_2^2, which lead to distortions of higher harmonic forms dependent on the rotational potential. Thus the coefficient ρ_{2l} in the expansion of the density will be found to be of order m^l.

Z is expanded in a Taylor series about the spherically symmetrical value of ρ, ρ_0:

$$Z(\rho) = Z(\rho_0) + Z'(\rho_0) \sum_{l=1} \rho_{2l} P_{2l} + Z''(\rho_0) \left(\sum_{l=1} \rho_{2l} P_{2l} \right)^2,$$

where

$$Z'(\rho_0) = \mathrm{d}Z_0/\mathrm{d}\rho.$$

To obtain the projection of the gravitational potential, the well-known expression of $|r - r'|^{-1}$ in a series of zonal harmonics is used. It is then found that

$$V = Q + \frac{2\pi G}{r} \sum P_{2l}(\mu) \frac{2}{4l+1} \left[\left\{ r^{-2l} \int_0^r \mathrm{d}r' \, r'^{2l+2} \rho_{2l}(r') \right\} \right.$$

$$\left. + \left\{ r^{2l+1} \int_r^b \mathrm{d}r' \, r'^{2l-1} \rho_{2l}(r') \right\} + b^{2-2l} r^{2l+1} SH_{l+1,m} \right].$$

b is the polar radius, and the first two terms in the square bracket give the contributions to the potential of material within a radius r and from the radius r to the radius b. The contribution from the remaining material outside the radius is given by the so-called shell integral $SH_{l+1,m}$. Here m denotes the surface value of μ and the integrals are defined by:

$$SH_{l+1,m} P_{2l} = T_{2l} \left[b^{2l-2} \int_b^{R(\mu)} \mathrm{d}r' \, r'^{1-2l} \rho(r', \mu) \right].$$

$R(\mu)$ is the value of the external radius for $\theta = \cos^{-1} \mu$; it is b when $\mu = 1$. The aim now is to solve the equations to give $R(\mu)$ in the form:

$$R(\mu) = b \left[1 + \sum_{l=0}^{\infty} d_{2l} P_{2l}(\mu) \right].$$

The d_{2l} are shape coefficients.

Now, in free space, for r greater than a, the equatorial radius,

$$V = Q + \frac{2\pi G}{r} \sum_{l=0}^{\infty} P_{2l}(\mu) \frac{2}{4l+1}$$

$$\times \left[r^{-2l} \int_0^b \mathrm{d}r' \, r'^{2l+2} \rho_{2l}(r') + r^{-2} b^{2l+3} SH_{l+1,0} \right],$$

with

$$SH_{l+1,0}P_{2l} = T_{2l}\left[b^{-3-2l}\int_{b}^{R(\mu)} dr'\, r'^{2l+2}\rho(r',\mu)\right].$$

The values of $J_2, J_4 \ldots$ may then be obtained by comparing the terms in the above expression for the external potential with the form

$$V = Q + \frac{GM}{r}\left[1 - \sum_{l=1}\left(\frac{a}{r}\right)^{2l} J_{2l}P_{2l}(\mu)\right].$$

The first few equations for the coefficients ρ_{2l} have the following forms:

(a) $l = 0$:

$$4\pi G\left[\frac{1}{r}\int_{0}^{r} x^2\rho_0\, dx + \int_{r}^{b} x\rho_0\, dx\right] = Z(\rho_0) + V_s.$$

The solution of this equation gives ρ_0, the density of a non-rotating planet, as a function of r when ρ is a known function of pressure. It may be noted that this relation is satisfied trivially when ρ is a constant. The left-hand side is then

$$4\pi G\rho\left[\frac{r^2}{3} + \frac{1}{2}(b^2 - r^2)\right] = 4\pi G\rho\,\frac{3b^2 - r^2}{6},$$

which is the potential at radius r within a sphere of radius b and is thus equal to the right-hand side.

(b) $l = 1$, order m:

$$\frac{4\pi G}{5}\left[\frac{1}{r^3}\int_{0}^{r} x^4\rho_2\, dx + r^2\int_{r}^{b} \rho_2\,\frac{dx}{x}\right] - \tfrac{1}{3}\varpi^2 r^2 = Z'(\rho_0)\rho_2.$$

This shows how ρ_2, the coefficient of P_2 in ρ, is dependent on ϖ^2.

(c) $l = 2$, order m^2:

$$\frac{4\pi}{9}G\left[\frac{1}{r^5}\int_{0}^{r} x^6\rho_4\, dx + r^4\int_{r}^{b} \rho_4\,\frac{dx}{x^3} + \frac{r^4}{b^2}SH_{3,1}\right] - \tfrac{18}{35}Z''(\rho_0)\rho_2^3$$

$$= Z'(\rho_0)\rho_4.$$

(d) $l = 3$, order m^3:

$$\frac{4\pi}{11}G\left[\frac{1}{r^7}\int_{0}^{r} x^8\rho_6\, dx + r^6\int_{r}^{b} \rho_6\,\frac{dx}{x^5} + \frac{r^6}{b^4}SH_{4,1}\right]$$

$$- \tfrac{10}{11}Z''(\rho_0)\rho_2\rho_4 - \tfrac{18}{77}Z'''(\rho_0)\rho_2^3 = Z'(\rho_0)\rho_6.$$

The shape coefficients d_{2l}, are needed to calculate the shell integrals, and Hubbard, Slattery and de Vito show how to find them. They also obtain analytical solutions for the case when the pressure is given by

$$p = \tfrac{1}{2}K\rho^2,$$

the particular case of the polytropic form

$$p = \frac{n}{n+1} K\rho^{1+(1/n)},$$

in which $n = 1$. For this form, it follows that

$$\frac{d\rho}{dr} = -\frac{g}{K} \rho^{1-(1/n)}$$

or

$$\frac{d\rho}{dr} = -\frac{g}{K}$$

if $n = 1$. Hubbard (1974b) considers that the material in Jupiter follows closely the polytrope with $n = 1$; the coefficients in the expansion of the density then have the form

$$\rho_{2l}(r) = j_{2l}(\kappa r),$$

where the j_{2l} are spherical Bessel functions and

$$\kappa = (2\pi G/K)^{1/2}.$$

It is instructive to compare the state of the theory of the internal gravity field with that of the external field. For a long time the latter was approached through an expansion of the potential in spherical harmonics, although it was realized that that might not be a correct procedure. The reason is that an expansion of the inverse distance in powers of r only converges in the space outside a spherical surface of radius a, equal to the equatorial radius, which contains no mass, whereas the expansion is used down to the surface of the planet and therefore in a region in which a sphere of radius less than a contains mass. This difficulty can be overcome for a perfect ellipsoid of revolution (to which planetary surfaces closely approach) by using oblate spheroidal co-ordinates, because the corresponding expansion for the inverse distance in spheroidal harmonics converges everywhere outside the co-ordinate surface which coincides with the planetary surface, and an exact theory of the external field of an ellipsoid of revolution is possible (Cook, 1959). If an attempt is made to apply spheroidal co-ordinates in the theory of the internal field, an apparently intractable difficulty arises in that the co-ordinate surfaces become flatter towards the centre of the planet, where they collapse to a disc, whereas the flattening of equipotential surfaces decreases towards the centre, where they become a sphere. Thus, it seems that spheroidal co-ordinates are less well adapted to the theory of the internal field than are spherical co-ordinates. The theoretical difficulty of the latter remains, however, namely that, in the region between the spheres of radii b and a, neither the expansion of the inverse distance in powers of r nor that in inverse powers of r converges. In so far as they use expansions in terms of spherical harmonics this objection applies equally to the Darwin type of theory and to the work of Hubbard, Slattery and de Vito.

8.4 Models of Jupiter and Saturn

All current models of Jupiter and Saturn are based on the ideas that these planets must be composed almost entirely of hydrogen and helium and that, at the pressures encountered in them, the hydrogen at least is in the metallic form in some part. The simplest models have been found to match Jupiter quite well, but Saturn less well. Models vary in the pressure at which the metallic transition in hydrogen is supposed to occur, but that does not have a large effect on the properties of the model because, as was seen in Chapter 7, the fact that the transition pressure is poorly determined reflects in part the similarity of the pressure–density relations of the metallic and molecular forms of hydrogen. Other sources of variation between models are the way in which the temperature is supposed to increase inwards and the mixing of helium with hydrogen. When a rule for the variation of density with pressure has been established, then the variation of density with position can be calculated either from the integral equations of Hubbard, Slattery and de Vito or by a process of successive approximations using the Darwin theory in which the pressure and J_2 and J_4 are calculated from an assumed variation of density with radius. The variation of density with radius is then recalculated from the pressure, and so on, at each stage the total pressure being that due to gravitational attraction and centrifugal acceleration. Models would be considered satisfactory when the values of the mass, J_2 and J_4 calculated from the distribution of density with radius and co-latitude agree with the observed values.

The first model derived on these principles was that of de Marcus (1958) who used the theory of the internal gravity field effectively in the form given by de Sitter. As for the equations of state, de Marcus recalculated the properties of the metallic form of hydrogen with results that have been discussed in Chapter 7. The densities of molecular hydrogen and of helium were based on measurements by Stewart (1956) and, as explained in Chapter 7, there are considerable difficulties in extrapolating these data, obtained at relatively low pressure, to a million atmospheres or so. An analytical expression was used to extrapolate the helium results, but a 'judicious' procedure was used for hydrogen.

de Marcus estimated that the transition pressure from the molecular form to the metallic form would lie between 1.9×10^{11} and 3.9×10^{11} Pa. He also considered that the molecular form would be mainly fluid up to the transition pressure and that the molecular fluid would have an equation of state close to that of solid molecular hydrogen at absolute zero. He considered that the thermal expansion of solid hydrogen would not be significant, and he also pointed out that the thermal conductivity of the molecular, as well as the metallic, form would be expected to be high. Thus, at a density of about 700 kg/m^3, he estimated the thermal conductivity of metallic hydrogen to be 420 W/m deg and that of molecular hydrogen at 172 K to be almost the same (copper at room temperature has a thermal conductivity of about 40 W/m deg). de Marcus found that he could obtain reasonable but not exact agreement with the mechanical data for Jupiter and Saturn available to him at the time if he supposed that the fraction of helium increased from zero at the surface to 1 at a radius of one-tenth the surface radius in Jupiter and one-quarter in Saturn. These detailed models thus confirm a conclusion arrived at earlier by de Marcus (1951) and by Ramsey (1951) that a model composed of hydrogen alone would not fit Jupiter or Saturn. It is also a straightforward matter to see that Jupiter and Saturn cannot have the same composition. Saturn is both smaller than Jupiter and has a lower mean density, so that the central pressure will be less. If, then, the compositions of the two planets are the same, the increase of density under self-compression will be less in Saturn than in Jupiter and consequently the dimensionless moment of inertia should be greater for Saturn than for Jupiter, whereas it is certainly less. It follows that the central parts of Saturn are composed of material of higher atomic weight than are the central parts of Jupiter. In de Marcus's models this is achieved by increasing the relative volume of pure helium.

A significant feature of de Marcus's work is that he somewhat arbitrarily took the density of molecular hydrogen at high pressures to be about 10 per cent less than a straightforward extrapolation of Stewart's (1956) results would indicate; he did so in order to get the best match between his model and the known properties of Jupiter. de Marcus's work was revised and extended by Peebles (1964) who took advantage of an electronic computer to calculate a wider range of models and, in particular, models with a wider range of helium abundance and with various assumptions about the atmosphere below the visible cloud layer, the atmosphere being defined as the region in which thermal expansion is significant. In particular, he considered models with thick atmospheres in the adiabatic state. Peebles used the same equation of state as de Marcus and fitted his models to the same dynamical data, but, whereas de Marcus allowed the proportion of helium to increase towards the centre in order to obtain agreement between the dynamical properties of the models and those of the actual planets, Peebles took the composition to be the same throughout. He admitted, however, the possible existence of a core of much denser material and, by taking the atmosphere to be thick and at a relatively high temperature, introduced an outermost zone of low density. Models of uniform composition appear not to be so strongly condensed toward the centre as the actual planets, even as Jupiter, and to obtain sufficient relative increase of density, the atomic weight of the central core must be increased, whether by increasing the proportion of helium or by postulating a core of terrestrial material and in addition, or alternatively, there must be supposed to be an atmosphere of particularly low density. de Marcus and Peebles both conclude that the abundance of hydrogen in Jupiter and Saturn is about 80 per cent, or somewhat greater than the value (75 per cent) now usually taken for the abundance in the Sun.

de Marcus and Peebles both effectively supposed the planets to be cold in that the temperature was low enough not to affect the density, and they also supposed the hydrogen and helium to be completely mixed. Both assumptions have been abandoned in more recent work (Hubbard and Smoluchowski, 1973; Stevenson and Salpeter, 1975; Zharkhov and Trubitsyn, 1975*a*). Low (1966) found, as a result of infra-red obser- vations of Jupiter at a wavelength of about 20 μm, that the atmosphere of the planet was at such a temperature (about 150 K) that it emitted nearly three times as much energy as it received from solar radiation; if the emission and absorption were in balance, the temperature would be expected to be 105 K. A similar result was found for Saturn, and both results were confirmed (with a slight downward revision of the re- radiated power) by Aumann, Gillespie and Low (1969) who observed Jupiter and Saturn from a high flying aircraft. Direct observations from the Pioneer 10 and 11 space craft (Kliore, Woiceskyn and Hubbard, 1976; Opp, 1980) further confirmed the general result, but reduced the re-radiated power from Jupiter to about 1.9 times the incident solar radiation and that from Saturn to about 2.2 times the solar radiation, or 2.4 ± 0.8 W/m^2.

Hubbard (1968) concluded from the initial result of Low (1966) that the additional power radiated from Jupiter came from an internal source of heat, which would have to be much more powerful than any source of heating by radioactive decay as in the Earth. It has usually been considered by those who follow this argument that the heat is generated by the contraction of an extended cloud of hydrogen to the present size of Jupiter. Hubbard wrote the free energy of metallic hydrogen in the form

$$F = NE_0 + NkT[3 \ln (1 - e^{-\Theta/T}) - D(\Theta/T)],$$

where N is the number of protons, E_0 is the binding energy at zero temperature, Θ is the Debye temperature and D is the Debye function. The expression follows from the straightforward Debye theory for the thermal energy of a lattice. For alkali metals, Hubbard takes Θ to be

$$1.74(2Z/A)\rho^{1/2}$$

and argues that, because A, the atomic mass, is 1, Θ will be of the order of 10^3 K. He then argues that the actual temperature of the metallic part of Jupiter cannot be much less than Θ for, otherwise, the heat capacity would be very small and the thermal energy would have been radiated away long since, instead of still being radiated. He considered that the heat capacity was, instead, 3 K, and he further argued that the temperature gradient would everywhere exceed the adiabatic gradient so that Jupiter and Saturn would be fully convecting throughout. On this basis, Hubbard (1969, 1970) constructed a number of models of Jupiter and Saturn. Hubbard's work has run parallel to that of Zharkhov and Trubitsyn in the U.S.S.R., with whom he has collaborated (see Zharkhov, Trubitsyn and Hubbard, 1978), and, like theirs, has been characterized by an intensive study to determine internal properties and equations of state from the external gravity field.

Hubbard's work has been based on his adaption of the James and Ostriker approach to the theory of a rotating planet in hydrostatic equilibrium, as described above and as recently summarized (Hubbard, 1974*b*). Hubbard has emphasized that the equation of state of the mixture of hydrogen and helium in Jupiter is close to that of a polytrope of index 1, being well represented by $p = 1.96\rho^2$, and has studied in some detail the properties of a planet made up of such material (Hubbard, 1974*a*). He finds the following results:

$$J_2 = 0.173\,273m - 0.197\,027m^2,$$

$$J_4 = -0.081\,092m^2,$$

$$J_6 = 0.056\,329m^3,$$

where m is $a^3\varpi^2/GM$; the consequent values for Jupiter are

$J_2 = 0.013\ 93$,

$J_4 = -0.000\ 52$,

$J_6 = 0.000\ 039$.

They are smaller than the observed values, but close enough to suggest that the polytrope of index unity is a good model from which to derive more realistic ones by small perturbations.

Further model calculations have been made by Slattery (1977), using the thermodynamic calculations for a solar mixture of molecular hydrogen and helium of Slattery and Hubbard (1976) in which a Monte Carlo procedure was used. Zharkhov and Trubitsyn have based their calculations on their extension of Darwin's theory of the figure of a rotating planet, now extended to the fifth order (Zharkhov and Trubitsyn, 1975*b*). They also use simple laws of density variation for which analytical solutions are possible, namely a linear law in which

$$\rho = \rho_0(1 - r/r_s),$$

where r_s is the surface radius and ρ_0 is about 4000 kg/m^3, or a quadratic form.

If the radius of an equipotential surface is

$$r = a[1 - f\cos^2\theta - (\tfrac{3}{8}f^2 + k)\sin^2 2\theta$$
$$+ \tfrac{1}{4}(\tfrac{1}{2}f^3 + h)(1 - 5\sin^2\theta)\sin^2 2\theta],$$

then the coefficients f, k and h are functions of a/a_s:

$f = f_0 + f_1(a/a_s)$,

$k = k_1(a/a_s)^3$,

$h = h_1(a/a_s)^4$.

They give formal methods for relating $J_2, J_4 \ldots$ to deviations from the linear law of density (Zharkhov and Trubitsyn, 1974).

A variety of models, with a range of compositions, but all with cores of high density materials has been calculated by Zharkhov, Makalkin and Trubitsyn (1974) who find that in all such models the central temperature is of the order of 20 000 K.

Zharkhov, Trubitsyn and Makalkin (1972) have calculated the gravitational parameters for certain models, using their theory of the figure carried to the third order, with the results shown in Table 8.9. The linear and quadratic models are those of Zharkhov and Trubitsyn mentioned above. They, in general, give the poorest fits to observation, while de Marcus's models are as good as any.

All the foregoing models of Hubbard, Zharkhov and Trubitsyn are based on the interior temperature being very high and the material entirely liquid. Hubbard (1968) himself remarked that the excess infra-red power radiated by Jupiter did not necessarily entail an internal source of heat and hence a high temperature, but that the atmosphere might be heated in some other way. Such a way has indeed recently been proposed by von Zahn and Fricke (1977) who find that data from the satellite ESRO 4 show that the solar wind contributes appreciably to the heating of the Earth's atmosphere, and they estimate that the effect of the solar wind on Jupiter would be still greater in relation to heating by ultra-violet radiation. However, they appear also to have revised the equilibrium temperature due to ultra-violet heating upward to about 150 K, which appears to agree well with values obtained from Pioneers 10 and 11 (Kliore, Woiceskyn and Hubbard, 1976) and there accordingly now seems little evidence for excess radiation from Jupiter from an internal source. Thus an essentially cold planet looks more plausible than it has done over the past ten years or more. At the same time, there is good evidence that Jupiter and Saturn are liquid at least in part. The existence of magnetic fields (Chapter 9) is powerful evidence for liquid zones, supposing, as it seems we must at present, that the fields originate by dynamo action in liquid cores or shells. Then again, Goldreich and Soter (1966) have studied mechanical damping in bodies of the solar system. They have estimated Q-factors, that is, the ratios of the total energy stored in an elastically vibrating body to the energy lost per cycle of oscillation, and find them to be greater than 10^5 for Jupiter and greater than 6×10^4 for Saturn. These values are some 100 times greater than for the solid Earth and strongly suggest that Jupiter and Saturn contain large proportions of liquid. Nonetheless, solid zones do not appear to be

Table 8.9. *Summary of calculated gravitational parameters*

Model	$10^3 J_2$ Jupiter	$10^3 J_2$ Saturn	$10^4 J_4$ Jupiter	$10^4 J_4$ Saturn	$10^6 J_6$ Jupiter	$10^6 J_6$ Saturn
de Marcus (1958)	14.8	16.80	−5.87	−12.90	3.91	—
Peebles (1964)	15.53	17.01	−6.50	−11.2	4.29	—
Hubbard (1969)	15.33	—	−6.34	—	4.11	—
linear	15.02	24.91	−6.11	−16.7	3.89	—
quadratic	17.76	29.50	−7.94	−10.3	5.41	—

excluded although, in the absence of an adequate theory of the melting of metallic hydrogen, it is not possible to estimate in any reliable way the maximum temperature at which solid metallic hydrogen could exist.

The earliest studies of models with pure hydrogen showed that the properties of the major planets could not be matched without some admixture of heavier elements and in particular of helium, whether in solar proportions or, as some models, in rather a lower proportion than in the Sun. It was supposed in earlier work that the hydrogen and helium would be completely mixed, but Smoluchowski (1967, 1973) has studied theoretically the behaviour of helium in hydrogen and has concluded that there will be zones in which helium will separate from hydrogen. At sufficiently high pressures, helium in the presence of metallic hydrogen will ionize and behave like a divalent metal. Hydrogen is, of course, a monovalent metal and Smoluchowski argued that, when ionized, helium would alloy with hydrogen in the same way as divalent metals do with alkali metals. On the other hand, if the helium is not ionized, it would not be miscible with metallic hydrogen and would separate out (see also, Stevenson and Salpeter, 1975); a similar behaviour has been found experimentally by Streett (1976). Thus it seems that Jupiter and Saturn may consist of layers rich in helium and layers depleted in helium, as indicated in Figure 8.1. So far no quantitative details have been calculated from which numerical models can be derived, but Smoluchowski

Figure 8.1. Zones of miscibility and immiscibility of hydrogen and helium in a major planet.

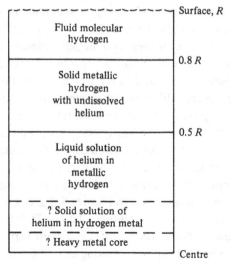

has suggested that the gravitational separation of immiscible liquids rich in hydrogen on the one hand and in helium on the other would provide an internal source of heat that would not entail a high internal temperature. The models of Jupiter and Saturn that have been generally investigated fall into three categories: models in which hydrogen and helium are fully mixed, but in which the composition changes with depth; models in which the ratio of helium to hydrogen is as in the Sun and in which the necessary central condensation is obtained with a core of denser material; and the models, so far just sketched, in which helium separates from hydrogen in some layers. Models with a core of denser material have been studied by Zharkhov, Makalkin and Trubitsyn (1974) and also by Podolak and Cameron (1974) who suppose the core to consist of metal silicates (rock) surrounded by an ice shell of methane, ammonium and water, and that, in turn, surrounded by a mixture of hydrogen and helium of solar composition.

Among the variety of models which may plausibly be constructed to fit the mass and gravitational parameters of Jupiter and Saturn, what may be picked out as common features of general validity? The first is that the major constituent is hydrogen which, on account of the high pressure, becomes metallic at some depth within the planet. Secondly, the metallic hydrogen is liquid somewhere within the planet, though not necessarily everywhere, so providing a means for the generation of the magnetic fields of Jupiter and Saturn. The third point is that both planets are more condensed than a uniform composition would predict, and Saturn is far more so than Jupiter. Objections have been raised that no mechanism has been suggested for changing the ratio of hydrogen to helium as required in the models of de Marcus and Peebles, and so models which have a denser core surrounded by a hydrogen and helium mixture in solar proportions may be thought to be the most likely. On this view, Saturn would have a relatively larger core than Jupiter. Furthermore, this model might be thought more reasonable than the immiscibility model of Smoluchowski, in that, to produce the greater central condensation in Saturn, the latter model would seem to entail greater separation of helium from hydrogen in Saturn than in Jupiter, whereas, with the greater range of pressure, and probably temperature, in Jupiter, the reverse would seem more likely. Indications of the variation of density and composition with radius for 'typical' models are given in Figures 8.2 and 8.3.

From time to time it has been suggested (see Chapter 7) that metallic hydrogen might be a superconductor (Ashcroft, 1968) with a critical temperature high enough for it to be superconducting in Jupiter (de Cesare,

1974). Evidently. this is more likely if Jupiter is relatively cold than if it is in the fully convecting state, and, in view of the reassessment of the atmospheric heating of Jupiter, indicated above, it may be that the possibility of a superconducting zone does require serious consideration and that attention should be paid to the effect of such a zone, which presumably would not extend to all the metallic hydrogen, upon the magnetic field. It is also worth speculating that protons in metallic hydrogen might be paired, giving rise to superfluid behaviour, and, if so, asking how that would affect the interpretation of the dynamical properties.

Figure 8.2. Typical model of Jupiter.

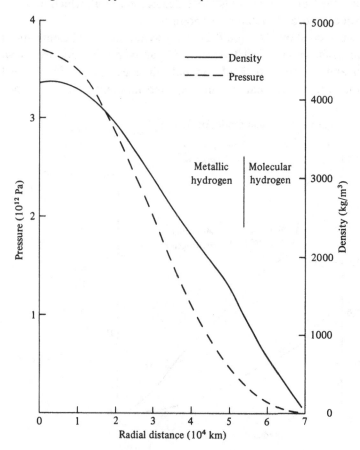

8.5 Model of Uranus and Neptune

Uranus and Neptune form a natural pair in that they are more similar to each other than they are to Jupiter and Saturn. The sizes and masses are both comparable, but Uranus is larger in volume, but less in mass, than Neptune, and so its mean density is less, and indeed less than that of Jupiter. The central pressures are of the order of those at the centre of the Earth rather than at the centres of Jupiter or Saturn and thus self-compression must be less than in Jupiter and Saturn. Consequently, the mean atomic mass of the material of Uranus and Neptune must be greater than that of hydrogen or helium (Ramsey, 1951) and it has generally been supposed, at least since the studies of Ramsey (1963), that those planets are composed of hydrogen and helium with admixtures of carbon, nitrogen and oxygen in greater than solar proportions. It is clear from the fact that Neptune is more massive than Uranus although smaller, that Neptune contains more heavier material, which may be in the form of a core of metallic compounds.

As was seen above (section 8.2), the gravitational coefficients and the dynamical and geometrical flattening of Uranus and Neptune are not well known and, indeed, some of the published estimates are physically impossible. It appears that the dimensionless moment of inertia, C/Ma^2,

Figure 8.3. Typical model of Saturn.

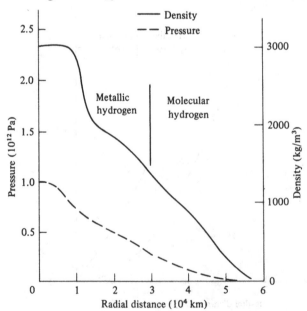

Radial distance (10^4 km)

is quite possibly of the order of 0.34 for Neptune and, less plausibly so, for Uranus. This is a value which can be reproduced with a suitable mixture of hydrogen, carbon, oxygen, nitrogen and neon, as shown by the calculations of Ramsey (1963). Ramsey assumed that Uranus and Neptune would be formed of saturated compounds of hydrogen with carbon, oxygen and nitrogen, together with neon, the mixture of methane, water, ammonia and neon being chosen so that the carbon, oxygen, nitrogen and neon atoms were in their cosmic proportions. He then calculated the pressure–density relations for these compounds and for the mixture, which he called CHONNE. Methane is a nearly spherical molecule which interacts only through repulsive forces and a weak van der Waals attraction. Ramsey assumed a Lennard-Jones type of potential and determined the constants from thermodynamic properties of the gas at low pressures (great separations) and for molecular scattering experiments at small separations. He then found that the indices m and n in the Lennard-Jones potential

$$\frac{A}{r^n} - \frac{B}{r^m}$$

were $n = 6$, $m = 13.75$, and, from those and the constants A and B, he could calculate the pressure–density relation for methane. The properties of neon were derived in a similar way from similar data. The properties of water were derived from shock-wave experiments (see Chapter 4).

Ramsey supposed that ammonia combined with hydrogen at high pressures to form ammonium, which, following the calculations of Bernal and Massey (1954), he took to be metallic. While Bernal and Massey's calculations of the metallic form are not unsatisfactory, Stevenson (1975) has calculated the properties of the molecular mixture $NH_3 + \frac{1}{2}H_2$ and found a considerable discrepancy; the energy is less than that of the metallic form as calculated by Bernal and Massey, which itself is somewhat less than that calculated by Stevenson, at pressures up to 10^{11} Pa. Stevenson speculates indeed that, if metallic ammonium exists, the transition pressure will be appreciably greater than 10^{11} Pa. Whether ammonia is in the metallic or molecular form does not, however, greatly affect the properties of CHONNE for it is not a large proportion (11 per cent) of the mixture. Indeed, the behaviour is dominated by water which forms 66 per cent by mass, and the properties of water are obtained from experiments up to 4×10^{10} Pa and extrapolated beyond that by the Bullen type of formula for the bulk modulus

$$K = K_0 + bp,$$

where, for water, $K_0 = 6.25 \times 10^9$ Pa and $b = 4.2$. Water is thus considerably less compressible at high pressures than is hydrogen, for which b is 2; values of b of the same order are also found for neon and methane (Ramsey, 1963; Cook, 1972). Indeed, water, neon and methane are less compressible at high pressures than the lower mantle and core of the Earth, for which b is about 3 (Table 8.10).

Now Ramsey's calculations for the CHONNE mixture give values of C/Ma^2 of about 0.36 for models that assume a radius of about 21 000 km. C/Ma^2 does not change much with radius, indeed it increases slightly, so that a CHONNE model with a radius of 25 000 km would have C/Ma^2 very close to 0.36 compared with a possible value of about 0.34. While Ramsey's CHONNE models give a plausible value of C/Ma^2, at least for Neptune, the densities are far too great, of the order of 2400 kg/m^3 for a radius of 21 000 km. A density much less than that of the CHONNE mixture is required to match the actual densities of Uranus and Neptune and, because the density of methane is about half that of water over the range of planetary pressures, it would seem that a model composed to a large extent of methane would fit the properties of Neptune to first order.

Uranus is yet less dense than Neptune so that a large proportion of methane is needed, while, if its value of C/Ma^2 is indeed as low as 0.25, a very dense core would be needed to give the high increase of density towards the centre. Apart from considerations of C/Ma^2, it is clear that Uranus and Neptune cannot be represented by the same type of model, for Uranus is the larger planet, yet has the lower mass and density. The

Table 8.10. *Densities of water and methane* (from Ramsey, 1963)

Pressure (10^{11} Pa)	Density (kg/m^3)	
	Water	Methane
0	1000	522
0.1	1650	1020
0.5	2320	1340
1.0	2730	1510
2.0	3220	1690
5.0	4000	1960
10.0	4710	2190
15.0	5190	2340
20.0	5560	2450

overall mean atomic weight of Neptune would therefore seem to be greater than that of Uranus, despite a possibly dense core in the latter. Podolak and Cameron (1974), as already mentioned in the previous section, considered models with a core of rock surrounded by a shell of solid methane, ammonia and water and that, again, by hydrogen and helium in solar proportions, while Podolak (1976) has constructed models of Uranus in which the composition is dominated by methane. With the dynamical properties as imperfectly known as they are, a wide variety of models is possible. It is possible to say two things with some assurance: first, the composition is dominated by materials heavier than hydrogen and helium and lighter than the materials of the terrestrial planets and, thus, considering cosmic abundances, comprising to a large extent some mixture of water, ammonia, methane and neon; secondly, because Neptune is smaller but more massive, it must have a larger proportion of dense material than Uranus, despite the presence of a larger rocky core in Uranus. On the other hand, it does not seem necessary to postulate the presence of a shell of hydrogen and helium, because the lower compressibility of methane and water (Table 8.10) is better matched to the values of C/Ma^2 as at present known.

8.6   Conclusion

The major planets form two pairs, Jupiter and Saturn, and Uranus and Neptune, as judged by size, mean density and, probably, central condensation. Jupiter and Saturn must be composed in large proportions of hydrogen and helium, Uranus and Neptune probably of methane.

Even before the detailed calculations on the properties of metallic hydrogen by de Marcus and subsequent authors it had been shown by Ramsey (1951) and Miles and Ramsey (1952) that the proportion of hydrogen in Jupiter exceeds that in Saturn, the exact amount of hydrogen depending on the way in which helium and heavier elements are distributed with radius. For this reason, it is difficult to decide whether any of the four planets has a rocky core. It is possible to produce satisfactory models of Jupiter and Saturn which differ in the distribution of helium with radius. If such a variation is thought to be implausible, then the differences between Jupiter and Saturn may be accounted for by a larger rocky or heavy metal core in Saturn than in Jupiter.

Similar considerations apply to Uranus and Neptune, although here there is more scope for speculation in view of the lack of definite knowledge of the flattening. If Uranus is predominantly composed of methane then Neptune almost certainly has a larger proportion of a heavier material such as water. Evidently more definite dynamical data for Uranus and Neptune are much to be desired.

While it is natural to group Jupiter and Saturn as a pair, and Uranus and Neptune, it is at the same time clear that there are sharp differences between them which make it impossible to represent the planets of a pair by the same model. There is one prediction that can be made if the foregoing ideas about the constitutions of the major planets are realistic. Jupiter and Saturn have magnetic fields which are currently thought to be generated by dynamo action in liquid metallic hydrogen. Now that it appears that ammonia, even if present in Uranus and Neptune, would not be in the metallic form at the pressures attained in those planets, the material of Uranus and Neptune would not seem to have metallic conductivity and thus there is no scope for dynamo action within them. Uranus and Neptune therefore would not be expected to have magnetic fields.

9

Departures from the hydrostatic state

9.1 Introduction

The models of the planets which have been adopted so far depend explicitly on the assumption that the planet is in the hydrostatic state, so that the density is a function only of radial distance (in a generalized sense when the planet is flattened by spin). That may be an appropriate first assumption to provide a starting point for further developments, but it is clearly not adequate: the gravity fields of the Earth, the Moon and Mars contain harmonic components that would be absent if the internal state were hydrostatic; the irregular surface features of the terrestrial planets are inconsistent with strict hydrostatic equilibrium; and the structure seen in the atmosphere of Jupiter reveals internal motions, if only superficial. A density distribution not in hydrostatic equilibrium requires a stress system to support it that departs from the simple normal pressure to which hydrostatic equilibrium corresponds. Such a stress system may be developed in two ways: statically, through strains of the planet, or dynamically, through movements of the material. According to which mode is effective, so the planet may be considered to be cold or hot (though, as has already been argued, no planet is hot in relation to the effect on the equation of state). If the planet is cold, the materials within it have high strengths, can support large stresses and so maintain statically non-hydrostatic distributions of density. If the planet is hot, then parts will be molten, as is the core of the Earth, or will be sufficiently hot to creep steadily under applied stress. In either case, motions will generate stresses that may support non-hydrostatic distributions of density. Such non-hydrostatic distributions are considered in this chapter from the point of view of what they may reveal about the internal structure of the planet. They are, of course, of great interest in their own right, whether we consider the features of the surface, the motions of the surface or the magnetic fields of such planets as possess them, but here we seek to use them as evidence for the state of the interior.

To recapitulate, the planets are cold in the sense that the density is primarily determined by pressure, together with chemical composition, and hardly at all by temperature, which, at planetary pressures, has only a minor effect upon the density. Thus, at a temperature of say 3000 K and a pressure of 10^{11} Pa, conditions corresponding to the boundary of the core in the Earth, the effect of temperature on the density is a few parts in a thousand at the most, whereas the compression is about 20 per cent. The pressures of self-gravitation and rotational acceleration thus determine the gross structure of planets and lead to a distribution of density which, overall, corresponds to the hydrostatic state. Superposed on the general hydrostatic state are variations of density, which presumably arise in one of two ways. Either initial irregularities in the formation of the planet have been maintained because the material is cold enough and, so, strong enough to support them, or internal sources of heat exist, which generate motions which lead to variations of density. In either case one would expect the departures from hydrostatic equilibrium to be small; in the first case, the strength of materials is small compared with planetary pressures, except for the Moon which is far from being in the hydrostatic state; in the second case, convective motions driven by differences of density arising from a temperature gradient are such as to reduce the variations of density.

9.2 Surface features

The surfaces of the Earth, the Moon, Mars and Mercury are now all well known and some indications of the surface structure of Venus have been obtained. Perhaps the most significant conclusion to be drawn from this information is that the Earth is quite different from all the others. All show considerable variations in relief, these being most pronounced in Mars, but, whereas processes generating relief are still active on the Earth, they must have ceased long since on the other planets and the Moon, for the surfaces of all of them are covered with craters produced by the impact of meteorites, some of these craters dating from a very long time ago. The other big difference between the Earth and the other bodies is that 70 per cent of the surface of the Earth is covered by the oceans, whereas liquid water is absent from the others.

On further inspection one sees that the character of the relief of the Earth is different from that of the Moon and other planets. The surface of the Earth, as we now appreciate, is divided into a few plates, which are relatively rigid so that displacements within them are small, but which more relative to one another at rates of a few centimetres per year. The plates, in general, comprise oceanic and continental sections. Plates are built up by volcanic activity along submarine ridges which, in terms of relief and length, are, after the continents and oceans, the major features of the topography of the Earth, although hidden by the waters of the oceans. The constancy of the surface area of the Earth requires that plates should also be destroyed, as takes place at the margins of continents with the formation of mountains. Earthquakes and volcanoes occur almost entirely along mid-oceanic ridges or along mountain chains. A rather important aspect of the tectonic activity of the Earth cannot at present be fitted naturally into the plate tectonic pattern, namely, vertical, or eustatic movements of continental structures on a large scale. The existence of such movements is well known, and some are related to accumulation or disappearance of ice caps, but others seem to be related to plate motion in ways not yet understood.

Two factors seem to determine the tectonic behaviour of the Earth. In the first place, temperature increases inwards relatively rapidly (30 deg/km near the surface) and in consequence only the outermost shell of the Earth, the lithosphere, is strong enough for sections of it to withstand internal stresses and move as a whole. At greater depths, in the asthenosphere, it is generally considered that the material is weak enough to deform in a quasi-liquid manner under stress, so that it allows sections of the lithosphere to move over it, or alternatively carries them with it as it moves itself. The cause of the motions of the plates is not known, except in so far as one can say that there must be a source of energy within the Earth that drives them through the flow of energy outwards through the Earth. It may be that the energy is heat energy, the sources of which are the heat generated by the radioactive nuclides in the Earth and that stored in the Earth since its formation.

The second factor which determines the tectonic behaviour of the Earth is the oceans. The rocks of the floors of the oceans form under some 3–5 km of the ocean and, undoubtedly, the characteristic of the ocean floor is in part determined by the cooling of volcanic rock under water; in particular, the thickness of the oceanic crust, that part which is chemically differentiated from the upper part of the mantle, is so determined. The oceans also play a part in the formation of mountains. In the first place, mountains form from sediments which have been accumulated under water. In the second place, mountains and the chemically differentiated continental crust are in isostatic equilibrium with the oceanic crust and, thus, in some indirect way, not as yet understood, the thickness of the continental crust is fixed by the thickness of the oceanic crust. Thus we may conclude that the major differentiation of the crust into continental and oceanic type is determined by the presence of the water cover on the Earth (see Cook, 1979).

The oceans affect the tectonics of the Earth in another important way, for they are the reservoir for the water which is taken up into and precipitates from the atmosphere and leads to erosion of the high land and the return of the material as sediment to the oceans where it can participate again in the formation of mountains and continental crust. It is clear that the continuation of tectonic activity, at least in the form in which it has occurred since the late pre-Cambrian, depends on the existence of the oceans.

To summarize, the tectonic activity of the Earth, at least as it is recorded for the past 2000 My, depends on two factors, the oceanic cover, on the one hand, and, on the other, the existence of a source of energy in the Earth which both provides the power to drive the tectonic activity and, through a fairly steep thermal gradient, ensures that the rigid surface layer, the part which forms the plates, is not more than about 100 km or so thick.

The surface appearances of the other planets are dominated by quite different structures from those that dominate the Earth. The Moon, Mars and Mercury are all covered by craters formed by the impact of meteorites, craters which on the Moon range in size from tens, and perhaps hundreds, of kilometres across down to a few micrometres. The form of the distribution of number of craters against size appears to be the same over a very wide range of size. Photographs of Mercury taken by Mariner 10 show much the same characteristics (see Strom, 1979). Those features of the planets which are of indigenous origin have to be looked for beneath the cover of craters. It is, of course, fairly obvious why craters are not evident on the Earth. In the first place, the smaller meteorites burn up through frictional heating in the atmosphere of the Earth, as no doubt also happens on Venus. Only the largest meteorites now penetrate the atmosphere with sufficient velocity to produce craters. In the past, the flux of meteorites is supposed to have been much greater, and the relics of some of the larger ancient craters have been detected, but, in general, 70 per cent of all meteorites will have fallen in the seas, and, of the craters formed on land, most are degraded by erosion or filled in with sediment.

Below the cover of craters, what appears on the Moon, Mars and Mercury? The dominant features appear to be volcanoes. The largest relief in the terrestrial planets is that of the Tharsis group on Mars (Mutch *et al.* 1976), which is apparently a great central volcano. Gravity measurements show that it is not isostatically compensated as is all terrestrial relief, but is supported by the strength of the material of Mars. Again, on the Moon, the major features are large areas of volcanic flows which lie above local concentrations of mass, or mascons, revealed by the disturbance of the gravity field and thus also apparently supported by the strength of the material. Mercury also appears to have volcanic structures below the crater cover. It is possible that all the tectonic features of the Moon and Mercury could have been formed by volcanic activity, but there is evidence from the extreme relief of Mars which is, in parts, unconnected with volcanic features, that other processes may have operated in that planet.

It has proved possible to estimate the ages of major features of the Moon and Mars by counting craters of different sizes upon them. All the major features of the Moon are of the order of 3×10^9 y old. Studies of Mercury have not as yet progressed so far as those of the Moon and Mars, but it seems that there, also, there is evidence of volcanic activity which ceased at an early stage in the history of the planet.

What may be said with assurance about the Moon, Mars and Mercury is that any substantial tectonic and volcanic activity ceased some considerable time ago, as shown by the craters covering all features of indigenous origin. Thus the source of energy for tectonic activity in those bodies expired long ago, whereas it is still present in the Earth. The other conclusion is that no plate structure is seen. Isolated volcanoes occur, but are not distributed in a systematic way as are those on the Earth. One may ascribe this, first, to the absence of oceans and, secondly, perhaps, to the temperature within the planets increasing inwards more slowly than in the Earth so that the rigid outer layer is much thicker. The conclusion to which one is led is that the temperatures within those bodies were never as great as in the Earth, although at an early period there was sufficient energy to drive volcanic activity. In so far as this conclusion may be related to the fact that the Earth is larger than the other bodies, it is unfortunate that, as yet, we know little about Venus. Is there evidence of continuing tectonic activity on that planet? There is some indication from infra-red photographs as well as from radar observations (Timber and Kirk, 1976) that the surface has craters and so may not be currently active. The question we would like to address is, whether the lower energy in the Moon, Mars and Mercury is due entirely to their small sizes, or is there a difference of composition and a lower radioactive content? If

Venus, of much the same size as the Earth, is inactive, it may be because the Earth had originally more radioactive material than the other planets. Of course, with no oceans, one would not expect Venus to show the same type of tectonic activity as the Earth, but, if it has a similar internal source of energy, one would expect it to have a thin lithosphere and a currently active surface.

Related to the question of sources of energy within the planets is the power of seismic activity. In the Earth, along mountain chains and the mid-oceanic ridges, there are continual earthquakes with, occasionally, some very large ones, all driven apparently by stresses that accumulate as the tectonic plates move. Seismic activity on the Moon, as revealed by the seismometers placed there in the Apollo programme, is very many orders of magnitude less than on the Earth, so much so that the stresses which control it are, to a large degree, of external tidal origin instead of being of wholly internal origin, as in the Earth. A single seismometer has been placed on Mars by the Viking Lander and has similarly shown very low activity, confirming, as with the Moon, that tectonic activity such as occurs on the Earth is negligible.

The surface features of the terrestrial planets, when compared with those of the Earth, thus lead us to conclude that the other planets have at the present time a much smaller internal source of energy than the Earth, and, also, that, being colder, the strength of the material can support larger stress differences and thus larger departures from hydrostatic equilibrium than in the Earth. We now look at the potential fields of the planets, the gravitational and magnetic fields, to see how far those conclusions are supported.

The major planets have fluid surfaces and so do not show permanent features as do the Moon and the terrestrial planets. However, Jupiter, in particular, shows features that bear witness to motions in the outer parts, that is, departures from the hydrostatic state, and thus to an internal source of energy driving the motions.

9.3 The gravitational fields of the planets

The gravitational field of the Earth is very well known indeed from analyses of orbits of artificial satellites as well as from measurements of gravity on the surface, and the field of the Moon and, to a lesser extent that of Mars, is well known from analyses of orbits of orbiting space vehicles, as already discussed to some extent in Chapters 3 and 6. It has also been seen that the knowledge of the fields of Mercury and Venus is confined to upper limits on the values of the harmonic coefficient, J_2.

Even for the Earth, the higher harmonic coefficients of the potential are not individually well determined, mainly because the number of possibly significant coefficients is greater than the number of distinct satellite orbits, and more or less arbitrary assumptions have to be made about their behaviour (as, for example, that those above a certain degree may be taken to be zero) for estimates of their magnitude to be made. The situation is worse for the Moon and Mars where the orbits of very few orbiting vehicles are available. Thus it is unlikely to be profitable to attempt any interpretation of individual harmonic coefficients, although it is known that the gravity field does reflect major topographic features, such as the large *maria* on the Moon (through the mascons) and the Tharsis volcanoes on Mars. There may be, however, some hope that statistical studies may be useful and, in particular, that reliable conclusions may be drawn from the way in which harmonic coefficients vary in general with degree.

It is well known that the magnitudes of the harmonic coefficients of the potential of the Earth, the Moon and Mars decrease with degree, and W. M. Kaula some while ago gave a rule which is closely followed by the terrestrial coefficients. If, in the usual notation, C_{lm} and S_{lm} are the non-dimensional coefficients of the normalized harmonics, $P_l^m (\cos \theta) \cos m\lambda$ and $P_l^m (\cos \theta) \sin m\lambda$, then Kaula's rule is (see Gaposchkin, 1973):

$$\frac{1}{2l+1} \sum_{m=0}^{l} (C_{lm}^2 + S_{lm}^2) = 10^{-5} l^{-2}.$$

The coefficients of the Moon and Mars have been fitted to a similar rule. An alternative comparison is, however, instructive, in which a specific hypothesis about the origin of the anomalies of the field is adopted. That hypothesis is that the anomalies are the potential of a random distribution of density variations, ρ, at some radius, r, less than a, the planetary radius. If ρ is expressed as a mass per unit area, then coefficients of the spherical harmonics in the potential are

$$\frac{a^2}{M} \left(\frac{r}{a}\right)^l \int_S \rho P_l^m (\cos \theta) \begin{array}{c} \cos \\ \sin \end{array} m\lambda \; dS,$$

where the integral is taken over the surface of the unit sphere and M is the mass of the planet.

If ρ is random and uncorrelated and if ρ^2 has a variance $\mathrm{var}(\rho)$ it follows that the expected values of C_{lm}^2 and S_{lm}^2 are given by

$$E(C_{lm}^2, S_{lm}^2) = \frac{a^4}{M^2}\left(\frac{r}{a}\right)^{2l+4} \mathrm{var}(\rho) \int_S [P_l^m(\cos\theta)]^2 \frac{\cos^2}{\sin^2} m\lambda \; dS$$

$$= \frac{4\pi a^4}{M^2}\left(\frac{r}{a}\right)^{2l+4} \mathrm{var}(\rho),$$

by the normalization convention for the spherical harmonics. Thus

$$\ln E(C_{lm}^2, S_{lm}^2) = (2l+4) \ln\left(\frac{r}{a}\right) + \ln\left(\frac{4\pi a^4 \mathrm{var}(\rho)}{M^2}\right)$$

so that a plot of $\ln E(C_{lm}^2, S_{lm}^2)$ against $2l$ should give a straight line from which r/a and $\mathrm{var}(\rho)$ may be determined.

Table 9.1. *Mean square coefficients of potential for the Earth, the Moon and Mars*

l		Values of $\dfrac{1}{2l+1} \sum_{m=0}^{l} (C_{lm}^2 + S_{lm}^2)$		
		Earth $\times 10^{16}$	Moon $\times 10^{12}$	Mars $\times 10^{12}$
2		—	1945	—
3		6805	233	419
4		2880	116	37.6
5		1536	38.3	10.1
6		583	159	3.40
7		372	111	—
8		245	26.8	—
9		79.7	12.9	—
10		67.2	3.6	—
11		61.1	18	—
12		31.4	3.6	—
13		24.8	3.4	—
14		17.3	8.4	—
15		15.3	9.8	—
16		17.7	19.4	—
17		12.1	—	—
18		8.52	—	—
r/a	a	0.68	0.8	0.46
	b	0.89	—	—
$\sigma(l)(\mathrm{kg/m^2})$	a	6.6×10^4	1.03×10^5	2.66×10^4
	b	4.0×10^3	—	—

a Low order terrestrial harmonics ($l \leqslant 9$).
b High order terrestrial harmonics ($l \geqslant 9$).

Values of $E(C_{lm}^2, S_{lm}^2)$ have been calculated from the solutions for the Earth, the Moon and Mars given, respectively, by Gaposchkin (1973), Ferrari (1977) and Gapcynski, Tolson and Michael (1977); as above

$$E = \frac{1}{2l+1} \sum_{m=0}^{l} (C_{lm}^2 + S_{lm}^2).$$

The results are listed in Table 9.1 and shown diagrammatically in Figure 9.1. The second degree harmonics have been excluded from the lists for the Earth and Mars because the zonal harmonic is determined by the spin of the planet and not by azimuthal irregularities of density. The second degree harmonics of the Moon are, however, listed because the hydrostatic spin contribution to the zonal harmonic is small.

Figure 9.1. Dependence of r.m.s. coefficient of harmonics in gravitational potentials upon the degree of the harmonic for the Earth, the Moon and Mars.

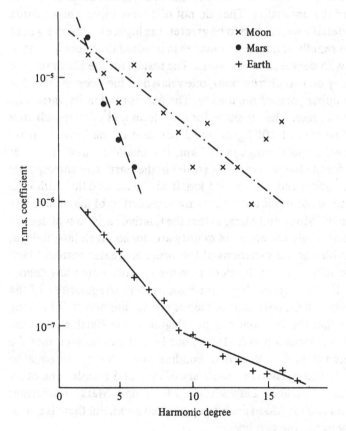

The data for Mars are, of course, much less extensive than for the Earth and the Moon. It must also be remembered that the uncertainties of many of the listed harmonic coefficients are large and themselves uncertain. Thus many of the lunar coefficients in Ferrari's solution are much less than the uncertainties that he quotes. Even so, all sets of data seem to follow linear laws according to the expression derived above. Two different regimes appear to hold in the Earth, one for harmonics of lower order with a large slope, implying the possibility of deep sources, and the other for harmonics of high order, indicating the need for supposing shallow sources. The results for the Moon are very scattered, possibly reflecting the unreliability of the data, but also possibly reflecting the inapplicability of the model to the Moon; that is, density variation may be distributed more generally throughout the Moon than would be implied by the assumed surface distribution.

Table 9.1 gives the relative radius r/a as found from the slope of the fitted line and also gives the density per unit area found from the magnitude of the anomalies. They do not of course represent a unique physical model. If r were taken to be greater, the higher harmonics would die away less rapidly and their amplitudes in ρ would therefore also have to decrease with degree, and vice versa. The results for the Earth suggest that there may be two distributions, one related to the deep mantle, the other to the upper parts of the mantle. The necessary density variations are rather moderate. The largest is for the Moon and corresponds to a variation of density of $100\,\text{kg/m}^3$ (or 3 per cent of the lunar density) extending over a radial range of $100\,\text{km}$; the smallest variation is that responsible for the harmonics of high order in the Earth and corresponds to $100\,\text{kg/m}^3$ over a radial range of $1\,\text{km}$. It will be noticed that, although all the terrestrial harmonic coefficients are some orders of magnitude less than those in the Moon and Mars, so that the relative variations of density are less, the absolute variations of density are not so much less; indeed, that responsible for the variations of low order is greater than in Mars. Two other conclusions are of interest. In the first place, the low degree line for the Earth suggests that contributions from irregularities of the boundary between the core and the mantle are not important. It has long been known that the harmonics of low degree in the Earth could arise from such irregularities (Cook, 1963), but here it will be seen that the lowest harmonics fit on a line corresponding to a value of r/a equal to 0.68, which is well outside the boundary of core and mantle. The other feature is the difference between the Moon and Mars. Similarities between Mars and the Moon are often remarked upon, but there is a clear difference between the two bodies.

A certain amount is known about the gravity field of Jupiter, from the fly-bys of the Pioneer 10 and 11 space craft as well as from long standing observations of the Jovian satellites, but so far the only harmonics definitely estimated are the even zonal harmonics, J_2, J_4 and J_6, which are consistent with hydrostatic equilibrium (Chapter 8).

9.4 The thermal states of the planets

While the temperature within the planets is the key to the questions discussed in this chapter and influences, to a much lesser degree, the structure of spherically symmetrical models, almost nothing is known of the thermal conditions within any planet. Indeed, it may be said that we know just the surface temperatures of the planets (determined by the balance between heating by solar radiation and cooling by radiation from the surface) and, besides those, just two other facts, namely, the rates at which heat flows out through the surface of the Earth and the Moon.

Attempts have been made for many years to estimate the flow of heat through the surface of the Earth by combining measurements of temperature gradients in boreholes with values of the thermal conductivity of the surrounding rocks. There are many difficulties in the way of that procedure, especially those connected with long term variations of the surface temperature and with heat convected past the borehole by flow of water, and it turns out that it is more straightforward to estimate the flow of heat through the floor of the ocean from measurements of temperature gradient and conductivity in the sediments at the sea-bottom.

Thus, despite the difficulties and expense of working at sea, the heat flow through the floor of the oceans is better known than that through the continental surface. The flow is not the same everywhere, and there is a strong relation to major tectonic features; in particular, the flow is larger from the mid-oceanic ridges. Nonetheless, a worldwide average is not too difficult to establish and is close to 0.06 W/m^2.

The heat flow through the surface of the Moon was estimated at two of the Apollo sites from measurements of heat flow and conductivity with probes driven into the surface. The two estimates are (Schubert, Young and Cassen, 1977)

Apollo 15 site: 0.022 W/m^2
Apollo 17 site: 0.016 W/m^2.

The average, so far as an average of two results can represent the global value for the Moon, is 0.019 W/m^2, about one-third of that for the Earth.

The surface flow of heat from the Earth is thus about thrice that from the Moon, but that is not the case on a volumetric basis. The ratio of the radius of the Earth to that of the Moon is about 3.7 to 1, and so that of the areas is about 13 to 1, and, consequently, the total rate of flow of heat out of the Earth is some 42 times that of the Moon. However, the volume of the Earth is about 49 times that of the Moon and, consequently, the loss of heat per unit volume from the Earth is effectively the same as that from the Moon. What does this conclusion entail?

Knowing nothing about the thermal state of the Earth or the Moon, we might adopt one of two extreme positions. On the one hand, the Earth and the Moon are in thermal equilibrium, so that the rate at which heat flows out of each is equal to the rate at which heat is produced in them by decay of radioactive nuclides. We should then conclude that the radio-active heat production in the Earth is, per unit volume, the same as that in the Moon. Alternatively we might suppose that the Earth and the Moon are cooling down from a state in which they were both at the same high temperature throughout. Suppose also, for we have no basis for any other supposition, that the thermal diffusivity, K, equal to $k/c\rho$ (k is the thermal conductivity, c the heat capacity and ρ the density) is the same for both bodies and has a value of about 7×10^{-7} m^2/s (Carslaw and Jaeger, 1959). The characteristic time for cooling of a sphere of radius a is a^2/K, which takes the following values:

Earth: 1.8×10^{12} y,
Moon: 1.4×10^{11} y.

Both times are much greater than the age of the Earth. For both bodies the solution to the equation of heat conduction therefore takes the approximate form (Carslaw and Jaeger, 1959)

$$T = 2aT_0 \sin(\pi r/a) \exp(-\pi^2 t/\tau),$$

where T_0 is the initial temperature and τ is equal to a^2/K. Now t, the time since the origin of the solar system, is about 5×10^9 y so that $\pi^2 t/\tau$ is about 0.02 for the Earth and about 0.35 for the Moon. In each case the heat flow through the surface (a rather better approximation for the Earth than for the Moon) is $2T_0/a$. Thus, the heat flow through the surface of the Earth would be expected to be less by a factor of 4 than that through the surface of the Moon. Allowance for the smaller value of a^2/K for the Moon would reduce the factor by about 2.5.

It seems clear from these comparisons that the thermal time constants of the Earth and the Moon are very much shorter than would be expected if heat transfer took place by conduction, and we conclude that some more effective process of transfer takes place throughout the major part of each body, so effective, in fact, that the flow of heat through the surface is currently in equilibrium with the rate of production in the interior. McKenzie and Weiss (1975) came to the same conclusion from consideration of the physics of the Earth by itself. It should be noted that the coincidence of the apparent volumetric rates of heat production in the Earth and the Moon does not preclude other solutions. What is excluded is that both bodies have cooled from the same initial temperature over the same time; they might have reached the present state by cooling from different initial temperatures over different times with different rates of internal heat generation. It is now commonly supposed that the mechanism of heat transfer is convective transport in a 'solid' material undergoing creep (see, for example, Schubert, Young and Cassen, 1977). Unfortunately, the theory of such processes is difficult and has been little explored, and it is not possible to make calculations of the rate of transport of heat from first principles. Tozer (1972) has pointed out that, in a sense, it is not necessary to do so. His argument is that the temperature within a planet will rise to such a value that the rate of creep under buoyancy forces is sufficient to transport the heat being produced by radioactive sources, or otherwise. While this argument has great force, it does not provide a specification for calculating the distribution of temperature within a planet.

For the purpose of understanding the dynamical state of the interior of a planet, it is necessary to know how much thermal energy is available to drive convective motions and also what the distribution of temperature is. If it is supposed that the Earth and the Moon are representative of all the terrestrial planets in having the same rates of heat production per unit volume, then it is possible to say that the heat production in Mars exceeds that of the Moon and is less than that in the Earth and so one might expect the tectonic activity to be intermediate between the Earth and the Moon, as indeed it appears to be; similarly, the heat production in Mercury would be expected to be close to that of the Moon and tectonic activity, correspondingly, at a low level. It must, however, be emphasized that these arguments take no account of how the radioactive sources are distributed through a planet; there is strong reason for thinking that in the Earth there is considerable concentration towards the surface and it may be that the same is true of other planets.

Although the temperature distribution within a planet cannot be calculated from just the known heat flow, it was seen in Chapter 5 that it is possible to estimate it to some extent from the distribution of electrical conductivity, the latter being found from the character of currents induced in the planet by external electromagnetic fields.

The heat production and distribution of temperature in the major planets may be expected to have some effect on their internal constitutions, but are even more uncertain than they are in the terrestrial planets. Observations of infra-red radiation from Jupiter and Saturn (Chapter 8) show that they radiate up to twice as much heat as they receive from the Sun as ultra-violet radiation, so that there are some other sources associated with them. If those sources are internal radioactive heating, the central temperatures are estimated to exceed 5000 K; however, as was seen in Chapter 8, the source of heat may be the solar wind, in which case no inference can be made about the internal temperature.

9.5 Magnetic fields

Among the striking properties of planets are their magnetic fields, the characteristics of planets with perhaps the widest spread, for the magnetic moments range from 1.4×10^{20} T m^3 for Jupiter to 2.4×10^{12} T m^3 for Mars. It seems clear that, for the Earth at least, and no doubt for Jupiter, the existence of a magnetic field is evidence for internal motions, for it is now no longer possible to suppose that the Earth's field arises from permanent magnetization of the interior of the Earth. Irrespective of arguments about the thermal state of the interior of the Earth, it is clear from the surface heat flow that the Curie temperatures of ferromagnetic minerals are attained at very modest depths within the Earth, no more than about 20 km, and that the amount of ferromagnetic material within the crust of that thickness is inadequate to provide the observed magnetic moment. Furthermore, the field is variable. It shows a secular variation with a time constant of about 500 y and reversals of polarity at a mean interval of about 2×10^5 y (Bullard, 1968), variations that are inconsistent with a source in a rigid crust. Thus the origin of the Earth's field is now sought in dynamo action in motions of the conducting fluid core of the Earth. The possibility that the Earth's field might be the remnant of an original field still decaying is excluded by the very short diffusion time of a field in the Earth compared with the age of the Earth, and by the fact of magnetic reversals, which are found in quite early rocks. It is not possible to make such definite statements about the fields of other planets, for as yet, we do not know how they vary with time, but supposing that they, like the Earth's field, are to be explained in terms of dynamo action, what can they tell us about the internal state of the planet?

The fields of all the planets, the Earth included, as well as the Moon, are known to us as a result of observations from artificial satellites and space probes. The space outside a planet is an insulator (except for any ionosphere or other plasmas) and so the field of internal origin can be expressed as the gradient of a potential that satisfies Laplace's equation in the space external to the planet:

$$\boldsymbol{B} = \nabla \Phi,$$

with

$$\nabla^2 \Phi = 0.$$

To the field so specified there must, in general, be added fields generated by currents in any external plasma, such as those that produce magnetic storms on the Earth; for our present purpose, the essential point is that the field of internal origin, the one with which we are here concerned, may be specified by a potential which satisfies Laplace's equation outside the planets and which, therefore, like the gravity field, may be expanded in spherical harmonics.

The usual notation is

$$\Phi = -\sum_{n=1}^{\infty} \left(\frac{a}{r}\right)^{n+1} \sum_{m=0}^{n} [g_{mn}(t) \cos m\lambda$$

$$+ h_{mn}(t) \sin m\lambda] P_n^m (\cos \theta),$$

where a is the radius of the Earth and the spherical polar co-ordinates of an external point are (r, θ, λ).

The units of g and h are those of magnetic field. Both g and h vary with time: the secular variation of the internal field.

Table 9.2 gives the values of g, h, $\dot{g}$ and $\dot{h}$ for the spherical harmonic expansion of the internal field.

Alternative descriptions of the main field are also useful. The harmonic part, proportional to $P_1(\cos \theta)$, represents an axial dipole, but the three first harmonic terms together represent a dipole inclined (at about 11 °) to the polar axis of the Earth.

The effects of the quadrupole terms may be represented by displacing the axis of the dipole from the centre of the Earth. Thus the Earth's field may be represented to first order by an off-axis inclined dipole. The terms of higher order in the potential may similarly be represented by dipoles (or current loops) now placed at the surface of the core of the Earth, and some six such dipoles suffice to give a good representation of the higher harmonics (Chapter 2).

Table 9.2. *Harmonic coefficients of the geomagnetic field* (Barraclough *et al.*, 1975)

n	m	g $(10^{-9}\,T)$	h $(10^{-9}\,T)$	$\dot{g}$ $(10^{-9}\,T/y)$	$\dot{h}(10^{-9}\,T/y)$
1	0	−30103.6	—	26.8	—
	1	−2016.5	5682.6	10.0	−10.1
2	0	−1906.7	—	−25.0	—
	1	3009.9	−2064.7	0.3	−2.8
	2	1633.0	−58.1	5.5	−18.9
3	0	1278.2	—	−3.8	—
	1	−2142.0	−329.8	−10.5	7.2
	2	1254.7	265.9	−4.7	2.8
	3	831.0	−227.0	−4.7	6.4
4	0	946.9	—	−0.9	—
	1	792.5	193.4	−2.2	5.4
	2	443.8	−265.8	−4.0	0.7
	3	−403.9	53.0	−2.1	2.6
	4	212.5	−285.2	−4.6	−0.7
5	0	−220.6	—	0.2	—
	1	351.4	24.5	−1.0	0.9
	2	262.3	148.4	1.3	2.6
	3	−63.8	−161.3	−2.1	−2.7
	4	−157.5	−83.4	−0.6	1.3
	5	−40.2	92.3	1.3	1.1
6	0	44.1	—	0.6	—
	1	69.9	−11.2	0.9	−0.3
	2	27.7	100.4	2.3	−0.2
	3	−194.4	77.6	3.5	0.2
	4	−0.9	−40.3	0.0	−1.6
	5	3.8	−7.9	0.8	0.4
	6	−108.7	15.6	−0.4	2.0
7	0	71.5	—	−0.4	—
	1	−53.3	−76.6	−0.2	−1.2
	2	2.3	−24.7	−0.5	−0.2
	3	13.4	−4.5	0.3	0.0
	4	−6.4	7.0	0.8	0.3
	5	3.2	24.5	0.6	−0.6
	6	17.0	−21.8	0.5	0.0
	7	−5.9	−12.9	−0.8	1.2
8	0	11.0	—	0.4	—
	1	5.1	4.9	0.3	−0.2
	2	−2.6	−13.9	0.0	−0.3
	3	−12.6	5.0	0.4	−0.3
	4	−13.8	−18.0	−0.2	−0.3
	5	−0.1	5.7	−0.4	0.5
	6	−2.4	14.5	0.6	−0.5
	7	12.3	−11.1	−0.3	−0.6
	8	4.9	−16.7	0.0	0.5

Normalization: The normalization of the spherical harmonics differs somewhat from that generally adopted for the gravity field and is

$$\int_{-1}^{+1} \{P_n^m(\mu)\}^2 d\mu = 1/(2n+1).$$

One can ask, just as for the gravity field, what is the greatest depth at which the sources of the field can be placed, and two procedures have been adopted to answer that question. Lowes (1974) followed the same method set out above (section 9.3) for the analysis of planetary gravitational fields, namely the mean square harmonic coefficients of given degree were plotted against degree, and Lowes found that the sources of different degree would be the same if located near the surface of the core. Hide (1978) has adopted a somewhat different approach, asking at what depth the flux through a sphere is no longer conserved, and again finds that the sources of the field must be located at the boundary of the core. Should harmonic expansions for the fields of other planets become known, it would be possible by similar arguments to determine the greatest depth at which sources could be placed, in effect the depth of the conducting core in which dynamo action takes place. At the present time, however, insufficient components have been determined for the other planets for which, in general, only the parameters of an off-axis tilted dipole representation of the field are known.

Lowe's analysis shows up another feature of the Earth's field, and no doubt of the fields of the other planets as well. Evaluated at the surface of the core, the field is no longer primarily a dipole field, for all multipole components of the magnetic potential have similar magnitudes. The field is not in fact an essential dipole field, it only appears so to us because the dipole field dies away the most slowly with radial distance.

The field of Jupiter has been determined from observations from the Pioneer 10 and 11 space craft, and a summary of these results is shown in Table 9.3. Acuna and Ness (1976) have made a more detailed analysis, reproduced in Table 9.4.

The existence of the Jovian field had been inferred, prior to direct space craft observations, from the properties of radio emission from Jupiter, and the value of the dipole moment, estimated from the radio observations, agrees well with the later space craft measurements, although the

Table 9.3. *Dipole field of Jupiter: summary of Pioneer 10 and 11 results*

	Pioneer 10	Pioneer 11
Moment (T m^3)	1.454×10^{20}	1.536×10^{20}
Tilt (degrees)	10.6	10.77
Offset (R_J)	0.111	0.101
Reference	Smith (1974)	Smith *et al.* (1975)

radio observations cannot provide the detail of the direct measurements (Warwick, 1967).

Radio emission from Saturn has been detected and the existence of a magnetic field, of the same order as that of Jupiter, has been inferred (Brown, 1975) and has now been confirmed by direct observation from Pioneer 11 (Smith *et al.*, 1980; Acuna and Ness, 1980) (see Table 9.5). In contrast to the Earth and Jupiter, Saturn has a dipole field with its axis very close to the spin axis and in the opposite sense to the other two.

The magnetic fields of the terrestrial planets and the Moon have been derived from a number of space craft observations. Mariner 10 observations give the field of Mercury. The initial analyses yielded a dipole moment of 5.1×10^{12} T m^3 (Ness, Behannon, Lepping and Whang, 1975), but a later study, with allowance for a quadrupole component, gave 2.41×10^{12} T m^3 (Jackson and Beard, 1977).

The magnetic moment of Venus was obtained from observations with the Venera-4 space craft. The first interpretation indicated that the moment was below the limit of detection (8×10^{11} T m^3), but Russell (1976) in a re-discussion, suggests a value of 6.5×10^{12} T m^3.

The magnetic moment of Mars is derived from observations with the Mars 2 and 3 space craft and is about 2.4×10^{12} T m^3 (Dolginov, Yerovshenko and Zhuzgov, 1973). Russell (1977) has discussed the morphology of the bow shocks formed by the solar wind flowing past the terrestrial planets, as observed by the Mariner and Venera space craft, and has noted that they fall into two sets: one the one hand, those around the Earth and Mercury, where the solar wind is deflected by the permanent magnetic dipole moment of the planet; and those around

Table 9.4. *Spherical harmonic coefficients of Jovian magnetic field* (Acuna and Ness, 1976)

n	m	$g(10^{-4}$ T)	h $(10^{-4}$ T)
2	0	−0.203	—
	1	−0.871	−0.037
	2	+0.331	−0.402
3	0	−0.233	—
	1	−0.357	−0.463
	2	+0.506	+0.096
	3	−0.202	+0.233

Magnetic moment: 1.56×10^{20} T m^3.

Venus and Mars where the deflection is by an ionosphere and where any permanent dipole moment is insignificant. The discussions of the direct observations of the field of Venus and Mars have been somewhat inconclusive, and it may be that there are no net dipole fields of those planets, but that they are magnetized in much the same way as the Moon.

The Moon has no significant net dipole field but the surface is magnetized in a somewhat random fashion (Chapter 5; Sonnett, 1977; Coleman and Russell, 1977). Any overall dipole field has a moment of less than $10^9 \, \text{T} \, \text{m}^3$, but the surface material is magnetized with field intensities ranging from about $\frac{1}{2}$ nT to $-\frac{1}{2}$ nT, and maintaining a given sign over areas of up to 300 km across. A past field in the range of 10^{-6} to 10^{-4} T is required to account for the remanent magnetization of lunar rocks, but no such field currently exists. Runcorn's suggestion (Chapter 5; Runcorn 1975, 1977) is accordingly that a field of internal origin and of that magnitude did once exist, and that it magnetized the outer shell of the Moon, and has since decayed away, leaving no net dipole moment. However, because of irregularities in the outer magnetized shell of the Moon, local fields appear at the surface. Viking observations have shown that rocks at the surface of Mars are magnetized, and it is tempting to apply Runcorn's model to Mars as well.

The information about the dipole moments of the planets and the Moon is summarized in Table 9.6. In order to use facts about the magnetic fields of the planets to elucidate their internal structure, it is necessary to understand how planetary fields are generated, and detailed understanding is lacking. For the reasons set out earlier, it appears that the Earth's field must be generated in a liquid core, and it is natural to say the same of the other planets, but as yet we have no evidence which requires the origin to lie in a liquid core. The evidence for the Earth is the

Table 9.5. *The magnetic field of Saturn* (Acuna and Ness, 1980; Smith *et al.*, 1980)

Magnetic moment $(\text{T} \, \text{m}^3)$	$(4.3 \pm 0.2) \times 10^{18}$
Tilt (degrees)	2 ± 1
Equatorial field (mT)	2×10^{-2}
N. polar field (mT)	6.3×10^{-2}
S. polar field (mT)	4.8×10^{-2}
Spherical harmonic coefficients (mT):	
$g_1^0 \quad 2.03 \times 10^{-2}$	
$g_2^0 \quad 0.15 \times 10^{-2} \qquad h_2^1 \quad 0.01 \times 10^{-2} \qquad h_2^2 \quad 0.02 \times 10^{-2}$	

1 mT = 10 gauss.

variation of the field in time, and we do not know how the fields of the Moon and planets change with time. There is, however, no theory other than a dynamo theory that shows any signs of accounting for the fields of the planets. The remainder of this section is therefore concerned to see what may be learnt about the interiors of the planets on the basis of dynamo theory.

The general idea of dynamo theory is that some source of energy causes motions in a fluid that is electrically conducting, that the motions of the liquid in the magnetic field induce electric currents and that those currents generate the magnetic field. The motion of the liquid is resisted by viscosity and by the forces corresponding to ohmic dissipation of the electric currents. From this brief statement we may infer that a planet with a magnetic field must have a liquid electrically conducting zone, not necessarily the innermost zone; the inner core of the Earth, for example, is solid, and so may be the innermost cores of Jupiter and Saturn. In the Earth, electrical conductivity of the core is consistent with a metallic composition, mainly iron, while dynamical and seismological evidence shows that the core is liquid. Metallic hydrogen, the major constituent of Jupiter and Saturn, must have a high electrical conductivity; there is no direct evidence that any part of Jupiter or Saturn is liquid, but we take the existence of magnetic fields as implying that there are liquid zones in those planets.

Table 9.6. *The magnetic fields of the planets*

	Surface radius (km)	Core radius (km)	Spin angular velocity (rad/s)	Dipole moment (T m^3)	Dipole fields at surface (mT)	at core (mT)
Moon	1738	<400	—	<5×10^9	<10^{-6}	—
Mercury	2442	—	1.22×10^{-6}	2.4×10^{12}	1.6×10^{-4}	—
Mars	3380	—	7.09×10^{-5}	2.4×10^{12}	6.2×10^{-5}	—
Venus	6053	—	−2.99×10^{-7}	6.5×10^{12}	2.9×10^{-5}	—
Earth	6378	3400	7.29×10^{-5}	8×10^{15}	3.1×10^{-2}	2×10^{-1}
Saturn	60 000	—	1.71×10^{-4}	4×10^{18}	4×10^{-2}	—
Jupiter	71 000	—	1.77×10^{-4}	1.5×10^{20}	4.2×10^{-1}	—

The bodies are arranged in order of radius.
The core radius is that of a possible fluid conducting core.
The 'surface field' is that magnetic moment divided by the cube of the surface radius and gives the order of magnitude of the dipole component at the surface. The field at the core is similarly evaluated; as mentioned in the text this may be far from a correct description of the actual field just outside the core.

Consider first the equation of motion in a rotating liquid (all planets are rotating). Let u be the velocity of the liquid, Ω the angular velocity of the body, p the pressure and ν the viscosity. Let F be the force acting on the liquid; it will include an electromagnetic term if a magnetic field is being generated. Then u satisfies the equation

$$\frac{\partial u}{\partial t} + u \cdot \nabla u + 2\Omega \wedge u = -\rho^{-1}\nabla p + F + \nu\nabla^2 u.$$

It is often permissible to ignore the convective acceleration $u \cdot \nabla u$ and the viscous term $\nu\nabla^2 u$, so that

$$\frac{\partial u}{\partial t} + 2\Omega \wedge u = -\rho^{-1}\nabla p + F.$$

Suppose that $\rho g = -\nabla p$, that the temperature within the body is T and that the coefficient of thermal expansion is α; there is then a buoyancy force per unit volume equal to $\alpha T g$. Further, let B be the magnetic induction and J the electric current; there is then an electromagnetic force per unit volume equal to

$$\rho^{-1} J \wedge B.$$

Let the liquid be confined to a volume V. Outside V, B satisfies

$$\nabla B = 0.$$

Within V,

$$\frac{\partial B}{\partial t} = -\nabla \wedge E \qquad \text{(Maxwell)},$$

where E is the electric field.

E has two parts, the first being the motional field arising from the liquid moving in the magnetic field, namely

$$E = -u \wedge B.$$

In addition, there is the ohmic term arising from the electric current, J, that is

$$E = J/\sigma,$$

where σ is the electrical conductivity. But

$$\mu_0 J = \nabla \wedge B$$

and thus

$$E = \frac{1}{\mu_0\sigma}\nabla \wedge B.$$

Accordingly

$$\nabla \wedge E = -\nabla \wedge (u \wedge B) - \frac{1}{\mu_0 \sigma} \nabla \wedge \nabla \wedge B.$$

Denote $(\mu_0 \sigma)^{-1}$ by λ. Then

$$\frac{\partial B}{\partial t} = \nabla \wedge (u \wedge B) + \lambda \nabla^2 B.$$

The boundary conditions on u and B are

$$u \cdot n = 0$$

on the boundary of V, and B must be continuous across the boundary.

Associated with the equations for $\partial u / \partial t$ and $\partial B / \partial t$ there should also be the equation of heat conduction. Together with the equation for the conservation of mass, they should allow the velocity u, induction B and temperature T to be found for given boundary conditions of, for example, heat flux or temperature. That is the problem one would like to solve, but no solutions have so far been obtained. The dynamo problem is, in fact, usually divided into two: the dynamical problem and the kinematic problem. The former problem is to determine u and B given a force field (including a temperature field), the latter problem is to determine B given u.

The kinematical problem is in the nature of an existence problem, it addresses the question whether there are *any* velocity fields which lead to the maintenance of a steady magnetic field, irrespective of whether there is a means of generating those velocity fields. A thorough account of the state of both problems has been given by Moffat (1978).

A very important general result is a theorem of T. G. Cowling who showed that dynamo action cannot occur if the velocities and magnetic fields have a common axis of symmetry. An important aspect of planetary fields is that they are generated in bodies that are rotating, in which the axial symmetry of the motions is removed; it may well be that the inclination of the dipole axis to the spin axis, as observed in the Earth and Jupiter, is a consequence of departure from axial symmetry, although no theoretical work has as yet shown that the inclination does naturally follow from dynamical requirements. The field of Saturn, on the other hand, may pose a problem of interpretation.

Any solenoidal vector field may be written in the form, for B for example,

$$B = B_T + B_P,$$

where B_T, the *toroidal* part, is given by

$$B_T = \nabla \wedge (xT(x))$$
$$= -x \wedge \nabla T,$$

where x is the position vector and T a scalar function.
B_P, the *poloidal* part, is given by

$$B_P = \nabla \wedge \nabla \wedge (xP(x)),$$

where P is another scalar function of position. It follows that $\nabla \wedge B_T$ is a poloidal field, and likewise $\nabla \wedge B_P$ is toroidal.

Kinematic dynamo theory is concerned with the interaction between toroidal and poloidal parts of velocity and magnetic fields, each being expanded in the form

$$f_{nm}(r) Y_{nm}(\theta, \phi),$$

where $f_{nm}(r)$ is a radial function and $Y_{nm}(\theta, \phi)$ a spherical harmonic.

It has been found possible to construct velocity fields that lead to self-sustaining dynamos. Dynamo action may be thought of in terms of the concentration of magnetic flux by motions of conducting fluids carrying the flux with it because of the high conductivity so that the field lines are 'frozen into' the fluid. Suitable motions will lead to greater concentration of field lines, but that concentration will be counterbalanced by diffusion according to the equation

$$\frac{\partial B}{\partial t} = \lambda \nabla^2 B.$$

Only if the rate of concentration by convection in the fluid is greater than diffusion will a magnetic field grow and be sustained. Thus, whether or not a field is generated is dependent on a magnetic Reynolds number which expresses that balance, namely

$$R_m = \frac{u_0 l_0}{\lambda} = u_0 l_0 \mu_0 \sigma,$$

where u_0 is a typical velocity and l_0 is a typical scale for variations of u.

The westerly drift of the secular variation field of the Earth suggests that R_m is about 150 for the core; values of R_m of about 20 to 100 are required to sustain dynamo action in some simple cases.

The work on kinematic dynamos briefly mentioned postulates that u and B can be specified as functions of position and time, but it is almost certain that motions in the core of the Earth, and no doubt in other planets, are turbulent with random components. Much attention has therefore been paid in recent years to kinematic dynamo theory in which random fields have an essential part.

The essential feature of such dynamos is that the interaction of the random components of velocity and magnetic field can produce a mean electromotive force, and have an electrical current parallel to the mean magnetic field.

Let the velocity U be written as the sum of mean, U_0, and fluctuating parts, u:

$$U = U_0 + u,$$

such that the mean value of u, averaged over an appropriate span of space and time, is zero.

Similarly, let B be separated:

$$B = B_0 + b.$$

Then the induction equation

$$\frac{\partial B}{\partial t} = \nabla \wedge (U \wedge B) + \lambda \nabla^2 B$$

similarly separates:

$$\frac{\partial B_0}{\partial t} = \nabla \wedge (U_0 \wedge B_0) + \nabla \wedge E + \lambda \nabla^2 B_0$$

and

$$\frac{\partial b}{\partial t} = \nabla \wedge (U_0 \wedge b) + \nabla \wedge (u \wedge B_0) + \nabla \wedge G + \lambda \nabla^2 b.$$

Here $E = \langle u \wedge b \rangle$ and $G = u \wedge b - E$, where the brackets $\langle \rangle$ denote a mean value.

Moffat (1978) shows that the components of E are linearly related to those of B_0 through the equation for $\dot{b}$ (which involves B_0 linearly), and consequently it is possible to write

$$E_i = \alpha_{ij} B_{0j} + \beta_{ijk} \frac{\partial B_{0j}}{\partial x_k} + \cdots ;$$

the coefficients α_{ij}, β_{ijk} are pseudotensors, changing sign on a change of parity because E is a polar vector and B_0 an axial vector.

In certain simple (isotropic) cases α_{ij} may be written as

$$\alpha_{ij} = \alpha \delta_{ij},$$

where α is a pseudoscalar. In that case

$$E = \alpha B_0$$

and

$$J = \sigma E = \sigma \alpha B_0.$$

The effect whereby a current parallel to B_0 can be established by interaction of turbulent motions and fields is known as the 'α-effect'. The effect of the term proportional to $\partial B_{0j}/\partial x_k$ is to change the magnetic diffusivity; in the simplest case λ is replaced by $\lambda + \beta$.

The numerical values of α and β depend on the type of motion that is established. It should be noted that, with suitable choice of the extent of smoothing, G can be taken to be negligible.

Dynamo theory is concerned with the form and magnitude of the α-tensor and β-tensor. Because α is a pseudotensor, changing sign with a change from a left-handed to a right-handed co-ordinate system, no motions which are symmetrical under such a change can produce dynamo action through the α-effect. It is here, then, that rotation plays an important part, for, if a spherical body of fluid is both rotating and convecting, the resulting motion will have an overall left-handedness or right-handedness and will be able to sustain an α-effect, whereas, if there is no rotation, the motions may be symmetrical under reflexion and no α-effect will occur.

Velocity fields which lack reflexional symmetry are said to be *helical*, helicity being defined as $u \cdot \omega$, where ω is the *vorticity*, equal to $\nabla \wedge u$. Whether velocity fields are laminar or turbulent, for them to be able to sustain dynamo action it is essential that they should have non-zero helicity.

If the α-effect operates, two types of dynamo are possible in a rotating fluid. Suppose that the mean velocity, magnetic and electric fields are all axisymmetrical, so that they may be written, dropping the suffix 0, in cylindrical polar co-ordinates (r, z, ϕ), as

$$U = r\omega(r, z)i_\phi + U_P,$$

$$B = B(r, z)i_\phi + B_P,$$

$$E = E_\phi i_\phi + E_P.$$

Here i_ϕ is the unit vector in the ϕ direction, $\omega(r, z)$ and $B(r, z)$ are functions of r and z, but not of ϕ, and U_P, B_P and E_P are poloidal fields.

B_P is derived from an azimuthal vector potential:

$$B_P = \nabla \wedge A(r, z)i_\phi$$

(for example, if A represents a simple ring current, B_P is a dipole field).

If $E = \alpha B - \beta \nabla \wedge B$, it follows that

$$\frac{\partial B}{\partial t} + r(U_P \cdot \nabla)\left(\frac{B}{r}\right) = r(B_P \cdot \nabla)\omega + [\nabla \wedge (\alpha B_P)]_\phi + \lambda_e(\nabla^2 - r^{-2})B$$

and

$$\frac{\partial A}{\partial t} + r^{-1}(U_P \cdot \nabla)(rA)\alpha = B + \lambda_e(\nabla^2 - r^{-2})A,$$

where $\lambda_e = \lambda + \beta$ is the effective value of λ (Moffat, 1978).

In the equation for $\dot{B}$, there are two source terms, namely

$$r(B_P \cdot \nabla)\omega \quad \text{and} \quad [\nabla \wedge (\alpha B_P)]_\phi,$$

the ratio of their orders of magnitude being

$$L^2 \omega_0'/\alpha_0.$$

L is a scale length over which mean quantities show significant variation and is large compared with the scale of turbulence, ω_0' is a typical value of $\nabla \omega$ and α_0 a typical value of α.

The ratio thus expresses the effect of differential rotation, $\nabla \omega$, in comparison to the strength of turbulent fields, α.

If differential rotation is negligible,

$$\frac{\partial B}{\partial t} + r(U_P \cdot \nabla)\left(\frac{B}{r}\right) = [\nabla \wedge (\alpha B_P)]_\phi + \lambda_e(\nabla^2 - r^{-2})B$$

and the term $[\nabla \wedge (\alpha B_P)]_\phi$ acts as the source of B, that is of the toroidal field.

On the other hand, if differential rotation dominates,

$$\frac{\partial B}{\partial t} + r(U_P \cdot \nabla)\left(\frac{B}{r}\right) = r(B_P \cdot \nabla)\omega + \lambda_e(\nabla^2 - r^{-2})B$$

and the source of B is then proportional to B_P and $\nabla \omega$.

In either case, the source of the vector potential A, corresponding to the poloidal part of B, is αB, through the equation

$$\frac{\partial A}{\partial t} + r^{-1}(U_P \cdot \nabla)(rA) = \alpha B + \lambda_e(\nabla^2 - r^{-2})A.$$

The scheme of dynamos depending on the α-effect is thus that a poloidal field, derived from a toroidal vector potential, is generated by the α-effect between a toroidal field and the turbulent motion, and then the toroidal field is regenerated from the poloidal field, either by interaction with the turbulent motion through the α-effect or through interaction with differential rotation. Dynamos in which the α-effect operates in both parts are known as α^2-*dynamos*, those in which differential rotation operates in producing toroidal field are known as $\alpha\omega$-*dynamos*. α depends in magnitude upon the magnitudes and spectrum of the turbulent velocities (Moffat, 1978), whilst $\nabla\omega$ depends in magnitude upon the spin angular velocity; accordingly one may anticipate that dynamos in slowly rotating bodies would be α^2-dynamos, but that those in fast spinning bodies would be $\alpha\omega$-dynamos or laminar dynamos.

Moffat (1978) shows that, in a turbulent fluid, the maximum value of the helicity is of order u_0^2/l_0, where l_0 is a characteristic scale of the turbulence. He also shows that if the helicity has its maximum value, the value of α is

$$l_0 u_0^2/\lambda.$$

Now l_0 and $\dot{u}_0$ are related by the Reynolds number

$$R_e = l_0 u_0/\nu,$$

and so $\alpha \sim R_e^2 \nu^2/l_0\lambda$.

Now, irrespective of the source of energy of a planetary dynamo, the typical velocity will increase with the rate of supply of energy and thus with the energy that has to be transported by the fluid motion; α, therefore, will increase with the energy generated within the dynamo regions.

While there is now a good understanding of certain kinematical features of dynamos, little has been achieved in solutions of the dynamical problem. The essence of the matter is that a force proportional to $J \wedge B$ acts upon the fluid. It may be expected to have two effects. In the first place, it will react back upon the fluid motion as a brake so setting a limit to the fluid motions and, hence, to the magnetic field, a limit not provided for in the kinematical theory. Secondly, it is expected to behave like the Coriolis force in modifying the onset and form of convective motions in the fluid. So far, however, insufficient has been done to make any general comments on either of these points.

The conditions for the onset of dynamo action have been discussed by a number of authors and it is usually considered that the magnetic Reynolds number must exceed some limiting value.

Gubbins (1974) has given the following values for the core of the Earth:

radius: 3483 km
electrical conductivity: 5×10^5 S/m
viscosity: $\not> 10^{12}$ Nm/s^2.

If the typical velocity is taken to be that of the westerly drift of the secular variation field and the typical scale that of the radius of the core, it will be found that R_m is of the order of 10^2, while the Reynolds number is about 3×10^8. It has been suggested that R_m should exceed about 10 for the dynamo action to take place.

The magnetic Reynolds number for Jupiter must certainly be greater than that for the Earth, probably very much greater, for the typical dimension is greater and the electrical conductivity of metallic hydrogen is some orders of magnitude greater than that of molten iron. It is not, of course, known what the typical velocities are in Jupiter, but they would have to be some orders of magnitude less than in the Earth for the magnetic Reynolds number to fall below the terrestrial value. In any event, we know that Jupiter has a magnetic field.

There is no general agreement on the source of power which drives the geomagnetic field. In principle, the necessary power can be estimated from the ohmic dissipation of currents needed to maintain the field. It is, of course, a fairly simple matter to estimate the currents needed to maintain the observed poloidal field; the difficulty is that, in the absence of reliable solutions of the dynamo equations, the magnitude of the currents that maintain the associated (and probably much larger) toroidal field is not known. Three sources have been suggested: thermal energy, whether from radioactive nuclides or from original heat; precessional couples; and energy derived from differentiation of the core from the mantle. Thermal energy is perhaps the natural one to think of, but has been criticized on the grounds that there are strong indications in the Earth that the radioactive nuclides are concentrated in the outer layers, so that the core would be expected to be depleted and so would not generate sufficient power to drive the dynamo. A somewhat related difficulty that has been put forward, and has been strongly criticized, is that the equation of state of the core of the Earth is such that it would be stable to thermal gradients and would not convect. Whether that is so or not, it is now believed that thermal convection, in the sense of motions in an unstable fluid, is not necessary to generate a magnetic field through the α-effect.

The core of the Earth is less oblate than the mantle, and the oblateness decreases towards the centre where it is zero. Thus the precessional couple, proportional to $(C - A)/C$, is not constant throughout the core, but also decreases towards the centre. It has been suggested that the differential couple would provide a torque to drive the geomagnetic dynamo, but the idea has been strongly criticized by Rochester, Jacobs, Smylie and Chong (1975).

Finally, it has been suggested that, on the assumption that the core of the Earth is still increasing in size by differentiation from the mantle, the gravitational energy released would provide the power to sustain the Earth's field.

If we compare Jupiter with the Earth, the precessional source seems even less likely, for the precessional couple on Jupiter is far less than that on the Earth, yet Jupiter's field is greater. As to Mercury, the Sun is much closer, but J_2 is much smaller than for the Earth and again the precessional idea seems less favourable.

It will be argued below (Chapter 10) that the core of the Earth was formed prior to and independently of the mantle around it and, if that is right, continued growth of the core would not now be going on, nor would it in Mercury, to which the argument for the prior formation of the core would apply as to the Earth. Smoluchowski, in particular, has however suggested that gravitational separation of helium from hydrogen in Jupiter may be a source of the energy which is radiated from Jupiter and so a possible source of the dynamo. It was seen in the previous chapter that there must be considerable doubt about the source of the energy radiated from the Jovian surface, and it is really quite uncertain how much energy has to be produced in the interior. To summarize, it may be said that in none of the planets with magnetic fields can the source of the energy be identified as yet, nor, in the absence of a reliable theory, can the magnitude and configuration of the field be used to estimate the power needed to sustain it.

9.6 Conclusion

In this chapter, I have drawn together a number of features of the planets which indicate how they deviate from simple models in hydrostatic equilibrium, in which the density may be supposed to depend only to an unimportant degree upon temperature.

The main features are the terms in the gravitational potential that would be absent for hydrostatic equilibrium, tectonic activity, heat flow through the surface and generation of a magnetic field. The first could be maintained in a static planet with sufficient strength, but the latter are connected through motions in the interior. In so far as the motions are driven by thermal convection, they will be in the sense of reducing differences of density and so the distribution of density will tend to the radial variation corresponding to the hydrostatic state. As knowledge of the planets increases, it is to be hoped that it will become possible to define such internal motions better than at present.

10

Conclusion

10.1   The lesser objects of the solar system

So far in our studies, no notice has been taken of the smaller bodies in the solar system, i.e. Pluto, the asteroids and the satellites of Mars and the major planets, for their properties are but poorly known on the whole and it is not very rewarding to apply to them the type of analysis that was applied in the foregoing parts of this book to the greater objects. Yet, in considering the solar system as a whole, their existence and such information as we have of them cannot be ignored.

Pluto, the outermost known planet, is in a highly eccentric orbit highly inclined to the ecliptic, and, in consequence, although it comes on occasion within the orbit of Neptune, it never approaches Neptune closely, and detailed studies have shown the outer solar system to be stable. Pluto is a very small object as seen from the Earth. Its diameter is estimated from the brightness and supposed reflectivity. The latter has recently been redetermined from infra-red spectroscopy and the diameter of Pluto is consequently now estimated to lie between 2800 and 3300 km (Cruickshank, Pilcher and Morrison, 1976). The mass of Pluto was originally estimated from the perturbations of the orbits of Uranus and Neptune, but a satellite has now been detected (Christy and Harrington, 1978) with a period of 6.4 d, from which the mass of Pluto is estimated to be about 0.002 times that of the Earth (Meadows, 1980). The density is consequently between about 650 and 1100 kg/m^3, similar to that of some of the smaller satellites (Table 10.1).

The asteroids form a cloud of small objects in planetary orbits about the Sun, lying between Mars and Jupiter at about 2.8 AU from the Sun, that is to say, closer to Mars than to Jupiter, which lies at about twice that distance from the Sun. Their number is unknown, but the orbits of some 1700 have been determined and their total mass is far less than that of the

302

satellites of planets as the following figures show:

mass of planets: 447.9 M_E
mass of satellites: 0.12 M_E
mass of asteroids: 0.0003 M_E,

where M_E is the mass of the Earth (Allen, 1963).

The mass and density of a few of them have been estimated; the largest, Ceres, has a radius of 350 km, a mass of 6×10^{20} mg and a density of 3340 kg/m^3.

The satellites of Mars and the major planets (Table 10.1) are, in general, small objects of low density, comparable with the asteroids, but four satellites – Ganymede and Callisto of Jupiter, Titan of Saturn, and Triton of Neptune – are larger than the Moon.

Table 10.1. *Properties of some satellites*

Planet	Satellite	Radius (km)	Mass (kg)	Density (kg/m^3)
Earth	Moon	1737	7.0×10^{22}	3340
Mars	Phobos	10	—	—
	Deimos	5	—	—
Jupiter	Io	1830	7.24×10^{22}	2820
	Europa	1550	4.71×10^{22}	3020
	Ganymede	2775	15.52×10^{22}	1730
	Callisto	2500	9.67×10^{22}	1480
	8 satellites less than 100 km in diameter			
Saturn	Mimas	180	3.8×10^{19}	1500
	Enceladus	300	7.4×10^{19}	700
	Tethys	520 ± 60	6.2×10^{20}	1100
	Dione	500 ± 120	1.05×10^{21}	2000
	Rhea	800 ± 100	2.2×10^{21}	1000
	Titan	2900 ± 200	1.35×10^{23}	1320
	Hyperion	112 ± 15	—	—
	Iapetus	725 ± 100	2.8×10^{21}	1800
	Phoebe	120	—	—
Uranus	Ariel	200	1.3×10^{21}	—
	Umbriel	200	5.4×10^{20}	—
	Titania	500	4.4×10^{21}	—
	Oberon	400	2.5×10^{21}	—
	Miranda	100	0.6×10^{20}	—
Neptune	Triton	1885	1.36×10^{23}	4800
	Nereid	100	3×10^{19}	—

Data from Allen (1963), Newburn and Gulkis (1973) and Anderson *et al.* (1980).

Phobos and Deimos are rocky irregular objects, pitted with craters and their diameters are only indications of the general dimensions. They have been photographed from the Viking orbiters.

The four innermost satellites of Jupiter were discovered by Galileo with his early telescope and are known as the Galilean or Medicean satellites.

The dynamical properties of many of the satellites of Jupiter and Saturn have been determined from space craft (Anderson *et al.*, 1980). They are small objects (Table 10.1) in which, as in the Moon, self-compression will be negligible. The densities suggest that the satellites are made up of rocky material, of density around $3000\,kg/m^3$, and ices of water, ammonia and methane, of densities between 500 and $1000\,kg/m^3$, but it must be said that these are only guesses based on general under-standing of the composition of the solar system. There may well be surprises in store. Many satellites have now been examined from space craft. Phobos and Deimos have been photographed and, like the Moon, Mars and Mercury, are heavily cratered as are Ganymede and especially Callisto (Stone and Lane, 1979). Io lies within the radiation belts of Jupiter, surrounded by an atmosphere of sodium and sulphur, which it injects into the radiation belts, the electrical properties of which it modulates. Io shows strong volcanic activity (Carr *et al.*, 1979) whereas Europa is covered with a thin layer showing a complex pattern of fractures (Stone and Lane, 1979).

No values have been quoted for the densities of the satellites of Uranus and Neptune, except for Triton, for, if calculated from the radii and masses given in Table 10.1, they would be implausibly high, as may be seen by comparing the masses of the satellites of Uranus with those of Saturn of supposedly the same radius. The fact appears to be that the diameters of the Uranian and Neptunian satellites are too small to be determined in any realistic way from the Earth.

Besides their satellites, three planets, Saturn, Jupiter (Stone and Lane, 1979) and Uranus (Nicholson *et al.*, 1979) have rings of dust about them.

The small objects tell us really very little about the solar system. They seem to be the debris, the inconsiderable remains of the principal processes. The asteroids are often supposed to represent a missing planet between Mars and Jupiter, but their total mass is so small – 3×10^{-4} of the mass of the Earth – that it is hard to see where the rest of any such planet can have gone; nonetheless, the average orbit of the asteroids fits neatly into the place provided for it in Bode's law of planetary distances.

10.2   The planets as a whole

The planets divide clearly into two groups, the terrestrial planets, a compact group, small and of rocky and metallic composition, and the major planets, far from the Earth, covering a wide range of distance and composed mainly of light materials, but probably with some core of rock and metal. Within each group there are differences of size, mean density and, by implication, of composition; Jupiter and Saturn are twins, but not identical twins, and so are Uranus and Neptune, for, while the planets of each pair comprise a light envelope in which a rocky core is probably present, the relative size of any core is greater in the outer planet of the pair. What can the internal constitutions, the similarities and dissimilarities of the planets tell us about the way in which the solar system may have come into being?

It is commonly supposed that the planets formed out of material that once was part of the Sun, that tidal disturbances produced by a second star caused irregularities of density in the outer parts of the solar atmosphere, which developed into condensations that were unstable and collapsed under self-gravitation into the planets. It is further considered likely that the planets formed during the so-called Hyashi phase of the formation of the Sun, when the solar atmosphere would have been extended out into the region occupied by the solar system as we now know it. The temperature within that extended atmosphere would, of course, fall away from the inner to the outer zones, so that in the inner parts materials such as hydrogen, helium, water or methane, would be gaseous, but metals and metal silicates could condense to form the terrestrial planets. As the temperature fell in the outer parts with the contraction of the early Sun, the light elements there could also condense, but would probably require some nucleus on which to do so; no doubt condensation would start when the escape velocity from the rocky or metallic nucleus was comparable with the thermal velocities in the surrounding gas. Among others, two questions arise from this general account: in what order of chemical composition did the condensation of the terrestrial planets proceed, and how did rocky or metallic nuclei of the major planets come into being so far from the Sun?

There are a number of broad classes of ideas about the condensation of the terrestrial planets. On the one hand, it has been supposed that they formed as liquid spheres, necessarily at high temperatures, in which the rocky and metallic constituents were mixed, and then, because they were immiscible, the metal separated from the rock as the planets cooled down and, being denser, formed the cores. On the other hand, it has often been thought that the material condensed as dust which accumulated into cold bodies that then heated up from the gravitational energy released in the condensation and from the energy of radioactive nuclides which have now decayed away; as the planet heated up, diffusion under gravitation in the solid led to the separation of rock from metal and to the formation of a core, with further release of gravitational energy. In yet a third scheme, (see Chapter 6) it is supposed that, on account of its lower vapour pressure, the metallic material of the core would first condense out of the solar gas and would be followed by the rocky material which would accumulate as mantles around the cores. The third process is distinct from the first two in that the core is seen as the essential primitive feature of the planet rather than as a secondary feature forming after the accumulation of the planet.

It seems that our understanding of the constitution of the terrestrial planets enables us to discriminate between the first two and the latter schemes of condensation. In the first two, the composition of the material would be expected to be the same for all the planets, or perhaps to change in a systematic way with increase of distance from the Sun according to changes of temperature and pressure in the inner atmosphere of the early Sun. Thus, the cores of the planets would be expected to be either a constant fraction of the planets or a proportion that changed systematically from Mercury to Mars. That is not found. No systematic dependence of the relative size of the core upon the position of the planet is to be seen in the data of Table 6.6, nor does the relative size of the core depend in any regular way upon the overall size of the planet: the Earth and presumably Venus have large cores and Mars and the Moon small ones or none at all, but Mercury presumably has a core large in relation to its size. This rather irregular state of affairs would, however, seem to be consistent with the formation first of cores, then of mantles around them, the two processes having no necessary relationship one to another, so that the relative size of the core would be arbitrary. A two-stage process of formation of the planets thus seems to be implied by the constitutions of the terrestrial planets.

When we look at the major planets, the question that arises there is how the cores of heavy material have come into existence so far from the Sun. The information we have at present about the major planets does not enable us to estimate the size of heavy cores unambiguously, and indeed, without seismic data of some sort, that may never be possible, but cores which are a few times larger than the Earth would be possible in

Jupiter and Saturn. How is it that heavy bodies, larger than any of the terrestrial planets, could form in the outer parts of the solar system in the lower densities and temperatures of the early Sun? The present masses of such presumed cores exceed by a factor of about ten the amount of rocky and satellite constituents appropriate to the present mass of hydrogen in composition.

A similar position is presented by the overall composition of the four major planets, in which the overall ratio of carbon, nitrogen and oxygen to hydrogen and helium exceeds the present solar proportion by a factor of about ten. It seems that a great deal of hydrogen has been lost from the outer parts of the solar system just as it has from the inner parts. Furthermore, as in the terrestrial planets, the presence of rocky or metallic cores of different proportions in the four major planets suggests that the cores condensed first and that the light shells of hydrogen and helium or methane and water accumulated around them.

10.3 The limits of knowledge

Our knowledge of the planets is limited in two ways: by the absence of data which it is conceivable we might be able to obtain and by the ambiguity of inference from conceivable data, however complete.

We would like above all to have seismic information about the terrestrial planets. If space craft are placed into orbit about Venus and Mercury, then it is possible that even a small value of J_2 will be measurable and some estimates of the dimensionless moments of inertia and, hence, of the central condensations of those planets will be feasible. The present information about the values of J_2 suggests, however, that such studies will be very difficult; the corresponding secular motions of any satellites would be small and would be dominated by the attraction of the Sun, especially for Mercury. Placing of seismometers on Venus and Mercury might, however, be of great value, although it must be recognized that not much has so far been harvested from the Viking seismometers on Mars and, also, that on all three planets there seems to be almost no current tectonic activity as we know it on the Earth and, so, no internal seismic sources. For no planet other than the Earth, therefore, may we anticipate any observations of free oscillations. Will seismic information ever be obtained for the major planets? Clearly seismometers cannot be placed upon them, nor travel times measured through them, but it is not entirely inconceivable that they may be excited into free elastic oscillations by processes of which we are ignorant, and that, as with the Sun, ways may be found to detect them in the motions of the surface of the atmosphere. More realistically, we may look forward to considerable improvements in knowledge of the dynamical properties of Uranus and Neptune. Observations from large telescopes in orbit about the Earth should produce firmer values for the periods of rotation of Uranus and Neptune and of the motions of the natural satellites, as well as of the geometrical figure, and in due course space craft will pass close to Uranus and Neptune. By all these means, far more reliable dynamical data should become available for Uranus and Neptune about which there is, as was seen in Chapter 8, much uncertainty.

Space craft passing by Uranus and Neptune should also settle the question of the existence of magnetic fields of those planets.

More and better data will help us to understand the planets better than we now do, but we have to recognize that there are other limitations to our possible knowledge. We return to where we began: the data we can obtain for the planets are functionals of distributions of density, elastic properties, and so on, with depth. The deductions we can make from them are subject to great ambiguity when only a few functionals can be measured and, even when the equations of state can be derived with some precision, as they can for the Earth, the identification of possible constitutions still extends over a considerable range of choice. The most reliable deductions about the nature of the planets will be those made from general properties not so dependent on detailed interpretation of data.

As we contemplate the planets as we see them in the light of observations from space craft, they retain in many ways their mystery. Much light has been shed on some problems, but others have been posed. The general ideas about the constitutions of the planets are indeed not so very different from those current before observations from space craft; there have been notable refinements, some of our ideas about properties of materials are more precise, but the internal states and mode of origin of the planets still attract our wonder and challenge us to understand them.

APPENDIX 1

Limits and conditions on planetary models

A.1 Introduction

The only observed mechanical data we have for any planet are the mass and moment of inertia, and infinite sets of models can be constructed consistent with such pairs of data. The sets of models are not, however, unbounded, and, further, certain models are in some sense more probable than others. It is the purpose of this appendix to set out the bounds on two particular models and to give some most probable models. The models considered are: that of two zones, each of constant density, and that in which the density is determined by hydrostatic compression alone. The terrestrial planets may be modelled by the former, and the major planets by the latter. Neither model can represent the complexities of actual planets but, given only two data, no more elaborate model is justified. Guided by the constitution of the Earth, and by such seismic data as are available for the Moon, it is natural to choose the two-zone model as an approximation to the structures of the terrestrial planets. In this model, the maximum pressure is such that changes of density under self-compression are less than differences of density arising from differences of chemical composition or crystal structure in different parts of the planet. Thus, a model comprising two zones of different density is chosen as a basis for study of the terrestrial planets. In the major planets, on the other hand, the mean density is so low that they must be composed predominantly of one material – hydrogen – while the central pressure is great enough for there to be a major increase of density under self-compression. The second type of model is therefore the one to choose for the major planets.

It is the purpose of this appendix to describe the main features of these simple models and the bounds which are imposed on them. There are bounds which are inherent in the models (for example, a density must not become infinite or vanish), and there are bounds set by our knowledge of

309

planetary materials (a density must not exceed a certain value); bounds are also set by considerations based on our principles of inference.

A.2 Two-zone models

Consider a planet of radius a. In a two-zone model it is divided into two by a surface of radius a_1; within a_1 the density is ρ_2, and between a_1 and a it is ρ_1. Given the mass and moment of inertia of the planet, what can be learnt about a_1/a and ρ_2/ρ_1?

The equations for the mass and moment of inertia are

$$\text{mass: } \frac{4\pi}{3}a_1^3\rho_2 + \frac{4\pi}{3}(a^3 - a_1^3)\rho_1 = M = \frac{4\pi}{3}a^3\bar{\rho}$$

$$\text{inertia: } \frac{8\pi}{15}a_1^5\rho_2 + \frac{8\pi}{15}(a^5 - a_1^5)\rho_1 = I = \frac{8\pi}{15}a^5\gamma\bar{\rho}.$$

In these expressions, the mass mean density, $\bar{\rho}$, is defined to be $3M/4\pi a^3$. An inertial mean density, $\gamma\bar{\rho}$, may also be defined as $15I/8\pi a^5$.

The observed data may be conveniently represented by $\bar{\rho}$ and γ. If the density of the planet is constant throughout, then γ is 1.

Let a_1/a be written as α and ρ_2/ρ_1 as β. Then the equations for the mass and moment of inertia read

$$\alpha^3\beta\rho_1 + (1 - \alpha^3)\rho_1 = \bar{\rho}$$

and

$$\alpha^5\beta\rho_1 + (1 - \alpha^5)\rho_1 = \gamma\bar{\rho}$$

or

$$1 - \alpha^3(1 - \beta) = \bar{\rho}/\rho_1$$

and

$$1 - \alpha^5(1 - \beta) = \gamma\bar{\rho}/\rho_1.$$

Thus α, β and γ are related by the expression

$$\beta - 1 = \frac{1 - \gamma}{\alpha^3(\gamma - \alpha^2)}.$$

A set of curves showing the dependence of β on α for different values of the observed ratio γ is given in Figure A.1. They are restricted to values of β less than 5 because that seems a reasonable limit on the basis of the constitution of the Earth.

In any hydrostatically stable planet, the density of the inner zone must exceed that of the outer, so that ρ_2 exceeds ρ_1 and β is greater than 1.

Thus α must not exceed $\gamma^{1/2}$. Also, β has a minimum value of

$$1 + \frac{5(1-\gamma)}{2(\frac{3}{5})^{3/2}\gamma^{5/2}}$$

at the value of α equal to $(\frac{3}{5}\gamma)^{1/2}$.

Figure A.2 shows the maximum possible value of α and the value of α for the least value of β, each as functions of γ, whilst Figure A.3 shows the least value of β as a function of γ.

It was seen in Chapter 1 that models may be required to satisfy some external condition. Parker's (1972) condition is obtained by supposing the density to be an element of a space of functions L^p; so that the norm of

Figure A.1. The dependence of β, the ratio of the densities in a two-zone planet, upon α, the relative radius of the zone boundary, for various values of γ, the ratio of the inertial mean density to the mass mean density.

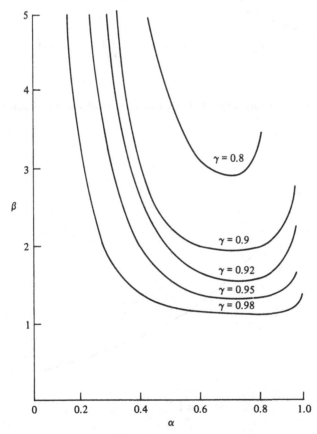

Figure A.2. Maximum possible value of α for a two-zone planet, and value of α for the least value of β, the ratio of densities, as functions of γ.

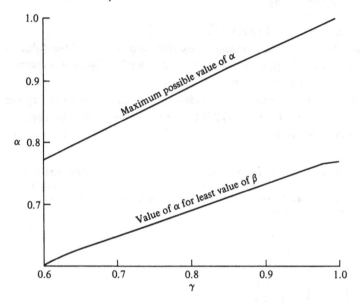

Figure A.3. Least value of β in a two-zone model as a function of γ.

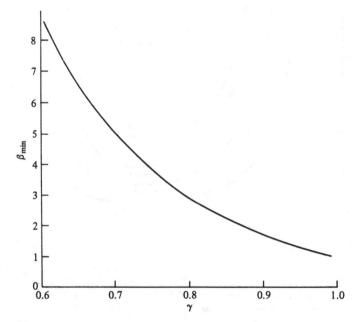

the space is

$$\|\rho\|_p = \left(\int_0^a |\rho(r)|^p dr/a \right)^{1/p}.$$

Parker also shows that if this norm is maximized subject to the constraints that the mass and moment of inertia are to be those observed, then a lower bound on the greatest density is obtained, and he finds that to be

$$\rho_m = \left(\frac{2}{5} \frac{Ma^2}{I} \right)^{3/2} \bar{\rho};$$

the corresponding value of α is

$$\left(\frac{5}{2} \frac{I}{Ma^2} \right)^{1/2}.$$

In this model, the density is ρ_m inside the radius a, and zero outside.

Consider, for example, the Moon, for which $I/Ma^2 = 0.392$, so that $\frac{5}{2}I/Ma^2 = 0.98$. Thus,

$$\rho_m = 1.03\bar{\rho}$$

and

$$\alpha = 0.99.$$

The inner density must exceed 1.03 times the mean density and α must be less than 0.99.

A different approach is to choose the function to be maximized or minimized on the basis of a probabilistic criterion. Following the principle of maximum entropy (Chapter 1) we should maximize

$$E = -\int_T k\rho \ln k\rho \, d\tau,$$

(where T is the volume of the planet) subject to the constraints of mass and moment of inertia, thus selecting the most random set of densities consistent with the mass and moment of inertia; k is a constant to be determined.

It is convenient to denote by σ_1 or σ_2 the ratio of ρ_1 or ρ_2 to $\bar{\rho}$, the mean density of the planet. Then it is convenient to take for the entropy

$$E = -\int_T \sigma \ln \sigma \, d\tau$$

and we wish to maximize E subject to constraints representing such data and *a priori* knowledge as we have. Suppose we know nothing about a planet. We then maximize E by putting

$$\frac{dE}{d\sigma} = 0$$

i.e.

$$\sigma = 1/e.$$

At first sight, this result determines ρ, but it does not, because we have no data nor *a priori* knowledge and so we do not know what $\bar{\rho}$ is and, therefore, we do not know ρ: if we know nothing about the planet, we can infer nothing.

Now suppose that we know the mass, i.e.

$$\int_T \rho \, d\tau = M = \frac{4\pi}{3} a^3 \bar{\rho},$$

where a is the mean radius of the planet. We then wish to maximize E subject to the condition that

$$\int_T \rho \, d\tau = \frac{4\pi}{3} a^3 \bar{\rho}.$$

According to the method of undetermined multipliers, we seek to maximize

$$E + \lambda \int_T \rho \, d\tau$$

with respect to variations of σ. With only one datum, we must suppose σ to be a constant throughout the planet, and so find

$$\frac{d}{d\sigma}\left[\frac{4\pi}{3} a^3 (-\sigma \ln \sigma + \lambda \bar{\rho}\sigma)\right] = 0$$

or

$$\ln \sigma + 1 - \lambda \bar{\rho} = 0$$

i.e.

$$\sigma = \exp(\lambda \bar{\rho} - 1).$$

As is usual in the method of undetermined multipliers, we determine λ by using the condition

$$\rho = \sigma \bar{\rho} = \bar{\rho} \exp[\lambda \bar{\rho} - 1] = \bar{\rho};$$

i.e.

$$\lambda \bar{\rho} = 1 \quad \text{or} \quad \lambda = 1/\bar{\rho}$$

and $\sigma = 1$. This trivial result reflects the fact that all we know about the planet is its mean density.

Consider now a model of two zones divided by a surface of radius a_1 and having relative densities σ_1 and σ_2 outside and inside a_1. As before, let $a_1/a = \alpha$. Once again, if we know only the mass of a planet, we can only suppose that the density is constant throughout. Now suppose that in addition to the mass we know the moment of inertia. The function to be maximized is

$$E + \frac{4\pi}{3} a^3 [\alpha^3 \sigma_2 + (1 - \alpha^3)\sigma_1]\bar{\rho}\lambda_1 + \frac{8\pi}{15} a^3 [\alpha^5 \sigma_2 + (1 - \alpha^5)\sigma_1]\bar{\rho}\lambda_2,$$

where a factor of a^2 has been included in λ_2. The variational equations are then

$$-3(\sigma_2 \ln \sigma_2 - \sigma_1 \ln \sigma_1) + 3(\sigma_2 - \sigma_1)\lambda_1 + 5\alpha^2(\sigma_2 - \sigma_1)\lambda_2 = 0$$

and

$$(1 - \alpha^3)(-\ln \sigma_1 - 1 + \lambda_1) + (1 - \alpha^5)\lambda_2 = 0,$$

while the equations of condition may be written

$$\sigma_2\alpha^3 + (1 - \alpha^3)\sigma_1 = 1 \quad \text{(mass)}$$
$$\sigma_2\alpha^5 + (1 - \alpha^5)\sigma_1 = \gamma. \quad \text{(inertia)}$$

It does not seem possible to obtain explicit solutions of these five equations, but some idea of the behaviour may be obtained by taking a different form for the entropy. The expression

$$-\sigma(\sigma - 1)$$

vanishes like $-\sigma \ln \sigma$ when $\sigma = 0$ and 1 and approaches $-\infty$ as σ approaches ∞, although at a different rate, and, if it is maximized, it is found that $\alpha = 0.752$, independent of the value of γ, provided it is close to 1. The same result is obtained if the function to be maximized is the variance of the relative density, that is

$$\alpha^3(\sigma_2 - 1)^2 + (1 - \alpha^3)(\sigma_1 - 1)^2.$$

Again, consider the Moon, for which $\gamma = 0.98$. With $\alpha = 0.75$, the corresponding values of σ_1 and σ_2 would be

$$\sigma_1 = 0.95, \quad \sigma_2 = 1.06.$$

Rietsch (1978) has also discussed the determination of the lowest possible value of the maximum density, ρ_u, in a two-zone model, and finds it to be given by

$$\rho_u = \rho_1 + (\bar{\rho} - \rho_1)^{5/2}/(\gamma\bar{\rho} - \rho_1)^{3/2}.$$

where ρ_u and ρ_l are the upper and lower bounds on the density. The corresponding value of α will be found to be

$$\alpha = \left(\frac{\gamma\bar{\rho}-\rho_l}{\bar{\rho}-\rho_l}\right)^{1/2}.$$

Suppose, for the Moon, for which $\bar{\rho} = 3340 \text{ kg/m}^3$, we take ρ_l to be 3600 kg/m^3. Then $\alpha = 0.9$, and $\rho_u = 3472 \text{ kg/m}^3$.

A.3 Continuous models

Instead of taking a two-zone model as a basis, models with a continuous variation of density have been studied according to the principles of maximum entropy. Gruber (1977) showed that if the entropy was taken to be of the form

$$k\rho \ln (k\rho),$$

that is, if the probability of obtaining a density ρ was taken to be proportional to the density, then the distribution of density has the form

$$\rho(r) = A \exp(-\lambda_2 r^2),$$

where A and λ_2 must be determined from the known mass and moment of inertia.

Rietsch (1977) took the entropy to be of the form

$$\int_T P(\rho) \ln [P(\rho)/\omega(\rho)] \, d\tau,$$

where $\omega(\rho)$ is a weighting factor associated with a continuous distribution of density and $P(\rho)$ is the probability of the occurrence of a density ρ and is normalized so that

$$\int_T P(\rho) \, d\tau = 1.$$

Rietsch calculates the expectation of the density, ρ_n, in an element of volume, T_n, namely

$$\tilde{\rho}_n = \int_{T_n} \rho_n P(\rho) \, d\tau.$$

The probability $P(\rho)$ is determined so as to maximize the entropy subject to the conditions on probability, mass and moment of inertia.

Rietsch finds that

$$\tilde{\rho}(r) = \rho_1 + 1/h(r, \lambda),$$

where

$$h(r, \lambda) = \lambda_1 + \tfrac{5}{3}\lambda_2(r/a)^2,$$

and λ_1 and λ_2 are the solutions of

$$\int_0^1 x^2 \, dx/h(xR, \lambda) = (\bar{\rho} - \rho_1)/3 \qquad \text{(mass)}$$

$$\int_0^1 x^4 \, dx/h(xR, \lambda) = (\gamma\bar{\rho} - \rho_1)/5 \qquad \text{(inertia)}.$$

APPENDIX 2

Combination of effects of small departures from a uniform distribution of density

The following results have been used to combine the effects of various features of lunar models in Chapter 5 and of Martian models in Chapter 6.

Suppose the density of a model departs by an amount $\rho_1(r)$ from a constant density ρ_0. Then the mass and moment of inertia are

$$M = M_0 + 4\pi \int \rho_1 r^2 \, \mathrm{d}r = M_0 + M_1$$

and

$$I = I_0 + \frac{8\pi}{3} \int \rho_1 r^4 \, \mathrm{d}r = I_0 + I_1.$$

Thus

$$\gamma = \frac{I}{Ma^2} = \frac{5}{2} \frac{I_0}{M_0 a^2}\left(1 + \frac{I_1}{I_0} - \frac{M_1}{M_0}\right) = \gamma_1,$$

where

$$\gamma_1 = 1 + \frac{I_1}{I_0} - \frac{M_1}{M_0}$$

since

$$I_0 = \tfrac{2}{5} M_0 a^2.$$

Now suppose there to be a second departure, $\rho_2(r)$. Then

$$\gamma = \frac{5}{2} \frac{I_0}{M_0 a^2}\left(1 + \frac{I_1}{I_0} - \frac{M_1}{M_0} + \frac{I_2}{I_0} - \frac{M_2}{M_0}\right) = \gamma_1 \gamma_2,$$

where

$$\gamma_2 = 1 + \frac{I_2}{I_0} - \frac{M_2}{M_0}.$$

Thus the effects of two small departures may be combined by calculating the separate values of γ and multiplying them.

318

APPENDIX 3

The physical librations of the Moon

This treatment follows that of Cook (1977). Figure A.4 shows the disposition of the Moon's axes of inertia and the co-ordinate systems used in the analysis of the physical librations; O is the centre of mass of the Moon; XY is the plane of the ecliptic; and Z the pole of the ecliptic. The axes $OXYZ$ rotate about OZ at the constant rate n, which is the average value of the motion of the Moon about the Earth. OX is in the

Figure A.4. Geometry of libration theory.

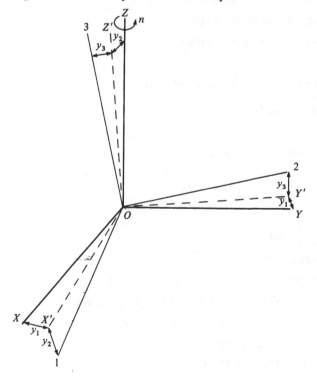

319

mean direction of the Earth and OY is approximately tangential to the Moon's orbit.

$O123$ coincide with the Moon's principal axes of inertia with the following correspondence:

$$O3{:}C, \ O2{:}B, \ O1{:}A.$$

Note that $C > B > A$.

The two systems of axes are related by the rotations, y_i:

y_1 about OZ: $OX \rightarrow OX'$
$OY \rightarrow OY'$

y_2 about OY': $OX' \rightarrow O1$
$OZ \rightarrow OZ'$

y_3 about $O1$: $OY' \rightarrow O2$
$OZ' \rightarrow O3$

The angular velocities are

$$n + \dot{y}_1 \text{ about } OZ$$

giving

$$(n + \dot{y}_1)\cos y_2 \text{ about } OZ',$$
$$-(n + \dot{y}_1)\sin y_2 \text{ about } O1,$$
$$(n + \dot{y}_1)\cos y_2 \sin y_3 \text{ about } O2$$

and

$$(n + \dot{y}_1)\cos y_2 \cos y_3 \text{ about } O3;$$
$$\dot{y}_2 \text{ about } OY$$

giving

$$\dot{y}_2 \cos y_3 \text{ about } O2$$

and

$$-\dot{y}_2 \sin y_3 \text{ about } O3;$$

and

$$\dot{y}_3 \text{ about } O1.$$

The resultants are

about $O1$: $-(n + \dot{y}_1)\sin y_2 + \dot{y}_3 = \Omega_1$

$O2$: $(n + \dot{y}_1)\cos y_2 \sin y_2 + \dot{y}_2 \cos y_3 = \Omega_2$

$O3$: $(n + \dot{y}_1)\cos y_2 \cos y_3 - \dot{y}_2 \sin y_3 = \Omega_3$.

The *kinetic energy*, T, is given by

$$2T = A\Omega_1^2 + B\Omega_2^2 + C\Omega_3^2.$$

Let l_1, l_2, l_3 be the direction cosines of the Earth relative to the principal axes of the Moon. The *potential energy*, V, is then given by

$$2V = -\frac{GM}{r^3}[A + B + C - 3(Al_1^2 + Bl_2^2 + Cl_3^2)].$$

M is the mass of the Earth and r its distance from the Moon.

If l is the vector (l_1, l_2, l_3), and l' is the corresponding vector of cosines relative to the ecliptic system $OXYZ$, then

$$l = M \cdot l',$$

where M is the product of three matrices:

$$M_1 = \begin{pmatrix} \cos y_1 & \sin y_1 & \cdot \\ -\sin y_1 & \cos y_1 & \cdot \\ \cdot & \cdot & 1 \end{pmatrix}$$

$$M_2 = \begin{pmatrix} \cos y_2 & \cdot & -\sin y_2 \\ \cdot & 1 & \cdot \\ \sin y_2 & \cdot & \cos y_2 \end{pmatrix}$$

$$M_3 = \begin{pmatrix} 1 & \cdot & \cdot \\ \cdot & \cos y_3 & \sin y_3 \\ \cdot & -\sin y_3 & \cos y_3 \end{pmatrix}.$$

l' is obtained from the theory of the lunar orbit; the leading terms are

$$l_1' = 1, \quad l_2' = 0, \quad l_3' = -2k \sin \nu t,$$

where $k = \sin \frac{1}{2}\theta$, θ being the inclination of the Moon's orbit to the ecliptic, and ν is $(n - n_0)g$, where n_0 is the mean motion of the Sun about the Earth, and g, a factor close to 1, represents the steady motion of the node of the Moon's orbit upon the ecliptic. The equations of motion are taken in the Lagrangian form:

$$\frac{d}{dt}\frac{\partial T}{\partial \dot{y}_i} - \frac{\partial T}{\partial y_i} = -\frac{\partial V}{\partial y_i}.$$

The variable part of the potential energy is proportional to $\sin \nu t$, so we take y_i to be proportional to $e^{i\nu t}$ and write $\dot{y}_i = i\nu y_i$, $\ddot{y}_i = -\nu^2 y_i$. The left-hand side of the equations of motion is then found to be

$$\begin{pmatrix} -\nu^2 & \cdot & \cdot \\ \cdot & \nu^2 - \beta & -i\nu(1 - \beta) \\ \cdot & i\nu(1 - \alpha) & \nu^2 - \alpha \end{pmatrix} \begin{pmatrix} y_1 \\ y_2 \\ y_3 \end{pmatrix} = N' \cdot y,$$

where $\alpha = (C - B)/A$ and $\beta = (C - A)/B$.

The right-hand side reduces to

$$(A, B, C)\frac{\partial V}{\partial y_i} = -3\begin{pmatrix}\gamma\\\beta\\0\end{pmatrix}\begin{pmatrix}y_1\\y_2\\y_3\end{pmatrix} + \begin{pmatrix}0\\6\beta k \sin \nu t\\0\end{pmatrix},$$

where $\gamma = (B - A)/C$. Then, since $\sin \nu t = (1/2i)(e^{i\nu t} - e^{-i\nu t})$, the equations of motion may be written as

$$\mathbf{N} \cdot \mathbf{y} = \frac{3\beta k}{i}\begin{pmatrix}0\\e^{\pm i\nu t}\\0\end{pmatrix},$$

where

$$\mathbf{N} = -\begin{pmatrix}\nu^2 - 3\gamma & \cdot & \cdot \\ \cdot & \nu^2 - 4\beta & -i\nu(1-\beta) \\ \cdot & i\nu(1-\alpha) & \nu^2 - \alpha\end{pmatrix}.$$

The solutions are then

$$\mathbf{y} = -\mathbf{N}^{-1}\frac{3\beta k}{i}\begin{pmatrix}0\\e^{\pm i\nu t}\\0\end{pmatrix},$$

where

$$\mathbf{N}^{-1} = -\begin{pmatrix}1/\nu^2 & \cdot & \cdot \\ \cdot & (\nu^2 - 1 - 3\beta)^{-1} & i\nu^{-1}(\nu^2 - 1 - 3\beta)^{-1} \\ \cdot & -i\nu^{-1}(\nu^2 - 1 - 3\beta)^{-1} & (\nu^2 - 1 - 3\beta)^{-1}\end{pmatrix},$$

with $+\nu$ written for ν when the forcing function is $e^{i\nu t}$, and $-\nu$ when it is $e^{-i\nu t}$. Hence

$$\mathbf{y} = -\frac{3k\beta}{i(\nu^2 - 1 - 3\beta)}\begin{pmatrix}0\\e^{i\nu t} - e^{-i\nu t}\\-i\nu^{-1}(e^{i\nu t} + e^{-i\nu t})\end{pmatrix}.$$

Now ν is $g(n - n_0)$ or $1 + g' - (n_0/n)$, where $g' = 0.0852$ and $n_0/n = 0.0808$. ν is very close to 1. Thus, the effect of the inclination of the Moon's orbit to the ecliptic is that there is no forced oscillation of y_1, but

$$y_2 = -\frac{6k\beta \sin \nu t}{2\{g' - (n_0/n)\} - 3\beta}$$

and

$$y_3 = \frac{6k\beta \cos \nu t}{[2\{g' - (n_0/n)\} - 3\beta]\nu}.$$

The solutions represent a rotation of the C axis about the pole of the ecliptic at the speed $(n - n_0)(1 + g)$ and with amplitude

$$3k\beta/[g' - (n_0/n) - \tfrac{3}{2}\beta]$$

(because ν is not exactly 1, the amplitudes of y_2 and y_3 differ slightly).

With $k = 0.045$, $\beta = 6 \times 10^{-4}$, and $g' - (n_0/n) \sim 0.005$, the angular amplitude of the motion is about $1°30'$.

The divisor $\nu^2 - 1 - 3\beta$ results in a large amplification.

To illustrate the theory, only one variable term in l' has been retained. In fact, l' contains a very large number of terms, the next most important being proportional to the lunar eccentricity. However, for the great majority of the terms, the divisor $\nu^2 - 1 - 3\beta$ is not very small. Other terms which must be included are the attraction of the Sun, and terms in the potential energy corresponding to the third and fourth harmonics in the gravity field of the Moon.

REFERENCES

Chapter 1

Backus, G. (1970a, b, c). Inference from inadequate and inaccurate data.
I, II, III. *Proc. Nat. Academy Sci.*, **65**, 1–7, **65**, 281–7, **68**, 282–9.
Backus, G. and Gilbert, F. (1967). Numerical applications of a formalism to
geophysical inverse problems. *Geophys. J. R. Astronom. Soc.*, **13**, 247–76.
Backus, G. and Gilbert, F. (1968). The resolving power of gross Earth data.
Geophys. J. R. Astronom. Soc., **16**, 169–205.
Backus, G. and Gilbert, F. (1970). Uniqueness in the inversion of inaccurate
gross Earth data. *Philos. Trans. R. Soc. A*, **266**, 123–92.
Cook, A. H. (1979). Geophysics and the human condition. *Q.J.R. Astronom.
Soc.*, **20**, 229–40.
Graber, M. A. (1977). An information theory approach to the density of the
Earth. *N.A.S.A. Tech. Memorandum 78034* (N.A.S.A.: Goddard Space
Flight Center, Greenbelt, Md.).
Parker, R. L. (1972). Inverse theory with grossly inadequate data. *Geophys.
J. R. Astronom. Soc.*, **29**, 123–38.

Chapter 2

Bullen, K. E. (1975). *The Earth's Density*. London: Chapman and Hall.
Cook, A. H. (1973). *Physics of the Earth and Planets*. London: Macmillan.
Dziewonski, A. M., Hales, A. L. and Lapwood, E. R. (1975). Parametrically
simple Earth models consistent with geophysical data. *Phys. Earth Planet.
Int.*, **10**, 12–48.
Gilbert, F. and Dziewonski, A. M. (1975). An application of normal mode
theory to the retrieval of structural parameters and source mechanisms from
seismic spectra. *Philos. Trans. R. Soc. A*, **278**, 187–269.
Moffat, H. V. (1978). *Magnetic field generation in electrically conducting
fluids*. Cambridge University Press.
Oldham, R. D. (1906). Constitution of the interior of the Earth as revealed
by earthquakes. *Q. J. Geol. Soc.*, **62**, 456–75.
Whittaker, E. T. and Watson, G. N. (1940). *A Course of Modern Analysis*,
4th edn. Cambridge University Press.
Williamson, E. D. and Adams, L. H. (1923). Density distribution in the
Earth. *J. Wash. Acad. Sci.*, **13**, 413–28.

Chapter 3

Cook, A. H. (1963). The contribution of observations of artificial satellites to the determination of the Earth's gravitational potential. *Space Sci. Rev.*, **2**, 355–437.

Cook, A. H. (1976). Measurements of distances to the Moon and artificial satellites. *Contemp. Phys.*, **17**, 577–98.

Cook, A. H. (1978). The estimation of gravitational potentials of planets. *Boll. geod. Sci. affine.*, **37** (no. 2–3), 325–6.

Darwin, Sir G. H. (1899). The theory of the figure of the Earth carried to the second order of small quantities. *Mon. Not. R. Astronom. Soc.*, **60**, 82–124.

Goldstein, R. M. (1971). Radar observations of Mercury. *Astronom. J.*, **96**, 1152–4.

Jeffreys, Sir Harold (1970). *The Earth*, 5th edn. Cambridge University Press.

Kaula, W. M. (1966). *Theory of Satellite Geodesy.* Blaisdell.

Klaasen, K. P. (1975). Mercury rotation period determined from Mariner 10 photography. *J. Geophys. Res.*, **80**, 3415.

Murray, J. B., Dollfus, A. and Smith B. (1972). Cartography of the surface of Mercury. *Icarus*, **17**, 576–84.

Newton, Sir I. (1687). *Philosphiae Naturalis Principia Mathematica.* Prop. XIX, Prob. II.

Radau, R. (1885). Sur la loi des densités à l'intérieur de la Terre. *C.R. Acad. Sci. Paris*, **100**, 972–4.

Shapiro, I. I. (1967). New method for the detection of light deflection by solar gravity. *Science*, **157**, 806–7.

Whittaker, E. T. and Watson, G. N. (1940). *A Course of Modern Analysis* 4th edn. Cambridge University Press.

Zharkhov, V. N. and Trubitsyn, V. P. (1970). Theory of the figure of rotating planets in hydrostatic equilibrium – a third approximation. *Soviet. Phys. Astronomy (Engl. Translation)*, **13**, 981–8

Chapter 4

Ahrens, T. J., Anderson, D. L. and Ringwood, A. E. (1969). Equations of state and crystal structures of high pressure phases of shocked silicates and oxides. *Rev. Geophys.* **7**, 667–707.

Altshuler, L. V., Krupnikov, K. K. and Brazhnik, M. I. (1958). Dynamic compressibilities of metals under pressures from 400,000 to 4,000,000 atmospheres. *J. Exptl. Th. Phys. (U.S.S.R.)*, **34**, 886–93, transl. *Sov. Phys. J.E.T.P.*, **7**, 614–19.

Altshuler, L. V., Krupnikov, K. K., Ledenev, B. N., Zhuchikhin, V. I. and Brazhnik, M. I. (1958a). Dynamic compressibility and equation of state of iron under high pressure. *J. Exptl. Th. Phys. (U.S.S.R.)*, **34**, 874–85, transl. *Sov. Phys. J.E.T.P.*, **7**, 606–14.

Anderson, D. L. (1967). A seismic equation of state. *Geophys. J. R. Astronom. Soc.*, **13**, 9–30.

Anderson, D. L. (1969). Bulk modulus–density systematics. *J. Geophys. Res.*, **74**, 3857–64.

Anderson, O. L. (1968). Some remarks on the volume dependence of the Grüneisen parameter. *J. Geophys. Res.*, **73**, 5187–94.

Anderson, O. L. and Liebermann, R. C. (1970). Equations for the elastic constants and their pressure derivatives for three cubic lattices and some geophysical applications. *Phys. Earth Planet. Int.*, **3**, 61–85.

Anderson, O. L., Schneider, E., Liebermann, R. and Soga, N. (1968). Some elastic constant data on minerals relevant to geophysics. *Rev. Geophys.*, **6**, 491–524.

Bassett, W. A. and Ming, L. (1972). Disproportionation of Fe_2SiO_4 to $2FeO$ and SiO_2 at pressures up to 250 kb and temperatures up to 3000 °C. *Phys. Earth Planet. Int.*, **6**, 154–60.

Berggren, K. E. and Fröman, A. (1969). Properties of compressed states of aluminium and iron from a spherical cellular model. *Ark. Phys.*, **39**, 355–81.

Bernal, J. D. (1936). Discussion Report, *Observatory*, **59**, 268.

Birch, F. (1947). Finite elastic strain of cubic crystals. *Phys. Rev.*, **71**, 809–24.

Birch, F. (1952). Elasticity and constitution of the Earth's interior. *J. Geophys. Res.*, **57**, 227–86.

Birch, F. (1972). The melting relations of iron and temperatures in the Earth's core. *Geophys. J. R. Astronom. Soc.*, **29**, 373–87.

Boschi, E. (1974a). On the melting curve at high pressures. *Geophys. J. R. Astronom. Soc.*, **37**, 45–50.

Boschi, E. (1974b). Melting of iron. *Geophys. J. R. Astronom. Soc.*, **29**, 327–34.

Boschi, E. and Caputo, M. (1969). Equations of state at high pressures and the Earth's interior. *Riv. Nuovo Cim. Ser.* 1, **1**, 441–513.

Bukowinski, M. S. T. and Knopoff, L. (1976). Electronic structure of iron and models of the Earth's core. *Geophys. Res. Lett.*, **3**, 45–8.

Bundy, F. P. (1963). Direct conversion of graphite to diamond in static pressure apparatus. *J. Chem. Phys.*, **38**, 631–43.

Carter, W. J., Marsh, S. P., Fritz, J. N. and McQueen, R. G. (1971). The equations of state of selected materials for high pressure reference. *Nat. Bur. Stds. Sp. Publ.*, **326** (Washington, D.C., U.S. Govt. Printing Office) 147–58.

Chan, T., Spetzler, H. A. and Meyer, M. D. (1976). Equation of state parameters for liquid metals and energetics of Earth's core. *Tectonophysics*, **35**, 271–83.

Cohen, M. L. and Heine, V. (1970). The fitting of pseudopotentials to experimental data. *Solid St. Phys.*, **24**, 37–248.

Cook, A. H. (1972). The dynamical properties and internal structures of the Earth, the Moon and the planets. *Proc. R. Soc. A*, **328**, 301–36.

Davies, G. F. and Anderson, D. L. (1971). Revised shock wave equations of state for high pressure phases of rocks and minerals. *J. Geophys. Res.*, **76**, 2617–27.

Dugdale, T. S. and MacDonald, D. K. C. (1953) The thermal expansion of solids. *Phys. Rev.*, **89**, 832–4.

Duvall, G. E. and Fowles, G. R. (1967). Shock waves. In *High Pressure Physics and Chemistry*, ed. R. S. Bradley, vol. 2, 209–291, London: Academic Press.

Feynman, R. P., Metropolis, N. and Teller, E. (1949). Equations of state of elements based on generalised Fermi–Thomas theory. *Phys. Rev.*, **75**, 1561.

Fröhlich, H. (1973). On the connection between macro and microphysics. *Riv. Nuovo Cim.*, **3**, 490–534.

Fürth, R. (1944). On the equation of state for solids. *Proc. R. Soc. A*, **183**, 87–110.

Gilvarry, J. J. (1954). Relativistic Thomas–Fermi atom model. *Phys. Rev.*, **95**, 71–2.

Gilvarry, J. J. (1969). Equations of state at high pressure from the Thomas–Fermi model. In *The Applications of Modern Physics to the Earth and Planetary Interiors*, ed. S. K. Runcorn, 313–403. London: Wiley.

Gilvarry, J. J. and Peebles, G. H. (1954). Solutions of the temperature perturbed Thomas–Fermi equation. *Phys. Rev.*, **99**, 550–2.

Goldschmidt, V. M. (1931). Zur Kristallchemie des Germaniums. *Nachr. Gessells. Wiss. Göttingen, Math. Phys. Kl.*, 184–90.

Hall, H. T. (1958). Some high-pressure high temperature apparatus design considerations: Equipment for use at 100,000 atmospheres and 3000 °C. *Rev. Sci. Instrum.*, **29**, 267–75.

Hall, H. T. (1960). Ultrahigh-pressure, high temperature apparatus, the "Belt". *Rev. Sci. Instrum.* **31**, 125.

Heine, V. (1970). The pseudopotential concept. *Solid St. Phys.*, **24**, 1–36.

Heine, V. and Weaire, D. (1970). Pseudopotential theory of cohesion and structure. *Solid St. Phys.*, **24**, 249–63.

Jeffreys, H. (1937). On the materials and density of the Earth's crust. *Mon. Not. R. Astronom. Soc. Geophys. Suppl.*, **4**, 50–61.

John, M. S. and Eyring, H. (1971). The significant structure theory of liquids. In *Physical Chemistry, an advanced treatise*, ed. D. Henderson, **VIIIa**. New York: Academic Press.

Kawai, N. (1971). Equipment for generating pressures up to 800 kb. *Nat. Bur. Stds. Sp. Publ.*, **326**, Washington, D.C., U.S. Govt. Printing Office, 45–8.

Kittel, C. (1968). *Introduction to Solid State Physics*, 3rd edn. London, New York, Sydney: Wiley.

Knopoff, L. (1963). Solids: Equations of state of solids at moderately high pressures. In *High Pressure Physics and Chemistry*, ed. R. S. Bradley, vol. 1, 227–45. London: Academic Press.

Knopoff, L. and MacDonald, G. J. F. (1960). An equation of state for the core of the Earth. *Geophys. J. R. Astronom. Soc.*, **3**, 68–77.

Knopoff, L. and Shapiro, J. N. (1969). Comments on the inter-relationships between Grüneisen's parameter and shock and isothermal equations of state. *J. Geophys. Res.*, **74**, 1439–50.

Kraut, E. A. and Kennedy, G. C. (1966). New melting law at high pressures. *Phys. Rev.*, **151**, 668–75.

Kumazawa, M. and Anderson, O. L. (1969). Elastic moduli, pressure derivatives and temperature derivatives of single crystal olivine and single crystal forsterite. *J. Geophys. Res.*, **74**, 5961–72.

Leppaluota, D. A. (1972). Melting of iron by significant structure theory. *Phys. Earth Planet. Int.*, **6**, 175–81.

Lindemann, F. A. (1910). *Phys. Zeit.*, **11**, 605.

Liu, L. G. (1975*a*). Post-oxide phases of forsterite and enstatite. *Geophys. Res. Lett.*, **2**, 417–19.

Liu, L. G. (1975*b*). Post-oxide phases of olivine and pyroxene and mineralogy of the mantle. *Nature*, **258**, 510–12.

Liu, L. G. (1976). The high pressure phase of $MgSiO_3$. *Earth Planet. Sci. Lett.*, **31**, 200–8.

McQueen, R. G. and Marsh, S. P. (1966). *Handbook of Physical Constants*, ed. S. P. Clark, Jr., *Mem. Geol. Soc. Amer.* **97**.

March, N. H. (1955). Equations of state of elements from Thomas–Fermi theory. *Proc. Phys. Soc. A.*, **68**, 726–34.

Merrill, L. and Bassett, W. A. (1974). Miniature diamond anvil pressure cell for single crystal X-ray diffraction studies. *Rev. Sci. Instrum.*, **45**, 290–4.

Ming, L. and Bassett, W. A. (1975). The post-spinel phases in the Mg_2SiO_4–Fe_2SiO_4 system. *Science*, **187**, 66–8.

Munro, D. C. (1967). Structural determinations by X-rays of systems at high pressure. In *High Pressure Physics and Chemistry*, ed. R. S. Bradley, vol. 2, 311–23. London: Academic Press.

Murnaghan, F. D. (1944). The compressibility of media under extreme pressures. *Proc. Nat. Acad. Sci.*, **30**, 244–7.

Murnaghan, F. D. (1951). *Finite deformation of an elastic solid*. New York: Wiley.

Navrotsky, A. and Kasper, R. B. (1976). Spinel disproportionation at high pressures; calorimetric determination of the enthalpy of formation of Mg_2SnO_4 and Co_2SnO_4 and some implications for silicates. *Earth Planet. Sci. Lett.*, **31**, 247–54.

Papika, J. J. and Cameron, Margyellen (1976). Crystal chemistry of silicate minerals of geophysical interest. *Rev. Geophys. Space Phys.*, **14**, 37–80.

Ramsey, W. H. (1950). On the compressibility of the Earth. *Mon. Nat. R. Astronom. Soc., Geophys. Suppl.*, **6**, 42–9.

Ramsey, W. H. (1963). On the densities of methane, metallic ammonium, water and neon at planetary pressures. *Mon. Not. R. Astronom. Soc.*, **125**, 469–85.

Ree, F. H. (1971). Computer calculations for model systems. In *Physical Chemistry, an Advanced Treatise*, ed. D. Henderson, vol. VIIIa. New York: Academic Press.

Rice, H. M., McQueen, R. G. and Walsh, J. (1958). Compression of solids by strong shock waves. *Solid St. Phys.*, **6**, 1–63.

Ringwood, A. E. (1962). A model for the upper mantle. *J. Geophys. Res.*, **67**, 857–66.

Ringwood, A. E. (1975). *Composition and petrology of the Earth's mantle*. New York: McGraw-Hill.

Ringwood, A. E. and Major, A. (1968). Apparatus for phase transition studies at high pressures and temperatures. *Phys. Earth Planet. Int.*, **1**, 164–8.

Shapiro, J. N. and Knopoff, L. (1969). Reduction of shock-wave equations of state to isothermal equations of state. *J. Geophys. Res.*, **74**, 1435–8.

Simon, F. E. (1937). On the range of stability of the fluid state. *Trans. Faraday Soc.*, **33**, 65.

Slater, J. C. (1940). Note on Grüneisen's constant for the incompressible metals. *Phys. Rev.*, **57**, 744–6.

Suito, K. (1972). Phase transformations of pure Mg_2SiO_4 into a spinel structure under high pressures and temperatures. *J. Phys. Earth*, **20**, 225–43.

Takeuchi, H. and Kanamori, H. (1966). Equations of state of matter from shock wave experiments. *J. Geophys. Res.*, **71**, 3985–94.

Teller, E. (1962). On the stability of molecules in the Thomas–Fermi theory. *Rev. Mod. Phys.*, **34**, 627–31.

Thomsen, L. and Anderson, O. L. (1971). Consistency in the high temperature equation of state of solids. *Nat. Bur. Stds. Sp. Publ.*, **326** (Washington, D.C., U.S. Govt. Printing Office) 209–17.

Tozer, D. (1967). Towards a theory of convection in the mantle. In *The Earth's Mantle*, ed. T. F. Gaskell. London: Academic Press.

Weaver, J. S., Takahashi, T. and Bassett, W. A. (1971). Calculations of the $p-V$ relation for sodium chloride up to 300 kb at 25 °C. *Nat. Bur. Stds. Sp. Publ.*, **326**, (Washington, D.C., U.S. Govt. Printing Office) 189–99.

Wildt, R. (1963). Planetary interiors. In *Planets and Satellites. The Solar System III*, ed. G. P. Kuiper and B. M. Middlehurst, 159–212. University of Chicago Press.

Wilson, A. H. (1966). *Thermodynamics and Statistical Mechanics.* Cambridge University Press.

Chapter 5

Ananda, M. P. (1977). Lunar gravity, a mass point model. *J. Geophys. Res.*, **82**, 3049–64.

Anderson, J. D., Efrom, L. and Wong, S. K. (1970). Martian mass and Earth–Moon mass ratio from coherent S-band tracking of Mariners 6 and 7, *Science*, **167**, 277–9.

Bender, P. L., Currie, D. G., Dicke, R. H., Eckhardt, D. H., Faller, J. E., Kaula, W. M., Mulholland, J. D., Plotkin, H. H., Poultney, S. K., Silverberg, E. C., Wilkinson, D. I., Williams, J. G. and Alley, C. O. (1973). The lunar laser ranging experiment. *Science*, **182**, 229–38.

Bessanova, E. N., Fishman, V. M., Ryaboyi, V. Z. and Sitnikova, G. A. (1974). The tau method for the inversion of travel times. I: deep seismic sounding data. *Geophys. J. R. Astronom. Soc.*, **36**, 377–98.

Bills, B. B. and Ferrari, A. J. (1977). A lunar density model consistent with topographic, gravitational, librational and seismic data. *J. Geophys. Res.*, **82**, 1306–14.

Blackshear, W. T. and Gapcynski, J. P. (1977). An improved value of the lunar moment of inertia. *J. Geophys. Res.*, **82**, 1699–1701.

Bryant, W. L. and Williamson, R. G. (1974). Lunar gravity analysis from Explorer 49. *AIAA Paper 74-810, AIAA Mechanics and Control of Flight Conference*, Anaheim, Cal., Aug. 5–9 1974.

Burnett, D. S. (1975) Lunar Science: the Apollo legacy. *J. Geophys. Res.*, **13** (No. 3), 13–34.

Cole, G. H. A. (1971). On inferring elastic properties of the deep lunar interior. *Planet. Space. Sci.*, **19**, 929–47.

Cook, A. H. (1972). The dynamical properties and internal structures of the Earth, the Moon and the planets. *Proc. R. Soc. A*, **328**, 301–36.

Cooper, M. R., Kovach, R. L. and Watkins, J. S. (1974). Lunar near surface structure. *Rev. Geophys. Space Phys.*, **12**, 291–308.

Dainty, A. M., Goins, N. R. and Toksöz, M. N. (1975). The structure of the Moon as determined from natural lunar seismic events. *Lunar Sci.*, **6**, 175–7. (Lunar Sci. Inst., Houston, Tex).

Duba, A., Heard, H. C. and Shock, R. N. (1974). Electrical conductivity of olivine at high pressure and under controlled oxygen fugacity. *J. Geophys. Res.*, **79**, 1667–73.

Dyal, P. and Parkin, C. W. (1973). Global electromagnetic induction in the Moon and planets. *Phys. Earth. Planet Int.*, **7**, 251–65.

Dyal, P., Parkin, C. W. and Daily, W. D. (1974a). Temperature and electrical conductivity of the lunar interior from magnetic transient measurements of the lunar tail. *Proc. V Lunar Sci. Conf.*

Dyal, P., Parkin, C. W. and Daily, W. D. (1974b). Magnetism and the interior of the Moon. *Rev. Geophys. Space Phys.*, **12**, 568–91.

Felsentreger, T. L. (1968). Classification of lunar satellite orbits. *Planet. Space. Sci.*, **16**, 285–95.

Ferrari, A. J. (1977). Lunar gravity: a harmonic analysis. *J. Geophys. Res.*, **82**, 3065–85.

Fuller, M. (1974). Lunar magnetism. *Rev. Geophys. Space. Phys.*, **12**, 23–70.

Gapcynski, J. P., Blackshear, W. T., Tolson, W. H. and Compton, H. R. (1975). A determination of the lunar moment of inertia. *Geophys. Res. Lett.*, **2**, 353–6.

Gilbert, F. and Dziewonski, A. M. (1975). An application of normal mode theory to the retrieval of structural parameters and source mechanisms from seismic spectra. *Philos. Trans. R. Soc. A*, **278**, 187–269.

Gubbins, D. (1974). Theories of geomagnetic and solar dynamos. *Rev. Geophys. Space. Phys.*, **12**, 137–54.

Hobbs, B. A. (1973). The inverse problem of the Moon's electrical conductivity. *Earth Planet. Sci. Lett.*, **17**, 380–4.

Keihm. S. J., Peters, K., Langseth, M. G. and Chute, J. L. (1973). Apollo 15 measurement of lunar surface brightness; thermal conductivity of the upper $\frac{1}{2}$ meters of regolith. *Earth Planet. Sci. Lett.*, **19**, 337–51.

King, R. W., Counselman, C. C. III and Shapiro, I. I. (1976). Lunar dynamics and selenodesy; results from analysis of V.L.B.I. and laser data. *J. Geophys. Res.*, **81**, 6251–6.

Kuckes, F. (1971). Lunar electrical conductivity profile. *Nature*, **232**, 249–51.

Kumagawa, M. and Anderson, O. L. (1969). Elastic moduli, pressure derivatives and temperature derivatives of single crystal olivine and single crystal forsterite. *J. Geophys. Res.*, **74**, 5961–72.

Lammlein, D. R. (1977). Lunar seismicity, structure and tectonics. *Philos. Trans. R. Soc. A.*, **285**, 451–61.

Lammlein, D. R., Latham, G. V., Dorman, J., Nakamura, Y. and Ewing, M. (1974). Lunar seismicity, structure and tectonics. *Rev. Geophys. Space Phys.*, **12**, 1–21.

Langseth, M. G., Clark, S. P., Chute, J. C., Keihm, S. J. and Wechsler, A. E. (1972). The Apollo 15 lunar heat flow measurement. *The Moon*, **4**, 390–410.

Langseth, M. G., Keihm, S. and Chute, J. C. (1973). Apollo 17 heat flow measurement. *Trans. Amer. Geophys. Un.* EOS, **54**, 349.

Lyttleton, R. A. (1963). On the origin of mountains. *Proc. R. Soc. A.*, **275**, 1–22.

McMechan, G. A. and Wiggins, R. A. (1972). Depth limits in body wave inversions. *Geophys. J. R. Astronom. Soc*, **28**, 459–73.

Michael, W. H. and Blackshear, W. T. (1972). Recent results on the mass, gravitational field and moment of inertia of the Moon. *The Moon*, **3**, 388–402.

Moffat, H. K. (1978). *Magnetic field generation in electrically conducting fluids*. Cambridge University Press.

Nakamura, Y., Latham, G., Lammlein, D., Ewing, M., Duennebier, F. and Dorman, J. (1974). Deep lunar interior inferred from recent seismic data. *Geophys. Res. Lett.*, **1**, 137–40.

Rietsch, E. (1977). The maximum entropy approach to inverse problems. *J. Geophys. Res.*, **42**, 489–506.

Runcorn, S. K. (1977). Interpretation of lunar potential fields. *Philos. Trans. R. Soc. A*, **285**, 507–16.

Schubert, G., Young, R. E. and Cassen, P. (1977). Solid state convection models of the lunar internal temperature. *Philos. Trans. R. Soc. A.* **285**, 523–36.

Sjogren, W. J. (1971). Lunar gravity estimate: independent confirmation. *J. Geophys. Res.*, **76**, 7021–6.

Sjogren, W. J. and Wollenhaupt, W. R. (1976). Lunar global figure from Mare surface elevation. *The Moon*, **15**, 143–54.

Sonnett, C. P., Colbµrn, D. S., Smith, B. F., Schubert, G. and Schwartz, K. (1972). The induced magnetic field of the Moon: conductivity profiles and inferred temperature. *Proc. III Lunar Sci Conf.* (Lunar Sci. Inst., Houston, Texas), 2309.

Toksöz, M. N. (1975). Lunar and planetary seismology. *J. Geophys. Res.*, **13** (No. 3), 306–11.

Toksöz, M. N., Dainty, A. M., Solomon, C. C. and Anderson, K. R. (1974). Structure of the Moon. *Rev. Geophys. Space Phys.*, **12**, 539–67.

Williams, J. G. (1976). Statement at Royal Society Discussion Meeting.

Chapter 6

Allen, C. W. (1963). *Astrophysical Quantities*, 2nd edn. University of London: Athlone Press.

Anderson, D. L., Miller, W. F., Latham, G. V., Nakamura, Y., Toksöz, M. N.,

Dainty, A. M., Duennebier, F. K., Lazarewicz, A. R., Kovach, R. L. and Knight, T. C. D. (1977). Seismology on Mars. *J. Geophys. Res.*, **82**, 4524–46.

Ash, M. E., Campbell, D. B., Dyce, R. B., Ingalls, R. P., Jurgens, R., Shapiro, I. I., Slade, M. A. and Thompson, T. W. (1968). The case for the radar radius of Venus. *Science*, **160**, 985–7.

Ash, M. E., Shapiro, I. I. and Smith, W. B. (1967). Astronomical constant and planetary ephemerides deduced from radar and optical observations. *Astronom. J.*, **72**, 338–50.

Binder, A. B. and Davis, D. R. (1973). Internal structure of Mars. *Phys. Earth Planet. Int.*, **7**, 477–85.

Born, G. H. (1974). Mars physical parameters as determined from Mariner 9 observations of natural satellites and Doppler tracking. *J. Geophys. Res.*, **79**, 4837–44.

Bullen, K. E. (1975). *The Earth's Density.* London: Chapman and Hall.

Cain, D. L., Kliore, A. J., Seidel, B. L. and Sykes, M. J. (1972). The shape of Mars from Mariner 9 occultations. *Icarus*, **17**, 517–24.

Christensen, E. J. (1975). Martian topography derived from occultation radar, spectral and optical measurements. *J. Geophys. Res*, **80**, 2909–13.

Cook, A. H. (1972). The dynamical properties and internal structures of the Earth, the Moon and the planets. *Proc. R. Soc. A*, **328**, 301–36.

Cook, A. H. (1977). The moment of inertia of Mars and the existence of a core. *Geophys. J. R. Astronom. Soc.*, **51**, 349–56.

Deane, J. A. (1976). Mariner 10 observations of Mercury. *Space Res.*, **16**, 965–8.

de Vaucouleur, G., Davies, M. E. and Sturms, F. M. Jr. (1973). Mariner 9 aerographic co-ordinate system. *J. Geophys. Res.*, **78**, 4395–404.

Dollfus, A. (1970). *Surfaces and Interiors of Planets and Satellites.* London and New York: Academic Press.

Dollfus, A. (1972*a*). New optical measurements of planetary diameters. Part II: Planet Venus. *Icarus*, **17**, 517–24.

Dollfus, A. (1972*b*). New optical measurements of planetary diameters. Part IV: Planet Mars. *Icarus*, **17**, 525–39.

Evans, J. V. and Hagfors, T. (eds) (1968). *Radar Astronomy.* New York: McGraw-Hill.

Gapcynski, J. P., Tolson, R. H. and Michael, W. H. Jr. (1977). Mars gravity field combined Viking and Mariner 9 results. *J. Geophys. Res.*, **82**, 4325–7.

Goldstein, R. M. (1971). Radar observations of Mercury. *Astronom. J.*, **76**, 1152–4.

Green, P. E. (1968). Radar measurement of target scattering properties. In *Radar Astronomy*, ed. J. V. Evans and T. Hagfors, 1–77. New York: McGraw-Hill.

Grossman, L. and Larimer, J. W. (1974). Early chemical history of the solar system. *Rev. Geophys. Space Phys.*, **12**, 71–101.

Howard, H. T., Tyler, G. L., Fjeldbo, G., Kliore, A. J., Levy, G. S., Brunn, D. L., Dickinson, R., Edelson, R. F., Martin, W., Postal, R. B., Seidel, B., Sesplankis, T. T., Shirley, D. L., Stelzried, C. T., Sweetnam, D. N.,

Zygielbaum, A. F., Esposito, P. B., Anderson, J. D., Shapiro, I. I. and Reasenberg, R. D. (1974a). Venus mass, gravity field, atmosphere and ionosphere as measured by the Mariner 10 dual frequency radio system. *Science*, **183**, 1297–1301.

Howard, H. T., Tyler, G. L., Esposito, P. B., Anderson, J. D., Reasenberg, R. D., Shapiro, I. I., Fjeldbo, G., Kliore, A. J., Levy, G. S., Brunn, D. L., Dickinson, R., Edelson, R. E., Martin, W. C., Postal, R. B., Seidel, B., Sesplankis, T. T., Shirley, D. L., Stelzried, C. T., Sweetnam, D. N., Wood, G. E., and Zygielbaum, A. F. (1974b). Mercury: results on mass, radius, ionosphere and atmosphere from Mariner 10 dual frequency radio signals. *Science*, **185**, 179–80.

Jeffreys, H. (1937). The density distribution of the inner planets. *Mon. Not. R. Astronom Soc., Geophys. Suppl.*, **4**, 62–71.

Johnston, D. H., McGetchin, D. R. and Toksöz, M. N. (1974). The thermal state and internal structure of Mars. *J. Geophys. Res.*, **79**, 3959–71.

Johnston, D. H. and Toksöz, M. N., 1977. Internal structure and properties of Mars. *Icarus*, **32**, 72–84.

Jordan, J. F. and Lorell, J. (1975). Mariner 9: an instrument of dynamical science. *Icarus*, **25**, 146–65.

Klaasen, K. P. (1975). Mercury rotation period determined from Mariner 10 photography. *J. Geophys. Res.*, **80**, 2415.

Kliore, A. J., Fjeldbo, G. and Seidel, B. L. (1971). Summary of Mariner 6 and 7 radio occultation results on the atmosphere of Mars. *Space. Res.*, **11**, 165–75.

Kovalevsky, J. (1970). Détérmination des masses des planètes et satellites. In *Surfaces and Interiors of Planets and Satellites*, ed. A. Dollfus, 1–77. London and New York: Academic Press.

Lewis, J. S. (1972). Metal/silicate fractionation in the solar system. *Earth Planet. Sci. Lett.*, **15**, 286–90.

Lorell, J., Born, G. H., Christensen, E. J., Esposito, P. B., Jordan, J. F., Laing, P. A., Sjogren, W. L., Wong, S. K., Reasenberg, R. D., Shapiro, I. I. and Slater, G. L. (1973). Gravity field of Mars from Mariner 9 tracking data. *Icarus*, **18**, 304–16.

Lyttleton, R. A. (1965). On the internal structure of the planet Mars. *Mon. Not. R. Astronom. Soc.*, **129**, 21–39.

Marov, M. Ya and Petrov, G. I. (1973). Investigations of Mars from the automatic stations Mars 2 and 3. *Icarus*, **19**, 163–79.

Melbourne, W. G., Muhlmann, D. O. and O'Handley, D. A. (1968). Radar determinations of the radius of Venus. *Science*, **160**, 987–9.

Murray, J. B., Dollfus, A. and Smith, B. (1972). Cartography of the surface of Mercury. *Icarus*, **17**, 576–84.

Null, G. W. (1969). A solution for the mass and dynamical oblateness of Mars using Mariner IV Doppler data. *Bull. Amer. Astronom. Soc.*, **1**, 356.

Rea, D. G. (1970). NASA Planetary Programme: Present and Future. *Space Res.*, **10**, 1028–35.

Reasenberg, R. D. (1977). The moment of inertia and isostasy of Mars. *J. Geophys. Res.*, **82**, 369–75.

Ringwood, A. E. and Clark, S. P. (1971). Internal constitution of Mars. *Nature*, 234, 89–92.

Shapiro, I. I. (1967). Resonance rotation of Venus. *Science*, 157, 423–5.

Shapiro, I. I. (1968). Spin and orbital motions of planets. In *Radar Astronomy*, ed. J. V. Evans and T. Hagfors, 143–85. New York: McGraw-Hill.

Sinclair, A. T. (1972). The motions of the satellites of Mars. *Mon. Not. R. Astronom. Soc.*, 155, 249–73.

Sjogren, W. L., Lorell, J., Wong, L. and Downs, W. (1975). Mars gravity field based on a short arc technique. *J. Geophys. Res.*, 80, 2899–908.

Smith, B. A. (1971). Mariner 6 and 7 television results. *Space Res.*, 11, 155–64.

Soffen, G. A. and Young, A. T. (1972). The Viking missions to Mars. *Icarus*, 16, 1–16.

Solomon, S. L. and Toksöz, M. N. (1973). Internal constitution and evolution of the Moon. *Phys. Earth Planet. Int.*, 7, 15–38.

Chapter 7

Anderson, D. L. (1977). Composition of the mantle and core. *Ann. Rev. Earth Planet. Sci.*, 5, 179–202.

Anderson, M. S. and Swenson, C. A. (1974). Experimental compressions for normal hydrogen and normal deuterium to 25 kbar at 4.2 K. *Phys. Rev. B*, 10, 5184–97.

Ashcroft, N. W. (1968). Metallic hydrogen: a high temperature superconductor? *Phys. Rev. Lett.*, 21, 1748–9.

Bardeen, J. (1938). An improved calculation of the energies of metallic lithium and sodium. *J. Chem. Phys.*, 6, 367–371.

Beck, H. and Strauss, D. (1975). On the lattice dynamics of metallic hydrogen. *Helv. Phys. Acta*, 48, 655–669.

Bowen, D. H. (1967). Superconductivity of solids. In *High Pressure Physics and Chemistry*, ed. R. S. Bradley, vol. 1, 355–373. London: Academic Press.

Brovman, E. G., Kagan, Yu and Kholas, A. K. (1972a). Structure of metallic hydrogen at zero pressure. *Sov. Phys. J.E.T.P.*, 34, 1300–15.

Brovman, E. G., Kagan, Yu and Kholas, A. K. (1972b). Properties of metallic hydrogen under pressure. *Sov. Phys. J.E.T.P.*, 35, 783–7.

Brovman, E. G., Khagan, Yu, Kholas, A. K. and Pushkarev, Zh. (1973). Role of electron–electron interaction in the formation of a metastable state metallic hydrogen. *J.E.T.P. Lett.*, 18, 160–2.

Caron, L. G. (1975). Metallic hydrogen. *Comm. on Solid St. Phys.*, 6, 103–15.

Carr, W. J. (1962). Ground state energy of metallic hydrogen—I. *Phys. Rev.*, 128, 120–5.

Critchfield, C. L. (1942). Theoretical properties of dense hydrogen. *Astrophys. J.*, 96, 1–10.

de Marcus, W. C. (1958). The constitution of Jupiter and Saturn. *Astronom. J.*, 63, 2–28.

Etters, R. D., Danilowicz, R. and England, W. (1975). Properties of solid and gaseous hydrogen based upon anisotropic pair interactions. *Phys. Rev.* A, **12**, 2199–222.

Gell-Mann, H. and Brueckner, K. A. (1957). Correlation energy of an electron gas at high density. *Phys. Rev.*, **106**, 364–8.

Grigorev, F. V., Kormer, S. B., Mikhailova, A. P., Tolohko and Urlin, V. D. (1972). Experimental determination of the compressibility of hydrogen at densities 0.5 to 2 g/cm³. *J.E.T.P. Lett.*, **16**, 201–4.

Hammerberg, J. and Ashcroft, N. W. (1974). Ground state energies of simple metals. *Phys. Rev. B*, **9**, 409–24.

Harris, F. E. and Monkhorst, H. J. (1969). Complete evaluation of the electronic energies of solids. *Phys. Rev. Lett.*, **23**, 1026–30.

Hawke, R. S. (1974). Experiments on hydrogen at megabar pressures; metallic hydrogen. *Adv. Solid State Phys.*; *Festkörperprobleme*, **14**, 111–18.

Herzfield, K. F. (1927). On atomic properties which make an element a metal. *Phys. Rev.*, **29**, 701–5.

Hubbard, W. B. (1970). Thermal structures of Jupiter. *Astrophys. J.*, **152**, 745–54.

Hubbard, W. B. and Lampe, M. (1969). Thermal conduction by electrons in stellar matter. *Astrophys. J. Suppl.*, **18**, 297–346.

Kamarad, J. (1975). Is metallic hydrogen a reality? *Cesk. Cas. Fis.* A, **25**, 631–2.

Kittel, C. (1968). *Introduction to Solid State Physics*. Wiley.

Kronig, R., de Boer, J. and Korringa, J. (1946). On the internal constitution of the Earth. *Physica*, **12**, 245–56.

Kuhn, W. and Rittman, A. (1941). Über die Zustand des Erdinnern and seine Entstehung aus einem homogenen Urzustand. *Geol. Rundschau.*, **32**, 215–56.

Leung, W. R., March, N. H. and Motz, H. (1976). Primitive phase diagram for hydrogen. *Phys. Lett.*, **56A**, 425–6.

Lindhard, J. (1954). On the properties of a gas of charged particles. *Kgl. Dan. Vidensk. Selsk. Mat. Fys. Medd.*, **28**, No. 8.

Mills, R. L. and Grilly, E. R. (1956). Melting curves of H_2, D_2 and T_2 up to 3000 kg/cm². *Phys. Rev.*, **101**, 1246–7.

Mlynek, R. (1974). Experiments with metallic hydrogen. *Cesk. Cas. Fis.* A, **24**, 614.

Neece, G. A., Rogers, F. J. and Hoover, W. G. (1971). Thermodynamic properties of compressed solid hydrogen. *J. Comput. Phys.*, **7**, 621–36.

Nozieres, P. and Pines, D. (1958). A dielectric formulation of the many-body problem: application to the free electron gas. *Nuovo Cim.* **9**, 470–90.

Pines, D. and Nozieres, P. (1966). *The theory of quantum liquids*—I. New York: Benjamin.

Ross, M. and McMahan, A. (1976). Comparison of the theoretical models for metallic hydrogen. *Phys. Rev. B.*, **13**, 5154–8.

Salpeter, E. E. and Zapolsky, H. S. (1967). Theoretical high pressure equations of state including correlation energy. *Phys. Rev.*, **158**, 876–86.

Stephenson, D. J. and Ashcroft, N. W. (1974). Conduction in fully ionized liquid metals. *Phys. Rev. A.*, **9**, 782–9.

Stewart, J. W. (1956). Compression of solidified gases to 20,000 kg/cm^2 at low temperature. *J. Phys. Chem. Solids*, **1**, 146–58.

Trubitsyn, V. P. (1966). Equation of state of solid hydrogen. *Sov. Phys. Solid St.*, **7**, 2708–14.

Trubitsyn, V. P. (1967). Equations of state of solid helium at high pressures. *Sov. Phys. Solid St.*, **8**, 2593–8.

Trubitsyn, V. P. (1971). Phase diagram of hydrogen and helium. *Sov. Astronomy A. J.*, **15**, 303–9.

Vashista, P. and Singwi, R. S. (1972). Electron correlation at metallic densities—V. *Phys. Rev. B*, **6**, 875–87.

Vereshchagin, L. F. (1973). Concerning the transition of diamond into the metallic state. *J.E.T.P. Lett.*; **17**, 301–2.

Vereshchagin, L. F., Yakolev, E. N. and Timofeev, Yu. A. (1975). Possibility of transition of hydrogen into the metallic state. *J.E.T.P. Lett.*, **21**, 85–6.

Wigner, E. (1934). On the interaction of electrons in metals. *Phys. Rev.*, **46**, 1002–11.

Wigner, E. P. and Huntington, J. B. (1935). On the possibility of a metallic modification of hydrogen. *J. Chem. Phys.*, **3**, 764–70.

Wigner, E. and Seitz, F. (1933). On the constitution of metallic sodium. *Phys. Rev.*, **43**, 804–10.

Wigner, E. and Seitz, F. (1934). On the constitution of metallic sodium—II. *Phys. Rev.*, **46**, 509–24.

Wilson, A. H. (1966). *Thermodynamics and Statistical Mechanics*. Cambridge University Press.

Woolley, H. W., Scott, R. B. and Brickewedde, F. G. (1948). Compilation of thermal properties of hydrogen in its various isotopic and ortho–para modifications. *J. Res. Nat. Bur. Stds.*, **41**, 379–475.

Chapter 8

Allen, C. W. (1963). *Astrophysical Quantities*, 2nd edn. London: Athlone Press.

Anderson, J. D., Null, G. W., Biller, E. D., Wong, S. K., Hubbard, W. B. and MacFarlane, J. J. (1980). Pioneer Saturn celestial mechanics experiment. *Science*, **207**, 449–53.

Anderson, J. D., Null, G. W. and Wong, S. K. (1974). Gravity results from Pioneer 10 Doppler data. *J. Geophys. Res.*, **79**, 3661–4.

Ash, M. E., Shapiro, I. I. and Smith, W. B. (1971). The 336 system of planetary masses. *Science*, **174**, 551–6.

Ashcroft, N. W. (1968). Metallic hydrogen, a high temperature superconductor. *Phys. Rev. Lett.*, **21**, 1748–9.

Aumann, H. H., Gillespie, C. M. Jr. and Low, F. J. (1969). The internal powers and effective temperatures of Jupiter and Saturn. *Astrophys. J.*, **157**, 169–72.

Bernal, M. J. M. and Massey, H. S. W. (1954). Metallic ammonium. *Mon. Not. R. Astronom. Soc.*, **114**, 172–9.

Brown, R. A. and Goody, R. M. (1977). The rotation of Uranus. *Astrophys. J.*, **217**, 680–7.

Carr, T. D. (1971). Jupiter's magnetospheric rotation period. *Astrophys. Lett.*, **7**, 157–62.

Cook, A. H. (1959). The external gravity field of a rotating planet to the order of e^3. *Geophys. J. R. Astronom. Soc.*, **2**, 199–214.

Cook, A. H. (1972). The dynamical properties and internal structures of the Earth, the Moon and the planets. *Proc. R. Soc. A*, **328**, 301–36.

Cruickshank, D. P. (1978). On the rotation period of Uranus. *Astrophys. J.*, **220**, L57–9.

Danielson, R. E., Tomasko, M. G. and Savage, R. D. (1972). High resolution imagery of Uranus obtained by Stratoscope II. *Astrophys. J.*, **178**, 887–900.

Darwin, G. H. (1899). The theory of the figure of the Earth carried to the second order of small quantities. *Mon. Not. R. Astronom. Soc.*, **60**, 82–124.

de Cesare, L. (1974). Superconductivity in astrophysics. *Mem. Soc. Astronom. Ital.*, **44**, 279–303.

de Marcus, W. C. (1951). Ph.D. Thesis, Yale University.

de Marcus, W. C. (1958). The constitution of Jupiter and Saturn. *Astronom. J.*, **63**, 2–28.

de Sitter, W. (1924). On the flattening and constitution of the Earth. *Bull. Astronom Inst. Neth.*, **55**, 97–108.

de Sitter, W. (1931). Jupiter's Galilean Satellites. *Mon. Not. R. Astronom. Soc.*, **91**, 706–38.

Dollfus, A. (ed.) (1970a). *Surfaces and Interiors of Planets and Satellites*. London and New York: Academic Press.

Dollfus, A. (1970b) Diamètres des planètes et satellites. In *Surfaces and Interiors of Planets and Satellites*, ed. A. Dollfus. pp. 45–139. London and New York: Academic Press.

Dollfus, A. (1970c). New optical measurements of the diameters of Uranus and Neptune. *Icarus*, **12**, 101–17.

Duncombe, R. L., Klepcynski, W. J. and Seidelman, P. K. (1971). A determination of the masses of the five outer planets. *Cel. Mech.*, **4**, 224–32.

Dunham, D. W. (1971). *The motions of the satellites of Uranus*, Ph.D Thesis, Yale University.

Elliot, J. L., Dunham, E. and Mink, D. (1979). The radius and ellipticity of Uranus from its occultations of SAO 158687. *Bull. Amer. Astronom. Soc.*, **11**, 568.

Elliot, J. L., Dunham, E., Wasserman, L. H., Millis, R. L. and Churns, J. (1978). The radii of Uranian rings α, β, γ, δ, ε, η, 4, 5 and 6 from their occultations of SAO 158687. *Astronom. J.*, **83**, 980–92.

Freeman, K. C. and Lyngå, G. (1970). Data for Neptune from occultation observations. *Astrophys. J.*, **160**, 767–80.

Garcia, H. H. (1972). The mass and figure of Saturn by photographic astrometry of its satellites. *Astronom. J.*, **77**, 684–91.

Gehrels, T., Baker, L. R., Beshore, E., Blenman, C., Burke, J. J., Castillo, N. D., da Costa, B., Degewu, J., Doose, L. R., Fountain, J. W., Gotobed, J.,

KenKnight, C. E., Kingston, R., McLaughlin, G., McMillan, R., Murphy, R., Smith, P. H., Stoll, C. P., Strickland, W., Tomasko, M. G., Wuesinghe, M. P. and Coffeen, D. L. (1980). Imaging photopolarimetry on Pioneer Saturn. *Science*, **207**, 434–9.

Gill, J. and Gault, B. (1968). A new determination of the orbit of Triton, the pole of Neptune's equator and the mass of Neptune. *Astronom. J.*, **73**, 595.

Goldreich, P. and Soter, S. (1966). Q in the solar system. *Icarus*, **5**, 375–89.

Hayes, S. H. and Belton, M. J. S. (1977). The rotational periods of Uranus and Neptune. *Icarus*, **32**, 383–401.

Hubbard, W. B. (1968). Thermal structure of Jupiter. *Astrophys. J.*, **152**, 745–54.

Hubbard, W. B. (1969). Thermal models of Jupiter and Saturn. *Astrophys. J.*, **155**, 333–44.

Hubbard, W. B. (1970). Structure of Jupiter: chemical composition, contraction and rotation. *Astrophys. J.*, **162**, 687–97.

Hubbard, W. B. (1974a). Inversion of gravity data for giant planets. *Icarus*, **21**, 157–65.

Hubbard, W. B. (1974b). Gravitational field of a rotating planet with polytropic index of unity. *Astronom. Zh.*, **51**, 1052–9 and *Sov. Astronom. A. J.*, **18**, 621–4.

Hubbard, W. B. (1977). de Sitter's theory flattens Jupiter. *Icarus*, **30**, 311–13.

Hubbard, W. B., Slattery, W. L. and de Vito, C. L. (1975). High zonal harmonics of rapidly rotating planets. *Astrophys. J.*, **199**, 504–16.

Hubbard, W. B. and Smoluchowski, R. (1973). Structure of Jupiter and Saturn. *Space Sci. Rev.*, **14**, 599–662.

Hubbard, W. B. and van Flandern, T. C. (1972). The occultation of Beta Scorpii by Jupiter and Io. 3: Astrometry. *Astronom. J.*, **77**, 65–77.

James, R. A. (1964). The structure and stability of rotating gas masses. *Astrophys. J.*, **140**, 552–82.

Jeffreys, H. (1954). Second order terms in the figure of Saturn. *Mon. Not. R. Astronom. Soc.*, **114**, 433–6.

Klepcynski, W. J., Seidelman, P. K. and Duncombe, R. L. (1971). The masses of the principal planets. *Cel. Mech.*, **4**, 253–72.

Kliore, A. J., Woiceskyn, P. M. and Hubbard, W. B. (1976). Temperature of the atmosphere of Jupiter from Pioneer 10 and 11 radio occultation studies. *Geophys. Res. Lett.*, **3**, 113–16.

Kovalevsky, J. (1970). Détermination des masses des planètes et satellites. In *Surfaces and Interiors of Planets and Satellites*, ed. A. Dollfus, 1–44. London and New York: Academic Press.

Kovalevsky, J., and Link, F. (1969). Diamètre, applatissement et propriétés optiques de la haute atmosphere de Neptune d'après l'occultation de l'étoile BD 17°4388. *Astron. Astrophys.*, **2**, 398–412.

Low, F. J. (1966). Observations of Venus, Jupiter and Saturn at λ 20 μ. *Astronom. J.*, **71**, 391.

Miles, B. and Ramsey, W. H. (1952). On the internal structures of Jupiter and Saturn. *Mon. Not. R. Astronom. Soc.*, **112**, 234–43.

Newburn, R. C. Jr. and Gulkis, S. (1973). A survey of the outer planets,

Jupiter, Saturn, Uranus, Neptune, Pluto and their satellites. *Space Sci. Rev.*, 3, 179–271.

Nicholson, P. D., Persson, S. E., Matthews, K., Goldreich, P. and Neugebauer, G. (1978). The rings of Uranus: results of 10 April 1978 occultation. *Astronom. J.*, 83, 1240–8.

Null, G. W., Anderson, J. D. and Wong, S. K. (1975). Gravity field of Jupiter from Pioneer 11 tracking data. *Science*, 188, 476–7.

Opp, A. G. (1980). Scientific results from Pioneer Saturn encounters: Summary. *Science*, 207, 401–3.

Ostriker, J. P. and Mark, J. W.-K. (1968). Rapidly rotating stars. I. The self consistent field method. *Astrophys. J.*, 151, 1075–88.

Peebles, P. J. E. (1964). The structure and composition of Jupiter and Saturn. *Astrophys. J.*, 140, 328–47.

Podolak, M. (1976). Methane rich models of Uranus. *Icarus*, 27, 473–7.

Podolak, M. and Cameron, A. G. W. (1974). Models of the giant planets. *Icarus*, 22, 123–48.

Ramsey, W. H. (1951). On the constitution of the major planets. *Mon. Not. R. Astronom. Soc.*, 111, 427–47.

Ramsey, W. H. (1963). On the densities of methane, metallic ammonium, water and neon at planetary pressures. *Mon. Not. R. Astronom. Soc.*, 125, 469–85.

Sinclair, A. T. (1976). Communication to the XVI IAU, Grenoble, 1976.

Slattery, W. L. (1977). The structure of the planets Jupiter and Saturn. *Icarus*, 32, 58–72.

Slattery, W. L. and Hubbard, W. B. (1976). Thermodynamics of a solar mixture of molecular hydrogen and helium at high pressure. *Icarus*, 29, 187–92.

Slavsky, D. and Smith, H. (1977). Rotation period of Neptune. *Bull. Amer. Astronom. Soc.*, 9, 512.

Smith, H. J. and Slavsky, D. B. (1979). Rotation period of Uranus. *Bull. Amer. Astronom. Soc.*, 11, 568.

Smoluchowski, R. (1967). Internal structure and energy emission of Jupiter. *Nature*, 215, 691–5.

Smoluchowski, R. (1973). Dynamics of the Jovian interior. *Astrophys. J.*, 185, L95–L99.

Stevenson, D. J. (1975). Does metallic ammonium exist? *Nature (Lond.)*, 258, 222–3.

Stevenson, D. J. and Salpeter, E. E. (1975). Interior models of Jupiter. In *Jupiter*, ed. T. Gehrels, 85–112. Tucson: University of Arizona Press.

Stewart, J. W. (1956). Compression of solidified gases to 20,000 kg/cm^3 at low temperature. *J. Phys. Chem. Solids*, 1, 146–58.

Streett, W. B. (1976). Phase equilibrium in gas mixtures at high pressures. *Icarus*, 29, 173–86.

Taylor, G. E. (1968). New determination of the diameter of Neptune. *Nature*, 219, 474–5.

Trafton, L. (1977). Uranus' rotational period. *Icarus*, 32, 402–12.

Trauger, J. T., Roesler, F. L. and Münch, G. (1978). A redetermination of the Uranus rotation period. *Astrophys. J.*, 219, 1079–83.

van Woerkom, A. J. J. (1950). The motion of Jupiter's fifth satellite, 1892–1949. *Astronom. Pap. Wash.*, **13**, 1–77.

von Zahn, U. and Fricke, K. H. (1977). Solar wind as a source of upper atmosphere heating on Earth and Jupiter. *J. Geophys. Res.*, **82**, 727–50.

Whitaker, E. A. and Greenberg, R. J. (1973). The eccentricity and inclination of Miranda's orbit. *Mon. Not. R. Astronom. Soc.*, **165**, 16P.

Zharkhov, V. N., Makalkin, A. B. and Trubitsyn, V. P. (1974). Models of Jupiter and Saturn, II. Structure and composition. *Astronom. Zh.*, **51**, 1288–1297, *Sov. Astron. A.J.*, **18**, 768–73.

Zharkhov, V. N. and Trubitsyn, V. P. (1969). Theory of the figure of rotating planets in hydrostatic equilibrium – a third approximation. *Sov. Phys. Astronomy*, **13**, 981–8 (transl. from *Astronom. Zh.*, **46** (1969) 1252–63).

Zharkhov, V. N. and Trubitsyn, V. P. (1974). Determination of the state of the molecular envelopes of Jupiter and Saturn from their gravitational moments. *Icarus*, **21**, 152–6.

Zharkhov, V. N. and Trubitsyn, V. P. (1975*a*). Structure, composition and gravitational field of Jupiter. In *Jupiter*, ed. T. Gehrels, 133–78. Tuscon: University of Arizona Press.

Zharkhov, V. N. and Trubitsyn, V. P. (1975*b*). Fifth approximation system of equations for the theory of figure. *Astronom. Zh.*, **52**, 599–614, *Sov. Astron. A.J.*, **19**, 366–72.

Zharkhov, V. N., Trubitsyn, V. P. and Hubbard, W. B. (1978). *Physics of Planetary Interiors.* Tucson: Pachart.

Zharkhov, V. N., Trubitsyn, V. P. and Makalkin, A. B. (1972). The high gravitational moments of Jupiter and Saturn. *Astrophys. Lett.*, **10**, 159–61.

Chapter 9

Acuna, M. H. and Ness, N. F. (1976). The main magnetic field of Jupiter. *J. Geophys. Res.*, **81**, 2917–22.

Acuna, M. H. and Ness, N. F. (1980). The magnetic field of Saturn: Pioneer 11 observations. *Science* **207**, 444–6.

Barraclough, L. D. R., Harwood, J. M., Leaton, B. R. and Malin, S.C. R. (1975). A model of the geomagnetic field at epoch 1975. *Geophys. J. R. Astronom. Soc.*, **43**, 645–59.

Brown, L. W. (1975). Saturn radio emission near 1 MHz. *Astrophys. J.*, **198**, L89–L92.

Bullard, E. C. (1968). Reversals of the Earth's magnetic field. *Philos. Trans. R. Soc. A.*, **263**, 481–524.

Carslaw, H. and Jaeger, J. C. (1959). *Conduction of heat in solids*, 2nd edn: Oxford University Press.

Coleman, P. J. and Russell, C. T. (1977). The remanent magnetic field of the Moon. *Philos. Trans. R. Soc. A*, **285**, 489–506.

Cook, A. H. (1963). Sources of harmonics of low order in the external gravity field of the Earth. *Nature*, **198**, 1186.

Cook, A. H. (1979). Geophysics and the human condition, *Q. J. R. Astronom. Soc.*, **20**, 229–40.

Dolginov, Sh. Sh., Yerovshenko, Ye. G. and Zhuzgov, L. N. (1973). Magnetic field in the very close neighbourhood of Mars according to data from the Mars 2 and 3 spacecraft. *J. Geophys. Res.*, **78**, 4771–89.

Ferrari, A. J. (1977). Lunar gravity: a harmonic analysis. *J. Geophys. Res.*, **82**, 3065–84.

Gapcynski, J. P., Tolson, R. M. and Michael, W. H. Jr. (1977). Mars gravity field, combined Viking and Mariner 9 results. *J. Geophys. Res.*, **82**, 4325–7.

Gaposchkin, E. M. (ed.) (1973). Smithsonian Standard Earth III. *Research in Space Science SAO Special Report* **353** (Smithsonian Astrophysical Observatory, Cambridge, Mass.).

Gubbins, D. 1974. Theories of the geomagnetic and solar dynamics. *Rev. Geophys. Space Phys.*, **12**, 137–54.

Hide, R. (1978). How to locate the electrically conducting core of a planet from external magnetic observations. *Nature*, **271**, 640–1.

Jackson, D. J. and Beard, D. B. (1977). The magnetic field of Mercury. *J. Geophys. Res.*, **82**, 2828–36.

Lowes, F. (1974). Spatial power spectrum of the main geomagnetic field and extrapolation to the core. *Geophys. J. R. Astronom. Soc.*, **36**, 717–30.

McKenzie, D. and Weiss, N. (1975). Speculations on the thermal and tectonic history of the Earth. *Geophys. J. R. Astronom. Soc.*, **42**, 131–174.

Moffat, H. K. (1978). *Magnetic field generation in electrically conducting fluids.* Cambridge University Press.

Mutch, T. A., Arvidson, R. E., Head, J. W. III, Jones, K. L. and Saunders, R. S. (1976). *The Geology of Mars.* Princeton and Guildford: Princeton University Press.

Ness, N. F., Behannon, K. W., Lepping, R. P. and Whang, Y. C. (1975). The magnetic field of Mercury-I. *J. Geophys. Res.*, **80**, 2708–16.

Rochester, M. G., Jacobs, J. A., Smylie, D. E. and Chong, K. F. (1975). Can precession power the geomagnetic dynamo? *Geophys. J. R. Astronom. Soc.*, **43**, 661–78.

Runcorn, S. K. (1975). An ancient lunar magnetic dipole field. *Nature*, **253**, 701–3.

Runcorn, S. K. (1977). Interpretation of lunar potential fields. *Philos. Trans. R. Soc. A*, **285**, 507–16.

Russell, C. T. (1976). The magnetic moment of Venus: Venera-4 measurements re-interpreted. *Geophys. Res. Lett.*, **3**, 125–8.

Russell, C. T. (1977). On the relative location of the bow shocks of the terrestrial planets. *Geophys. Res. Lett.*, **4**, 387–90.

Schubert, G., Young, R. E. and Cassen, P. (1977). Solid state convection models of the lunar internal temperature. *Philos. Trans. R. Soc. A*, **285**, 523–36.

Smith, E. J. (1974). The planetary magnetic field and magnetosphere of Jupiter: Pioneer 10. *J. Geophys. Res.*, **79**, 3501–13.

Smith, E. J., Davis, J., Jones, D., Coleman, P., Colburn, D. S., Dyal, P. and Sonnett, C. P. (1975). Jupiter's magnetic field, magnetosphere and interaction with the solar wind: Pioneer 11. *Science*, **188**, 451–5.

Sonnett, C. P. (1977). Some consequences of solar wind and induction in the Moon. *Philos. Trans. R. Soc. A*, **285**, 537–47.

Strom, R. G. (1979). Mercury: a post-Mariner 10 assessment. *Space Sci. Rev.*, **24**, 3–70.

Timber, M. E. and Kirk, D. B. (1976). Impact craters on Venus. *Icarus*, **28**, 351–7.

Tozer, D. (1972). The present thermal state of the terrestrial planets. *Phys. Earth Planet Int.*, **6**, 182–97.

Warwick, J. (1967). Radiophysics of Jupiter. *Space Sci. Rev.*, **6**, 841–91.

Chapter 10

Allen, C. W. (1963). *Astrophysical Quantities*. University of London: Athlone Press.

Anderson, J. D., Null, G. W., Biller, E. D., Wong, S. K., Hubbard, W. B. and MacFarlane, J. J. (1980). Pioneer Saturn celestial mechanics experiment. *Science*, **207**, 449–53.

Carr, M. H., Masursky, H., Strom, R. G. and Terrile, R. J. (1980). Volcanic features of Io. *Nature*, **280**, 729–33.

Christy, J. W. and Harrington, R. S. (1978). The satellite of Pluto. *Astronom. J.*, **83**, 1005–8.

Cruickshank, D. P., Pilcher, C. D. and Morrison, D. (1976). Pluto: evidence for methane frost. *Science*, **194**, 835–7.

Meadows, J. (1980). Lecture to the Royal Astronomical Society, March 1980.

Newburn, R. L. Jr. and Gulkis, S. (1973). A survey of the outer planets, Jupiter, Saturn, Uranus, Neptune, Pluto and their satellites. *Space Sci. Rev.*, **3**, 179–271.

Nicholson, P. D., Persson, S. E., Matthews, K., Goldreich, P. and Neugebauer, G. (1978). The rings of Uranus: results of the 10 April occultation. *Astronom. J.*, **83**, 1240–8.

Stone, E. C. and Lane, A. L. (1979). Voyager 1 encounter with the Jovian system. *Science*, **204**, 945–8.

Appendix 1

Parker, R. L. (1972). Inverse theory with grossly inadequate data. *Geophys. J. R. Astronom. Soc.*, **29**, 123–38.

Gruber, M. A. (1977). An information theory approach to the density of the Earth. *NASA Tech. Mem. 78034* (Greenbelt, Md.: N.A.S.A Goddard Space Flight Centre).

Rietsch, E. (1977). The maximum entropy approach to inverse problems. *J. Geophys.*, **42**, 489–506.

Rietsch, E. (1978). Extreme models from the maximum entropy formulation of inverse problems. *J. Geophys.*, **44**, 273–5.

Appendix 3

Cook, A. H. (1977). Theories of lunar libration. *Philos. Trans. R. Soc. A*, **284**, 573–85.

INDEX

α-effect 297
Adams–Williamson method 27ff, 37
adiabatic gradient 23
age of lunar surface 277
ALSEPS 141
aluminium quantum calculations 99
ammonium 269, 271, 272
Apollo
  missions 132
  seismometers 142, 143
  sites 283
Ariel 245, 303
asteroids 302, 304
asthenosphere 275
atomic units 201

β-phase 125
'belt' apparatus 107, 108
Birch–Murnaghan equation 103
Bode's Law 304
bulk modulus, 'jellium' 207
  major planets 238
  pressure dependence 36, 37, 96, 97,
    103, 104, 105, 106, 119, 120
Bullen Earth models 29ff
  incompressibility hypothesis 29

canonical constants 63, 64
canonical variables 63
Callandreau 250
CHONNE 269, 270
Clairaut 250
Clairaut's equation 80, 83
Clapeyron equation 123
Clausius's equation 129
condensation of planets 305, 306
continuous models 316, 317
core 306

composition 35
  Earth 17, 28, 33, 197, 199
  electrical properties 299
  equation of state 36
  inner 29, 33, 38
  Jupiter and Saturn 261
  and magnetic field 292
  major planets 271
  Moon 151, 152
  oblateness 300
  shadow zone 24, 25
  terrestrial planets 197
correlation energy 214, 215, 222
Cowling's theorem 294
craters 276
creep, solid state 93, 285
crust 33
  of Moon 160, 169, 170
crystal structure determination 110

Darwin 250, 251
  relation 84, 86
Deimos 175, 178, 181
degeneracy 204
  temperature 203
density
  light elements 6
  major planets 249
  and composition 89, 118
  pressure dependence 89, 90, 117ff
  temperature dependence 89, 91, 274
  (see also metallic hydrogen, molecular
    hydrogen, silicates)
diameters of planets 54
diamond press 109, 110
dielectric function 212
  metallic hydrogen 213, 222–6
Dione 303

343